Lecture Notes in Civil Engineering

Volume 298

Lecture Notes in Civil Engineering (LNCE) publishes the latest developments in Civil Engineering—quickly, informally and in top quality. Though original research reported in proceedings and post-proceedings represents the core of LNCE, edited volumes of exceptionally high quality and interest may also be considered for publication. Volumes published in LNCE embrace all aspects and subfields of, as well as new challenges in, Civil Engineering. Topics in the series include:

- Construction and Structural Mechanics
- Building Materials
- Concrete, Steel and Timber Structures
- Geotechnical Engineering
- Earthquake Engineering
- Coastal Engineering
- Ocean and Offshore Engineering; Ships and Floating Structures
- Hydraulics, Hydrology and Water Resources Engineering
- Environmental Engineering and Sustainability
- Structural Health and Monitoring
- Surveying and Geographical Information Systems
- Indoor Environments
- Transportation and Traffic
- Risk Analysis
- Safety and Security

To submit a proposal or request further information, please contact the appropriate Springer Editor:

- Pierpaolo Riva at pierpaolo.riva@springer.com (Europe and Americas);
- Swati Meherishi at swati.meherishi@springer.com (Asia—except China, Australia, and New Zealand);
- Wayne Hu at wayne.hu@springer.com (China).

All books in the series now indexed by Scopus and EI Compendex database!

Kasinathan Muthukkumaran · Deendayal Rathod ·
Evangelin Ramani Sujatha · M. Muthukumar
Editors

Transportation and Environmental Geotechnics

Proceedings of the Indian Geotechnical Conference 2021 Volume 4

Springer

Editors
Kasinathan Muthukkumaran
Department of Civil Engineering
National Institute of Technology
Tiruchirappalli, Tamil Nadu, India

Deendayal Rathod
Department of Civil Engineering
National Institute of Technology
Tiruchirappalli, Tamil Nadu, India

Evangelin Ramani Sujatha
School of Civil Engineering
SASTRA University
Thanjavur, Tamil Nadu, India

M. Muthukumar
Department of Structural and Geotechnical Engineering
Vellore Institute of Technology
Vellore, Tamil Nadu, India

ISSN 2366-2557 ISSN 2366-2565 (electronic)
Lecture Notes in Civil Engineering
ISBN 978-981-19-6773-3 ISBN 978-981-19-6774-0 (eBook)
https://doi.org/10.1007/978-981-19-6774-0

This Springer imprint is published by the registered company Springer Nature Singapore Pte Ltd.
The registered company address is: 152 Beach Road, #21-01/04 Gateway East, Singapore 189721, Singapore

Preface

The Indian Geotechnical Society, Trichy (IGS-Trichy) Chapter, and National Institute of Technology (NIT) Tiruchirappalli, India, organized the Indian Geotechnical Conference (IGC-2021) at Trichy during 16–18 December 2021. The main theme of the conference was "GEO-INDIA"—GEOTECHNICS FOR INFRASTRUCTURE DEVELOPMENT AND INNOVATIVE APPLICATIONS.

The sub-themes of the conference included:

1. Soil Behaviour and Characterization of Geomaterials
2. Geotechnical, Geological and Geophysical Investigation
3. Foundation Engineering
4. Ground Improvement Techniques
5. Geo-environmental Engineering
6. Soil Dynamics and Earthquake Geotechnical Engineering
7. Earth Retaining Structures, Dams and Embankments
8. Slope Stability and Landslides
9. Transportation Geotechnics
10. Geosynthetics Application
11. Computational, Analytical and Numerical Modelling
12. Rock Engineering, Tunnelling, Deep Excavations and Underground Constructions
13. Forensic Geotechnical Engineering and Case Studies
14. Others: Behaviour of Unsaturated Soils, Offshore & Marine Geotechnics, Remote Sensing & GIS, Instrumentation & Monitoring, Retrofitting of Geotechnical Structures, Reliability in Geotechnical Engineering, Geotechnical Education, Codes & Standards, & any other relevant topic.

The proceedings of this conference consists of selected papers presented at the conference. The proceedings is divided into six volumes. A special issue on IGC-2021 keynote and theme lecture presentations were published by Indian Geotechnical Journal.

We sincerely thank all the authors who have contributed their papers to the conference proceedings. We also thank all the theme editors and reviewers who have been instrumental in giving their valuable inputs for improving the quality of the final papers. We greatly appreciate and thank all the student volunteers for their unwavering support that was instrumental in preparation of this proceedings. Finally, thanks to Springer team for their support and full cooperation for publishing six volumes of this IGC-2021 proceedings.

Trichy, India

Kasinathan Muthukkumaran
Chairman, IGC-2021

Contents

About the Editors

Dr. Kasinathan Muthukkumaran is currently Professor in Civil Engineering at National Institute of Technology, Tiruchirappalli, India. He obtained Ph.D. in Soil-Structure Interaction and Marine Geotechnical Engineering from Indian Institute Technology Madras. He has published more than 150 papers in international and national journals and conferences. He has completed five R&D (including ISRO—**Chandrayaan-2 Mission project**) and 70 major consultancy projects in Geotechnical Engineering and published two patent including "**Moon Soil**" (A Method for Manufacture of Highland Lunar Soil Simulant). He has guided 10 Ph.D. scholars and six more are in progress, five MS (by research) and more than 40 M.Tech. students in Geotechnical and allied research areas. He is the **Founder Chairman** of Indian Geotechnical Society (IGS-Trichy) Trichy Chapter. He is a member of Technical Committee (TC-301 on "**Preservation of Historic Sites**") of International Society for Soil Mechanics and Geotechnical Engineering. Professor Muthukkumaran area of research is in geotechnical engineering, which includes pile foundation, soil-structure interaction, marine geotechnics and foundations, field instrumentation, geotechnical physical modeling, ground improvement and forensic geotechnical engineering. He has received DST Young Scientist Award, IGS—Smt. Indra Joshi Biennial Award and Keynote Paper Award—GEOMATE Conference 2015 at Osaka, Japan. He is an Associate Editor of *Australian Journal of Civil Engineering and Serving* editorial board member of several journals. Professor Muthukkumaran has significant administrative contribution as Estate Officer, Associate Dean (Planning and Development), Member of Buildings and Works Committee and Member of **Board of Governors (BoG)** of National Institute of Technology, Tiruchirappalli and Member of Buildings and Works Committee, IIM Trichy. He has received **NIT Trichy Achiever Awards** for research publications, research projects, maximum citation and consultancy projects.

Dr. Deendayal Rathod has obtained his B.Tech. (Civil Engineering) from JNTU College of Engineering (University Campus), Hyderabad, and M.E. and Ph.D. both from Indian Institute of Science (IISc) Bangalore. He is currently working as professor at National Institute of Technology (NIT) Trichy since 2006. He has published more than 15 papers in the referred international journals and 12 papers in the reputed national/international conferences. He has more than 15 years of teaching, research and consultancy experience at NIT Trichy, and two months of abroad working experience at the University of Warwick (UK) as Visiting Fellow. Dr. Rathod has involved in all the major consultancy work carried out by the Department of Civil Engineering, NIT Trichy, for various organizations like Tamil Nadu Housing Board, Southern Railway, NLC-Neyveli, TNEB, PWD Trichy, etc.

Dr. Evangelin Ramani Sujatha currently works as Professor in the School of Civil Engineering, SASTRA Deemed University, India. She holds a Master's degree in soil mechanics and foundation engineering and a Ph.D. in slope stability. Her research interests include sustainable ground improvement, landslide analysis and GIS applications, landfill liner applications, biopolymer and nano-additive characterization and applications. She previously served as Assistant Engineer in Tamil Nadu Public Works Department and has an academic career spanning 21 years. Also, she is the member secretary of the Indian Geotechnical Society, Thanjavur chapter. She published over 50 technical articles in many peer-reviewed national and international journals. She has also given over 30 conference presentations and has 12 peer-reviewed book chapters to her credit. She successfully investigated two extra-mural and one intra-mural funded project. At present, she is investigating a project funded by TNSCST on biopolymer-based brick development for economical housing. She is an active reviewer for many peer-reviewed journals.

Dr. M. Muthukumar is Associate Professor and Head of the Department of Structural and Geotechnical Engineering Division in the School of Civil Engineering, Vellore Institute of Technology (VIT), Vellore, India. He has over 20 years of professional experience in the field of geotechnical engineering. His primary areas of research include ground improvement techniques, helical piles, granular pile anchors, clay liners and soil reinforcement. He is also member in various professional societies. Professor Muthukumar has published several research articles in peer-reviewed international journals and reputed conferences. He is also a recipient of Tamil Nadu young scientist fellowship award in 2017. He is also a reviewer for various journals of national and international repute in geotechnical engineering.

Chapter 1
Study of Bearing Capacity of Shallow Foundation on Bamboo Geotextile-Reinforced Clayey Bed

Sudeepta Malakar and Parbin Sultana

Introduction

With increase in population, the demand of livable places also increases day by day. But livable place or the places where we can construct any structure without any help of engineer is in specific amount. We have other places like marshy land, sand area, clayey soil, and so on. Construction on these delicating lands will definitely take engineers advice. As we know India is having 2nd largest population in the world, so it has been necessary to construct structure on the marshy land. But constructing on delicate soil is challenging for engineers. In such situation, improving the bearing capacity of soil will be the main objective. Many researchers and engineers have already used many ways for improvising the bearing capacity of soil. Also, they have used geosynthetics for improvising the bearing capacity of soil. As geosynthetics are costly, we can use bamboo as an alternative to improve the bearing capacity of soil. This paper is concerned with the economical as well as biodegradable solution for improvising the bearing capacity of delicate soil. As the main concern about bamboo is they are not going to sustain like other reinforcing materials, so bitumen coating is used over bamboo. Also, bamboos have been tested on four months of weathered bamboo with bitumen coating and without bitumen coating. Bitumen-coated bamboo mat will also degraded after sometime, but due to consolidation process, the soil will improve its bearing capacity by itself with time.

S. Malakar (✉)
Indian Institute of Technology Guwahati, Assam, India
e-mail: sudeepta.m23@gmail.com

P. Sultana
National Institute of Technology Silchar, Assam, India
e-mail: parbin@civil.nits.ac.in

K. Muthukkumaran et al. (eds.), *Transportation and Environmental Geotechnics*,
Lecture Notes in Civil Engineering 298, https://doi.org/10.1007/978-981-19-6774-0_1

Background

Reinforcing earth means combination of earth with any reinforcing components. Since start of human progress, man has utilized soil with variety of reinforcing material to increase the capacity of soil according to their needs. With the passing year, we started using those earlier methods in a different or better way but with same mechanism. Sawwaf and Nazir [1] concentrated the conduct of strengthened and over and again stacked model footings on both unreinforced and fortified sand. Huat [2] examined on peat soil and strengthened the soil by utilizing bond segment strategy. Nugroho [3] states that we can increase the bearing capacity of peat soil by reinforcing soil with geotextile grid and bamboo. Farsakh et al. [4] examined the sandy soil with and without reinforcement. He also stated that geosynthetics can increase the bearing capacity of soil. The improvement in the bearing capacity of silty clay soil reinforced with geogrids at various depths is examined by Kolay et al. [5]. The findings demonstrate that when the number of geogrid layers increases, bearing capacity increases significantly. Waruwu [6] investigated the peat soil reinforced with bamboo and observed an improvement in the bearing capacity of peat soil. The bearing capacity is increased by 140% for one layer of bamboo reinforcement, 224% for two layers of bamboo reinforcement, and 279% for three layers of bamboo. Hegde et al. [7] investigated the possibility of using naturally available bamboo to increase the bearing capacity of soft soil through experimental and analytical studies. The results of the laboratory plate load tests indicated that the ultimate bearing capacity of the clay bed reinforced with a combination of bamboo cells and the bamboo grid was about 1.3 times higher than the geocell and geogrid reinforced clay beds. The findings of laboratory model testing of a surface strip footing on unreinforced and reinforced sand beds are presented by Cicek et al. [8]. Additionally examined how the type and number of reinforcement affected the optimum reinforcement length. According to Waruwu et al. [9], using bamboo grids as reinforcement on peat layers can minimize settlement caused by embankment loads. Mathew et al. [10] obtained maximum bearing capacity in three layers of bamboo grids, and it has better performance than geonet reinforcement. In conclusion, we can say that with reinforcing soil we can use the land having problem of taking major loads. We have many alternatives for reinforcing the soil, but as a civil engineer, we should always approach the method having safety as well as economic. And, bamboo is cheaper than any other reinforcing material.

Material Used

There are three important materials which have been used for completion of this test. The main three materials are soil from marshy land, bamboo as a reinforcing material, and bitumen for decreasing the degradation rate of bamboo.

Table 1.1 Properties of soil

Properties of soil	Value
Specific gravity	2.73
Plastic limit	36%
Liquid limit	71%
Plasticity index	35%
Optimum moisture content	31%
Maximum dry density	1.41 g/cm^3
Unconfined compression strength	2.55 kg/cm

Soil Used for the Experiment

The soil used for the tests was clayey soil. The soil was collected from marshy land of Cachar district of Assam, which has been used for the making of foundation bed. The soil is air-dried and then pulverized for the laboratory tests. Index properties of soil have been determined for the experiments which have been mentioned in Table 1.1. By the method of ASTM D 854, specific gravity has been determined. Liquid limit and plastic limit were determined by the method of ASTM D 4318. Plasticity index was determined by the formula of liquid limit and plastic limit. Optimum moisture content and maximum dry density were determined by the method of proctor test as per ASTM D 698. Unconfined compression strength was determined by the method of ASTM D 2166.

Bamboo Used in the Experiment

For the experiment, bamboo mats have been used for reinforcing the soil. Each strip has width, thickness of 10 mm and 2 mm, respectively. Bamboo mats can be easily available in market. Northeast of India is famous of growing and utilization of bamboo in various method. So, it can be easily available in Assam as a construction or reinforcing material. Bamboo is the fastest-growing plant, and availability of bamboo in Cachar district of Assam is not a big issue. Figure 1.1 demonstrates the photograph of the bamboo mats utilized in the research. For reinforcing the soil, the selection of bamboo should be done accurately. Some points are given below of selecting bamboo as reinforcement:

i. The color of the bamboo should be of brown color which will ensure that the bamboo is of at least more than 3 years;
ii. Always select the bamboos from the maximum height of bamboo trees;
iii. The green part of the bamboo will impart weakness to the strength. So, try to avoid using the green part of the bamboo;
iv. Due to high moisture content in the spring season, the bamboos posses weakness in the strength. So, avoid cutting the bamboos in the spring season.

Fig. 1.1 Bamboo mats

Bitumen Used on the Bamboo Mats

For non-degradation of bamboo mats, bitumen was used. Using natural bamboo or without applying any kind of coating on the bamboo will be degraded easily in the soil due to presence of moisture content in the soil. Therefore, it should be waterproof to reduce the degradation of bamboo mats and also reduce the swelling of bamboo. There are many ways to give the waterproof coating to bamboo mats like by paint, by native latex, by sodium silicate method. Coating with bitumen is also an alternate solution for coating the bamboo mats to reduce the degradation. Here, reason of choosing bitumen as coating material is mainly because of its simple availability and economic reason. There is no need of any guidance while coating with bitumen on the bamboo mat; bitumen can be available in any place or market easily, and the cost of bitumen is very affordable. In this research, bitumen coating is used to prevent the bamboo mat from rotting shown in Fig. 1.2.

Fig. 1.2 Process of bitumen coating

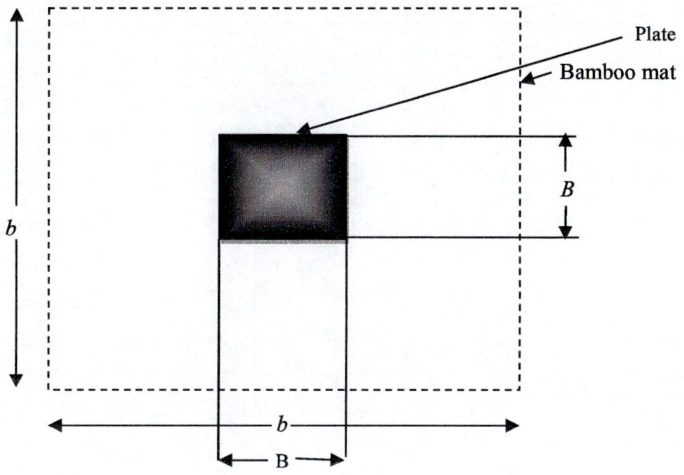

Fig. 1.3 Schematic diagram of test setup from top view

Laboratory Model Tests

The laboratory test was conducted in geotechnical laboratory of National Institute of Technology Silchar, Assam.

Experimental Setup

In the laboratory, a series of plate load test were conducted. The tank used for experiment has dimension of 1000 × 1000 × 1000 mm. Iron plate of square size has been taken as the model square footing for the tests. Size of the plate was taken 150 × 1500 mm with 2 mm thickness. Load has been applied from above to the model footing using hydraulic jack which was placed at the center of footing. The applied load was measured by pre-calibrated proving ring. Figures 1.3 and 1.4 represent the test setup, and Fig. 1.5 represents experimental setup, respectively.

Preparation of Foundation Bed

The soft soil used in preparing the foundation bed has been air-dried and pulverized. A moisture content of 44% has been mixed with the air-dried soil to make the foundation bed. The tank has been filled with ten layers of soil, and each layer has been compacted. Polyethylene was used in the sides as well as on the bottom of the model tank so that friction can be reduced. After completion of foundation bed, plate load tests have been conducted. The settlements corresponding to each load have been measured using two dial gauges.

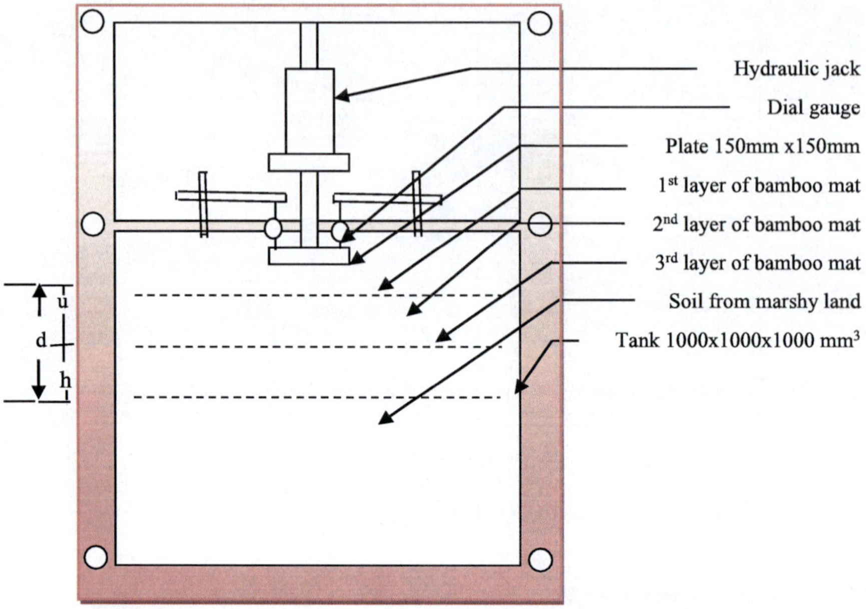

Fig. 1.4 Test setup

Fig. 1.5 Experimental setup

Placed Bamboo Mats Under the Soil for Weathering

Before placing the bamboo mats under the soil, bitumen coating has been applied to bamboo mats. When bitumen was dry, one bamboo mat was placed under the soil about 1 foot depth for weathering process on November 1, 2018. And, one bamboo mat without bitumen coating was placed under the soil at same depth and time. To compare the degradation process, two of the bamboo mats have been placed under the soil with and without bitumen coating. After four months of weathering process, it was found that bitumen-coated bamboo mat was still in good position while the other one which has no coating of bitumen was demolished and mix with the soil. In conclusion, bitumen coating can be the best alternative for decreasing the rate of degradation of bamboo.

Test Procedure

The plate load tests have been conducted on model footing surface of dimensions of 1000 $\times$ 1000 $\times$ 1000 mm with a plate size of 150 $\times$ 150 mm. Totally, eight tests have been done in the tank and then compared the results of the test. The first test has been done in the soil without reinforcement. The next three tests have been done in the soil having different layers of bamboo reinforcement without bitumen coating. The second test has been done in the soil having one layer of bamboo mat, the third test has been done in the soil having two layers of bamboo mats, and the fourth test has been done in the soil having three layers of bamboo mats. Again, the next three tests have been done in the soil having different layers of bamboo mats but with coating of bitumen. Also, the last eighth test has been performed on the reinforcement of degraded bamboo mat in the soil. The reinforcements have been placed in a test bed layer. The test has been performed by varying the multi-layers reinforcement to compare the results. It has been done to get the effective depth of reinforcement.

Flowchart of Tests

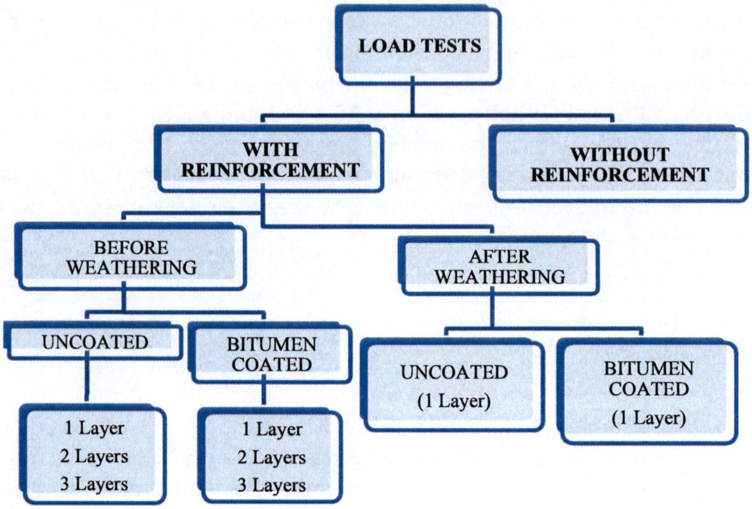

Results and Discussion

The results of all eight tests are mentioned below in tabular form. It can be seen from Table 1.2 that settlement on reinforced soil is effective than unreinforced soil.

Table 1.2 Results of all plate load tests on clayey bed

q_u	*s*1	*s*2	*s*3	*s*4	*s*5	*s*6	*s*7	*s*8
6.7	0.86	0.15	0.11	0.10	0.15	0.11	0.10	0.15
13.33	2.37	1.08	0.81	0.77	1.09	0.82	0.78	1.14
20	5.26	2.76	2.07	1.70	2.78	1.65	2.01	2.79
26.7	9.67	5.67	3.31	2.90	5.22	3.35	3.93	6.10
33.3	18.74	10.21	6.46	5.81	9.31	6.54	6.38	10.75
40	30.55	15.94	11.73	10.38	16.08	11.83	11.47	19.46
44.7	43.18	24.73	21.06	19.91	24.96	21.25	20.09	30.91
53.3	74.04	59.99	56.46	52.94	55.21	50.68	48.15	63.55

q_u: bearing pressure (kN/m^2)
$s1$: settlement on unreinforced soil
$s2$: settlement on reinforcement soil with one layer of bamboo mat (mm)
$s3$: settlement on reinforcement soil with two layers of bamboo mats (mm)
$s4$: settlement on reinforcement soil with three layers of bamboo mats (mm)
$s5$: settlement on reinforcement soil with one layer of bitumen coated bamboo mat (mm)
$s6$: settlement on reinforcement soil with two layers of bitumen coated bamboo mat (mm)
$s7$: settlement on reinforcement soil with three layers of bitumen coated bamboo mat (mm)
$s8$: settlement on reinforcement soil with degraded bitumen-coated bamboo mat (mm)

Interpretation of Test Outcomes in Graphical Form

Based on the experiments conducted in the laboratory, it concluded that bamboo mats also an alternate of reinforcing the soil. As most of the researchers are using geosynthetics for stabilizing the soil which are costly and non-biodegradable in nature, so we can use bamboo as an alternative which is eco-friendly and economical. India is a country where bamboo is found abundantly however, we haven't explored the use of bamboo as much as bamboo has the potential for example by using it to reinforce the soil. Figure 1.6 showing the improvisation of bearing capacity of soil after using bamboo mats as reinforcement in comparison with unreinforced soil. This experiment also concluded that by applying bitumen coating the degradation problem of bamboo can be solved. Ultimately bitumen-coated bamboo mats will also have been degraded. Till the time bamboo decomposes, clayey soil will gain sufficient strength due to the process of consolidation. In certain circumstances, reinforcement might not be required over time. The effect of bamboo may be lesser than geosynthetics, but it is not a bad choice for improving the quality of clayey soil for the shallow foundation, slope area, road area, and so on. The results of all tests in graphical form are represented below. Figure 1.6 represents settlement versus. bearing capacity of the soil.

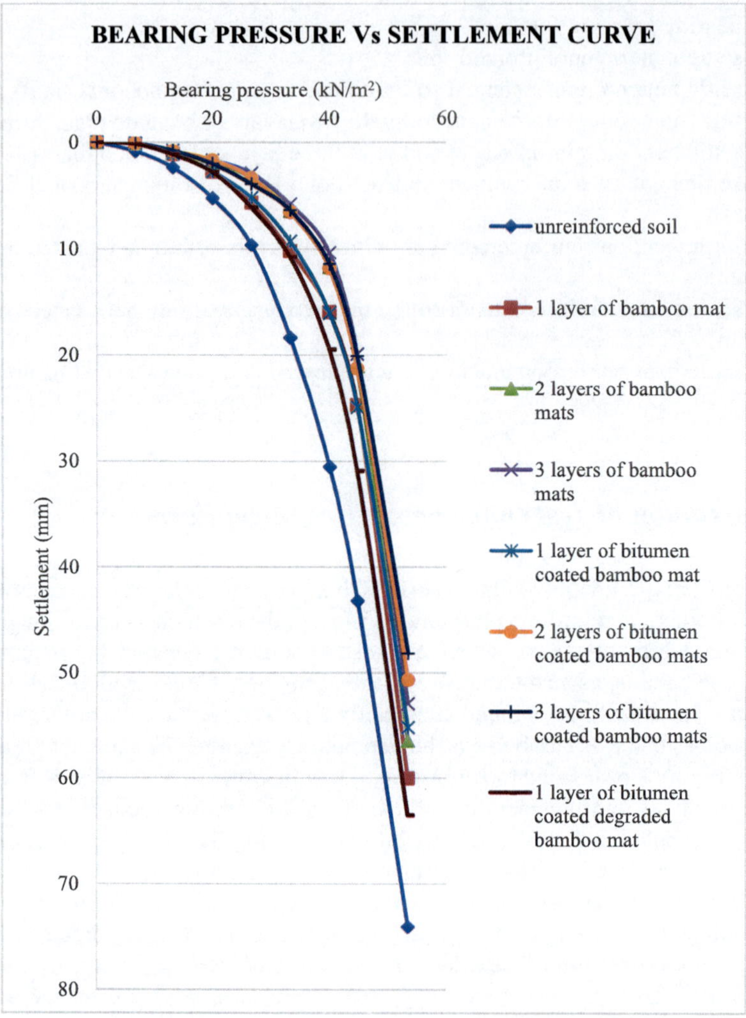

Fig. 1.6 Comparison of all eight tests outcomes

Conclusions

(i) By comparing all the eight results of the test, the conclusion has been made that using bamboo mats as reinforcement in the soil can work where there is no need of heavy reinforcement like embankments, in a slope area, on a road, and so on.

(ii) Stabilizing of soil in an economic way can be possible by placing bamboo mats in between the delicate soil. It is concluded that bamboo can be used to increase the bearing capacity of delicate soil.
(iii) The problem of degradation of the bamboo can be overcome by applying the bitumen coating over the bamboos. The test on weathering bitumen coating bamboo concluded that bitumen coating is effective in case of moisture soil which decreases the rate of degradation of bamboo.
(iv) Results of all eight tests concluded that settlement on reinforced soil is effective than unreinforced soil.
(v) It can also be noticed from the result that using two layers of bamboo mat give a similar settlement result as three layers of the bamboo mat. Therefore, soil stabilization and increased bearing capacity can be achieved by employing just two layers of bamboo mat.
(vi) Similar findings are obtained from bitumen-coated and uncoated bamboo mat settlements as shown in Fig. 1.6. As a result, applying bitumen on bamboo mats does not decrease the strength of the bamboo. In conclusion, the degradation of bamboo and the bearing capacity of soil both can improve simultaneously.

Scope of Future Work

The present thesis concerns the study of the bearing pressure-settling of the foundation without soil strengthening as well as with soil strengthening. Other aspects relating to the analytical approach could not be studied due to time constraints. Future research work should address the following points:

- The present work can be extended to cohesion less soil foundations,
- Work can be compared with numerical modeling,
- Work can be extended to inclined loaded reinforced soil condition,
- This work can be extended by using the different sand densities.

References

1. Sawwaf M, Nazir AK (2011) Behavior of repeatedly loaded rectangular footings resting on reinforced sand. Alexandria Eng J
2. Huat (2011) A study of the compressibility behavior of peat stabilized by DMM: lab model and FE analysis. Acad J
3. Nugroho (2011) Studies on the carrying capacity of shallow foundations with the combination of peat soil geotextile and bamboo grid. J Civil Eng
4. Farsakh M, Chen Q, Sharma R (2013) An experimental evaluation of the behavior of footings on geosynthetic-reinforced sand. Soils Found 53(2): 335–348
5. Kolay PK, Kumar S, Tiwari D (2013) Improvement of bearing capacity of shallow foundation on geogrid reinforced silty clay and sand. Hindawi Publishing Corporation J Constr Eng

6. Waruwu A (2014) Bamboo reinforcement in shallow foundation on the peat soil. J Civil Eng Res 4(3A):96–102
7. Hegde A, Sitharam TG (2015), Experimental and analytical studies on soft clay beds reinforced with bamboo cells and geocells. Int J Geosynthetics Ground Eng 1(2):1–11
8. Cicek E, Guler E, Yetimoglu T (2015) Effect of reinforcement length for different geosynthetic reinforcements on strip footing on sand soil. Soils Found
9. Waruwu AM, Halim H (2017) Settlement estimation of peat reinforced with bamboo grid under embankment. Int Rev Civil Eng
10. Mathew A, Sashikumar A (2017) Performance of Soft Soil Reinforced with Bamboo and Geonet. Int Res J Eng Technol

Chapter 2
Strip Footing on Reinforced Granular Bed Using Geogrid and Bearing Reinforcement

S. Aditya and R. Shivashankar

Introduction

The construction industry is noticing huge growth over time. Construction requires sites consisting of soils with good bearing capacity and other properties. There is a decrease in the availability of proper construction sites owing to which there has been increased use of marginal sites, where the soil has very low bearing capacity. There are a lot of places that are low lying and soil consisting of silt and clay which have low bearing capacity, hence rejected for constructional use. The soil is improved by various techniques like ground replacement, soil stabilization using chemical agents, soil reinforcement, etc. The modern concept of reinforced soil was introduced by Henry Vidal in 1960. The technique of soil reinforcement gained significant popularity due to its cost-effectiveness and various applications like slope stabilization, increasing bearing capacity, building retaining walls, and embankments on soft soil, pavements, etc. Several studies and experiments were carried out for evolving methodologies to determine the improvement in bearing capacity in terms of bearing capacity ratio (BCR). The load bearing capacity can be further be increased by providing granular fill over soft soil which acts as a strong base and distributes the load over a wider area, thus increasing the load to be applied on the soft soil. The addition of geo-synthetics at the interface of granular fill and soft soil can further improve the bearing capacity. Geo-synthetic layers resist destabilizing forces by mobilizing tensile forces in the reinforcement. The concept of soil reinforcement is improved with bearing reinforcement for increasing the bearing capacity and reducing the settlement of soft soils. The concept of bearing reinforcement was introduced by [1] as a cost-effective inextensible reinforcement for mechanically stabilized earth (MSE). Bearing reinforcement (Fig. 2.1) is an inextensible reinforcement type that consists of

S. Aditya (✉) · R. Shivashankar
Department of Civil Engineering, National Institute of Technology, Surathkal, Karnataka 575 025, India
e-mail: aditnitk8@gmail.com

K. Muthukkumaran et al. (eds.), *Transportation and Environmental Geotechnics*, Lecture Notes in Civil Engineering 298, https://doi.org/10.1007/978-981-19-6774-0_2

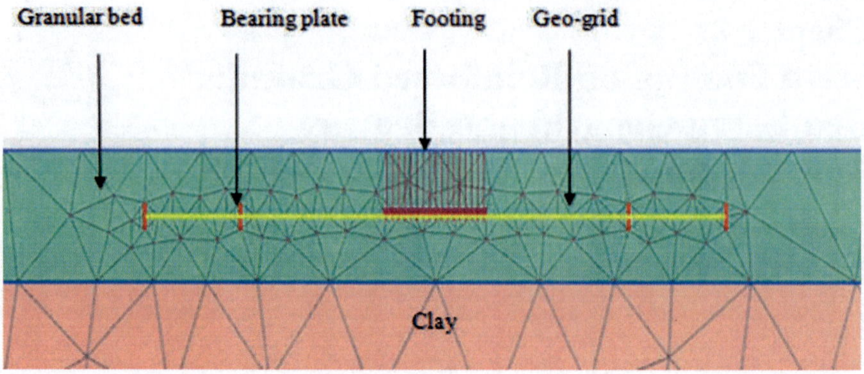

Fig. 2.1 Bearing reinforcement with two bearing placed at the ends of the reinforcement resting on clay overlying granular layer

a horizontally placed longitudinal element with a set of transverse members welded strongly. This study is of strip footings on reinforced granular bed over clays using bearing capacity by numerical analysis using PLAXIS 2D. The values of moduli of elasticity of soil, length of reinforcement, the ratio of the height of granular fill to the breadth of strip footing (H/B) are varied for a strip footing resting on granular bed overlying clays using bearing reinforcement.

Problem Definition

Strip footing resting on top of granular layer underlain by soft clay and stiff clay is analyzed. A granular bed is provided over soil, and reinforcement is provided at the interface or sandwiched between the granular bed. The aim of this study is to analyze and compare the bearing capacities of the soil with and without reinforcement and to study the bearing capacity when provided with bearing reinforcement. The cases studied are strip footing resting on clay, granular bed overlying clay with and without reinforcement at the interface, and reinforcement sandwiched between granular bed, and strip footing with granular bed overlying clay provided with bearing reinforcement sandwiched between the granular bed.

Numerical Modelling

Numerical modeling for the analysis in this study is done using PLAXIS 2D. It is assumed that the footing is resting on the ground surface and dry soil is considered. The groundwater is situated at great depth causing no interference with soil strata. The problem defined is modeled as a plain strain model since the length of the footing is greater than its width and triangular elements with 15 nodes were used. The reinforcement is designed as a geogrid element with suitable properties, and the bearing members are designed as plate elements with suitable properties (Tables 2.1, 2.2 and 2.3).

Table 2.1 Soil properties

Soil parameters	Soft clay	Stiff clay	Sand 1	Sand 2
Material model	Mohr–Columb	Mohr–Columb	Mohr–Columb	Mohr–Columb
Material type	Undrained *B*	Undrained *B*	Drained	Drained
Unsaturated unit weight, γ (kN/m^2)	14	18	18	18
Saturated unit Weight, γ_{sat} (kN/m^2)	17	20	20.8	20.8
Young's modulus E (kN/m^2)	20,000	80,000	80,000	12,000
Poisson's ratio, υ	0.4	0.3	0.3	0.3
Cohesion, C (kN/m^2)	20	80	0	0
Angle of internal friction, ϕ (°)	0	0	38	38
Interface reduction factor	0.8	0.8	0.67	0.67

Table 2.2 Properties of reinforcement

Parameter	Value
Material property	Geogrid element
Material type	Elastic
Normal stiffness, EA (kN/m)	20,000

Table 2.3 Properties of bearing plates

Parameter	Value
Material type	Elastic; isotropic
Prevent punching	Yes
Normal stiffness, EA (kN/m)	1.2×10^7
Flexural rigidity, EI (kNm^2/m)	1.2×10^5
Weight, w (kN/m/m)	8.3
Width of bearing plate (mm)	100
Poisson's ratio, υ	0.15

Results and Discussion

Soft Clay Condition

Effect of change in length of reinforcement The width of the strip footing is kept as 1 m, and the modulus of elasticity of the clay layer and granular base are assigned a value of 20 MPa and 80 MPa, respectively. The length of reinforcement (L_{rf}) is kept as 2*B* for the following cases. Reinforcement is placed at the interface between the granular bed and the clay layer which are the two cases considered. Figure 2.2a is plotted for parameter $H/B = 1$. The case where reinforcement is provided at the interface fails early, but when the reinforcement is provided between the granular bed, there is the development of load intensity and soil also does not fail early. Due to the presence of a clay layer below the reinforcement, the friction between the soil and reinforcement is not developed. Figure 2.2b is plotted when $H/B = 2$, it shows an increase in the load intensity in case where reinforcement is provided between granular bed. The performance of reinforcement sandwiched between granular bed is better for large settlements. The improvement is not significant for permissible settlement.

Effect of location of reinforcement. A strip footing of width 1m is considered, and the modulus of elasticity of the clay layer and granular base are assigned a value of 20 MPa and 80 MPa, respectively. The length of reinforcement (L_{rf}) is kept as 2*B* for the following cases. Reinforcement placed at the interface between the granular bed and the clay layer is the two cases considered. Figure 2.3a is plotted for parameter H/B=1. The case where reinforcement is placed at the interface fails early but when the reinforcement is provided between the granular bed there is the development of load intensity and soil also does not fail early. Due to the presence of a clay layer below the reinforcement, the friction between the soil and reinforcement

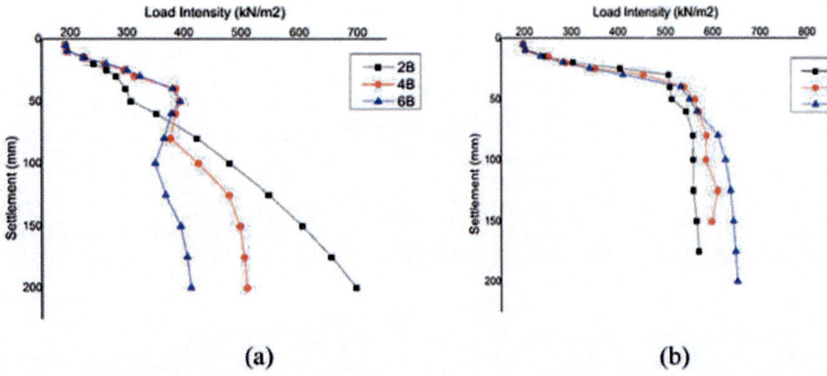

Fig. 2.2 Effect of change in length of reinforcement length of 2B, 4B, and 6B when **a** $H/B = 1$, $E2 = 80$ MPa and **b** $H/B = 2$, $E2 = 80$ MPa for reinforced granular bed

is not developed. Figure 2.3b is plotted when H/B=2; it shows an increase in the load intensity in case where reinforcement is provided between granular bed. The performance of reinforcement sandwiched between granular bed is better for large settlements. The improvement is not significant for permissible settlement.

Effect of change in thickness of granular bed The graphs Fig. 2.4a and b are plotted by varying the H/B ratio from 1 to 2. The increase in H/B from 1 to 2 shows a significant increase in load intensity with respect to settlement when the modulus of elasticity of sand is 80 Mpa (Fig. 2.4a). The same trend is followed when the modulus of elasticity of sand is 120 MPa (Fig. 2.4b). This shows that as the thickness of the granular bed is increased the soil properties are improved and the soil can take more load intensity.

Effect of change in the modular ratio Modular ratio ($E2/E1$) is varied from 4 to 6, and the following graphs (Fig. 2.5a and b) are plotted. There is little increase in load intensity when the modular ratio is varied and H/B=1 (Fig. 2.5a). When H/B=2,

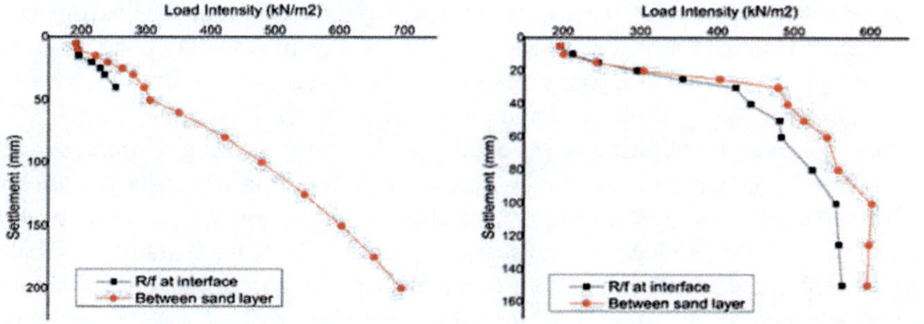

(a) (b)

Fig. 2.3 Effect of location of reinforcement when **a** $H/B = 1$, $L_{rf}/B = 2$, $E2 = 80$ MPa and **b** $H/B = 2$, $L_{rf}/B = 2$, $E2 = 80$ MPa for reinforced granular bed

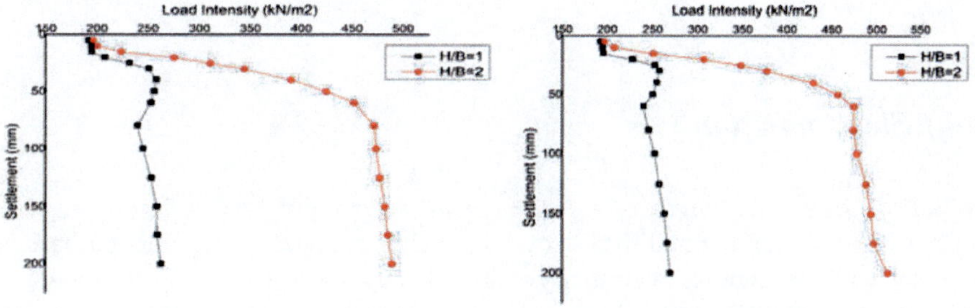

(a) (b)

Fig. 2.4 Effect of change in thickness of granular bed when **a** the modulus of elasticity of sand $E2 = 80$ MPa for granular bed and **b** the modulus of elasticity of sand $E2 = 120$ MPa for granular bed

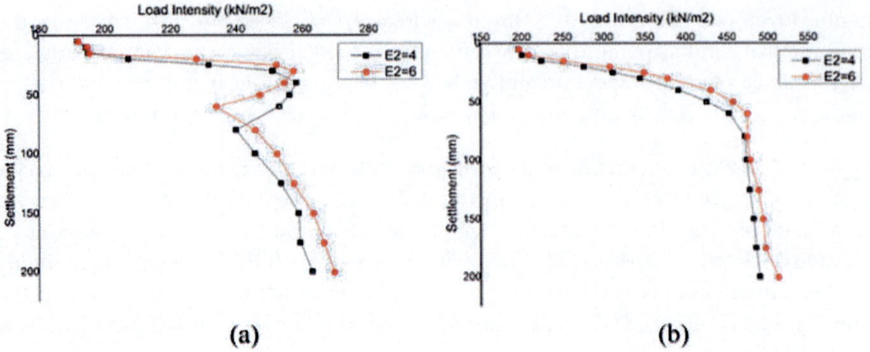

Fig. 2.5 Effect of change in modular ratio when **a** $H/B = 1$ and **b** $H/B = 2$ for granular bed

the same trend is followed were a little increase in load intensity with respect to settlement but the variation is not significant (Fig. 2.5a).

Comparison of variation of bearing capacities Fig. 2.6a and b indicate load intensity vs settlement curves for the case when the footing is resting on clay layer, granular bed overlying clay layer, reinforced granular bed overlying clay with reinforcement sandwiched between granular bed and on bearing reinforced granular bed (BRGB) with one plate and two bearing plates overlying clay where bearing reinforcement is in between the granular bed. For Fig. 2.6a, the length of reinforcement provided is $6B$, $E2 = 80$ MPa, and H/B=1. The graph indicates an increase in load intensity with respect to settlement for each of the cases considered. Reinforced granular bed and bearing reinforced granular bed with two plates spaced at a distance ‘B’ from the end of the reinforcement show similar load intensity with respect to settlement up to permissible values of settlement. In Fig. 2.6b, the length of reinforcement provided is 6B, $E2$= 120 MPa, and H/B=2. The graph indicates up to the permissible limit of settlement RGB and bearing reinforced granular bed (BRGB) with one and two plates welded at the end show an increase in load intensity with respect to the settlement. In this case, the bearing reinforced granular bed shows the most improvement.

Stiff Clay Condition

Effect of change in length of reinforcement Strip footing of width 1m resting on a reinforced granular bed (RGB) overlying clay is modeled; the granular base is assigned with the modulus of elasticity of 80 and 120 MPa. Soft clay with the modulus of elasticity of 80MPa is considered. The length of the reinforcement is varied by $2B$, $4B$, and $6B$. Graphs of load intensity versus settlement are plotted in each of the cases to follow. Figure 2.7a is plotted when H/B=1; there is improvement in the load intensity for settlement in the case where the length of reinforcement provided is

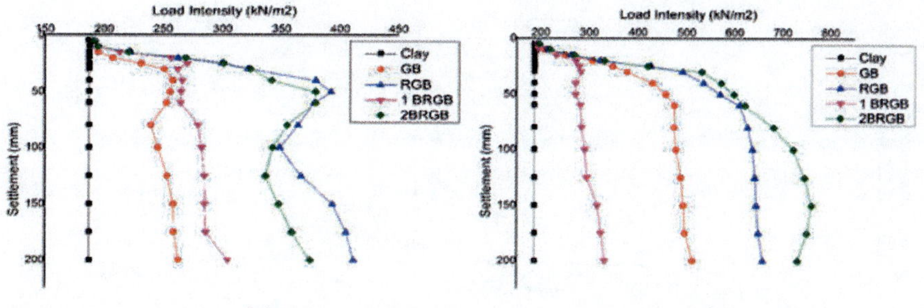

Fig. 2.6 Comparison when **a** the modulus of elasticity of sand $E2 = 80$ MPa, $L_{rf}/B = 6$, and $H/B = 1$ and **b** the modulus of elasticity of sand $E2 = 120$ MPa, $L_{rf}/B = 6$, and $H/B = 2$ for reinforcement sandwiched between the granular bed

4*B* compared to 2*B*. The improvement is not very significant for permissible values of settlement. Figure 2.7b is plotted when the *H*/*B* ratio is equal to 2; there is no significant improvement observed.

Effect of location of reinforcement The width of the strip footing is kept as 1m, and the modulus of elasticity of the clay layer is 80MPa, and the granular base is assigned a value of 80MPa and 120MPa, respectively. The length of reinforcement (L_{rf}) is kept as 2*B* for the following cases. The layer of reinforcement is placed at the interface between the granular bed and the clay layer which are the two cases considered. Figure 2.8a is plotted for parameter *H*/*B*=1. Figure 2.8b is plotted to assign *H*/*B*=2. There is some improvement in the case where the reinforcement is provided between the sand layer compared to when the reinforcement is placed at the interface. When *H*/*B*=2, there is no significant improvement but the soil fails early when reinforcement is provided at the interface. The soil fails at large settlements when the reinforcement is sandwiched between granular bed.

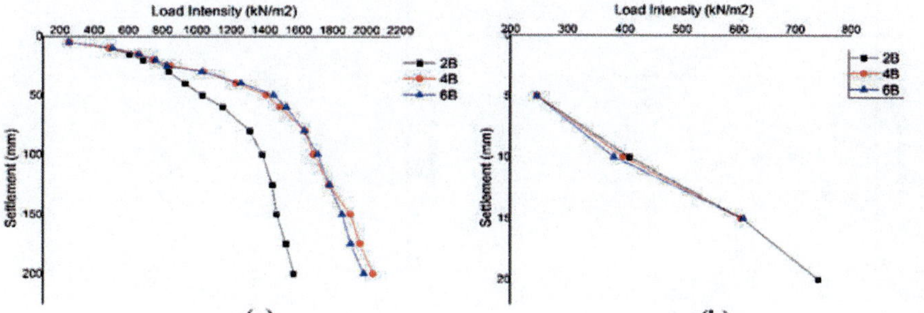

Fig. 2.7 Effect of change in length of reinforcement length of 2*B*, 4*B*, and 6*B* when **a** $H/B = 1$, $E2 = 80$ MPa and **b** $H/B = 2$, $E2 = 80$ MPa for reinforced granular bed

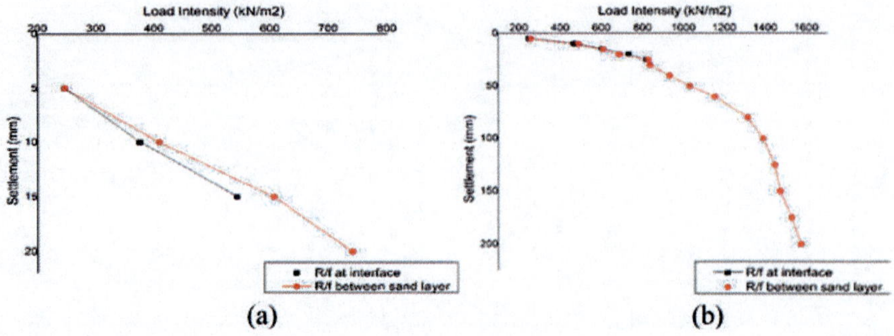

Fig. 2.8 Effect of location of reinforcement when **a** $H/B = 1$, $L_{rf}/B = 2$, $E2 = 80$ MPa and **b** $H/B = 2$, $L_{rf}/B = 2$, $E2 = 80$ MPa for reinforced granular bed

Effect of change in thickness of granular bed The graphs in Fig. 2.9a and b are plotted by varying the H/B ratio from 1 to 2. The increase in H/B from 1 to 2 shows no significant increase in load intensity for settlement when the modulus of elasticity of sand is 80 Mpa for permissible values of settlement (Fig. 2.9a). The same trend is followed when the modulus of elasticity of sand is 120 MPa (Fig. 2.9b). The soil fails for large settlements when H/B = 2 compared to when $H/B = 1$; also, there is an increase in load intensity with respect to settlement for large settlements.

Effect of change in the modular ratio Modular ratio ($E2/E1$) is varied from 1 to 1.5, and the following graphs (Fig. 2.10a and b) are plotted. There is little increase in load intensity when the modular ratio is varied and H/B=1 (Fig. 2.10a). When H/B=2, the same trend is followed were a little increase in load intensity with respect to settlement but the variation is not significant (Fig. 2.10b).

Comparison of variation of bearing capacities Figure 2.11a and b indicate the load intensity versus settlement curves for the case when the footing is resting on

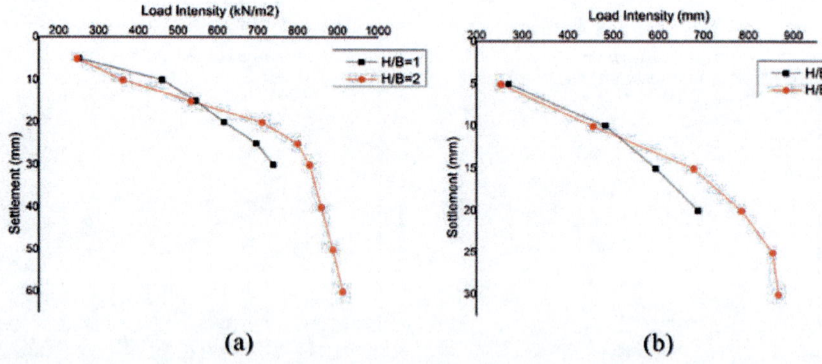

Fig. 2.9 Effect of change in thickness of granular bed when **a** the modulus of elasticity of sand $E2 = 80$ MPa; **b** the modulus of elasticity of sand $E2 = 120$ MPa for granular bed

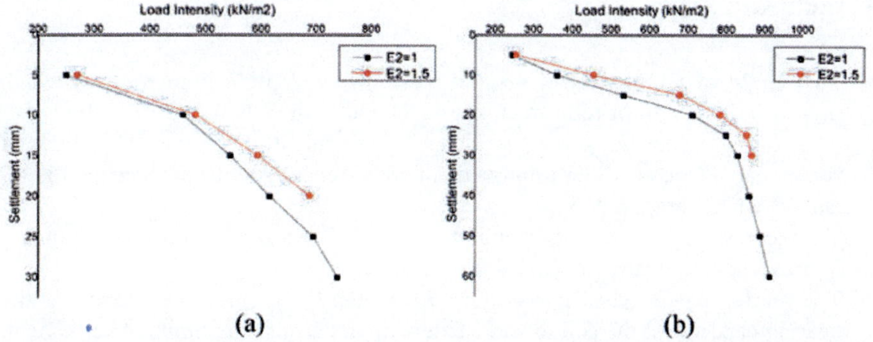

Fig. 2.10 Effect of change in modular ratio when **a** $H/B = 1$ and **b** $H/B = 2$ for granular bed

clay layer, granular bed overlying clay layer, reinforced granular bed overlying clay with reinforcement sandwiched between granular bed and on bearing reinforced granular bed (BRGB) with one plate and two bearing plates overlying clay where bearing reinforcement is in between the granular bed. For Fig. 2.11a, the length of reinforcement provided is 6*B*, *E*2= 80 MPa, and *H*/*B*=1. The graph indicates an increase in load intensity with respect to settlement for each of the cases considered. Reinforced granular bed and bearing reinforced granular bed with two plates spaced at a distance '*B*' from the end of the reinforcement show similar load intensity with respect to the settlement. Modulus of elasticity of sand *E*2=120MPa, L_{rf}/B=6, and *H*/*B*=2 for reinforcement are sandwiched between granular bed for Fig. 2.11b. Reinforced granular bed case and granular bed case show maximum improvement in soil.

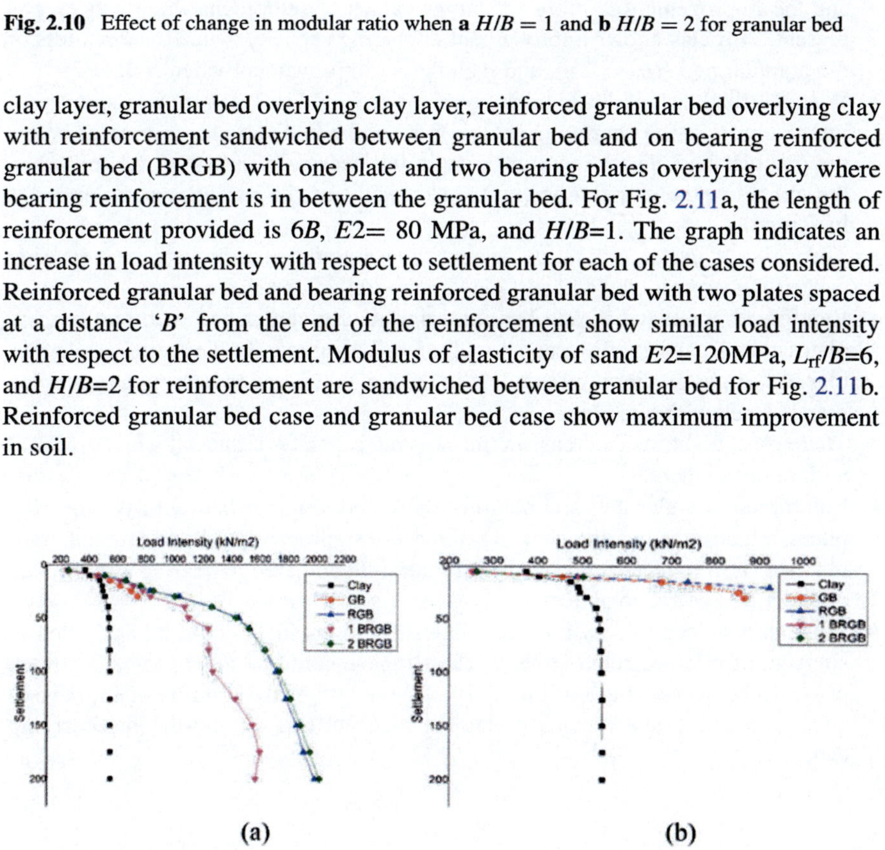

Fig. 2.11 Comparison when **a** the modulus of elasticity of sand $E2 = 80$ MPa, $L_{rf}/B = 6$, and $H/B = 1$; **b** the modulus of elasticity of sand $E2 = 120$ MPa, $L_{rf}/B = 6$, and $H/B = 2$ for reinforcement sandwiched between the granular bed

Conclusion

- The length of reinforcement $4B$ and $6B$ shows very little improvement which indicates the length of reinforcement $4B$ is optimum. When $H/B = 2$, the variation of length does not show significant improvement for permissible values of settlement. There is a little improvement when the length of reinforcement is $4B$ and $6B$ when compared to $2B$.
- The increase in thickness of the granular bed ($H/B = 2$) does not show significant improvement in bearing capacity for stiff clay.
- The reinforcement when it was placed between the granular bed enhances the bearing capacity of the soil to some extent up to permissible limits of settlement, and the improvement is more for larger values of settlement when soft clay is present. Stiff clay shows improvement in bearing capacity, when the thickness of the granular bed $H/B = 1$ m, and it shows no improvement when $H/B = 2$.
- When the thickness of the granular bed is varied from 1 to 2 m, it shows a significant improvement in the bearing capacity of the soil when soft clay is present below the granular bed. There is an increase in the bearing capacity of the soil when the soil below the granular bed is stiff clay as the thickness of the granular bed is increased.
- The variation in modular ratio does not show significant improvement in bearing capacity when soft clay and stiff clay are underlying granular beds.
- Reinforced granular bed and bearing reinforced granular bed with two bearing plates enhance the bearing capacity of the soil significantly. The reinforced granular bed shows greater improvement in the soil compared to bearing a reinforced granular bed with two plates.
- There is a 2 to 3 times increase in the carrying capacity of the soil when provided with reinforcement.
- Reinforced granular bed and bearing reinforced granular bed with two bearing plates enhance the bearing capacity of the soil significantly. The reinforced granular bed shows greater improvement in the soil compared to bearing a reinforced granular bed with two plates.
- There is a sudden decrease in load intensity observed; it could be attributed to slippage of reinforcement at ends. The slippage could be due to the adjustment of reinforcement and the soil particles. The length of the reinforcement beyond edges of the footing is reduced due to slippage, and this reduces the load carrying capacity.

Reference

1. Horpibulsuk S, Niramitkornburee A (2010) Pull-out resistance of bearing reinforcement embedded in sand. Jap Geotech Soc Soils Found 50(2):215–226

Chapter 3
Geosynthetics—Past Findings, Present Trends, and Future Prospects in Road Pavements

Deendayal Rathod and P Devahi

Introduction

In the early decades of the 1970s, the usage of fabrics has been introduced for reinforcing the earth structures in the USA. During the 1980s, woven and non-woven fabrics have been witnessed in the market. The term "geosynthetics" was coined in 1986, which addresses every synthetic material used in earth structures. The typical application of geosynthetic reinforcement emphasizes the reflective cracks identified on asphalt overlays, enhancement of road bases, strength improvisation in the soft subgrade materials. In addition, the installation of a geosynthetic layer in flexible pavement restrains the lateral movements between aggregates and subgrade layers, which improvise the strength and stiffness of road structure.

Pavement failure may occur either by traffic load or by environmental conditions. Traffic loads include repeated load cycles, which may cause structural or functional failure. Environmental conditions refer to environmental factors such as temperature variation, moisture differences in the subgrade, or sequences of freezing and thawing, leading to cracking of the subgrade. This paper elaborates on the efficient application of geosynthetics in the stream of flexible pavements and their recent advances in the field. It focuses on how the failure has been restrained by the incorporation of Geosynthetics in paved and unpaved roads.

Global Scenario

Soft ground improvement technique using geosynthetics is entitled to evolve as a well-known theme in the past four decades. In Germany, a high embankment for a

D. Rathod (✉) · P. Devahi
National Institute of Technology, Tiruchirappalli, India
e-mail: deendayal@nitt.edu

K. Muthukkumaran et al. (eds.), *Transportation and Environmental Geotechnics*,
Lecture Notes in Civil Engineering 298, https://doi.org/10.1007/978-981-19-6774-0_3

depth of 16 m has been developed over-saturated soil with a two-layer system [1]. In Japan, a lattice enclosed along with polyester fabric sheet and woven textile tubes has been utilized in combination with pumped mortar [2]. The lattice incorporation benefits the settlement reduction in comparison with the conventional fabric materials over the soft ground region. Besides, prefabricated vertical drains have proven to be effective in the enhancement of filter porosity and flow rates in the core region.

Stage-constructed full-height rigid has been replaced for a conventional retaining wall in Japan [3]. Their wall length is more than 100 km, and the process involves a cast in situ stage construction with short reinforcements. This particular fashion of construction exhibits a high connection strength, further providing high confining strength, stiffness, and strength of backfill [2] (Fig. 3.1).

Figure 3.2 illustrates the percentage of a research study carried using different types of geosynthetic materials. From the research findings, it can be stated that the introduction of geosynthetic reinforcement in the unpaved and paved construction has always benefitted the environment in terms of reducing the swelling, settlement, restrains the lateral movement, stresses in different levels of pavement, and mainly strengthens the soil by enhancing their CBR values.

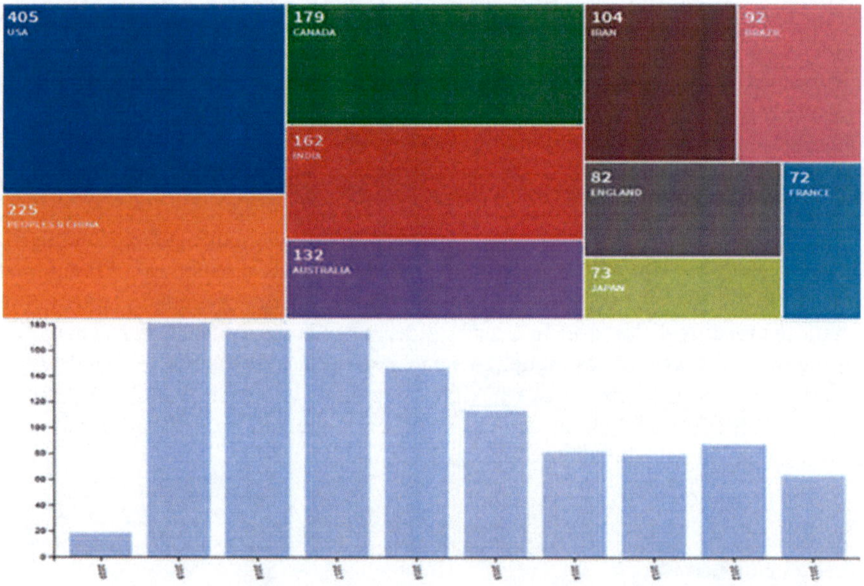

Fig. 3.1 **a** Country-wise publications in geosynthetics domain and **b** year-wise publication using geosynthetics

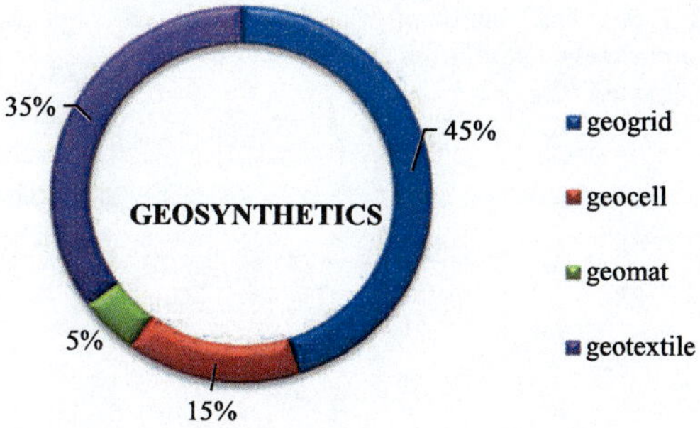

Fig. 3.2 Types of geosynthetic application in the pavement

Experimental Investigations on Different Types of Geosynthetics

Several studies were conducted on the usage of geosynthetics (various forms) on pavements, and it was observed that geotextiles and geogrids were the most commonly used ones. Many tests were done to explore the properties of geosynthetic-stabilized pavements. Table 3.1 illustrates the findings of laboratory results by using geogrid materials in different sizes and fashions.

Experimental investigations are conducted on full-scale laboratory tests and field tests. It includes the CBR test, plate load tests, standard proctor compaction test, cyclic plate load test, swell test, consolidation test, triaxial test, vane shear tests, sand cone tests, etc. Figure 3.3 shows an overview of total tests that were performed on pavement samples.

CBR test is most commonly adopted to analyze the geosynthetic-stabilized pavements (paved and unpaved). By CBR test index of strength has been determined easily. In CBR, test laboratory setup is more reliable than field analyzing. From the

Table 3.1 Summary of literatures on geogrid materials

Size dimension of geogrid used	Inferences made	References
Diameter—152.4 mm	Reduction in surface penetration and enhancement of stress distribution; 20% increment in the CBR value	[4]
A square aperture of 40 mm size	Better lateral movement resistance as compared to geotextile reinforcement	[5]
The aperture size of 1.47 and 0.27 mm thickness	Jute geotextile exhibits excellent reinforcing efficiency comparative to the geogrid material	[6]
Two square aperture of 40 and 30 mm size	Maximum CBR was observed with minimal fines content	[7]
Biaxial and triaxial geogrids with rectangular and triangular aperture shape	Permanent surface deformation was restricted, and better redistribution of the load was observed	[8]

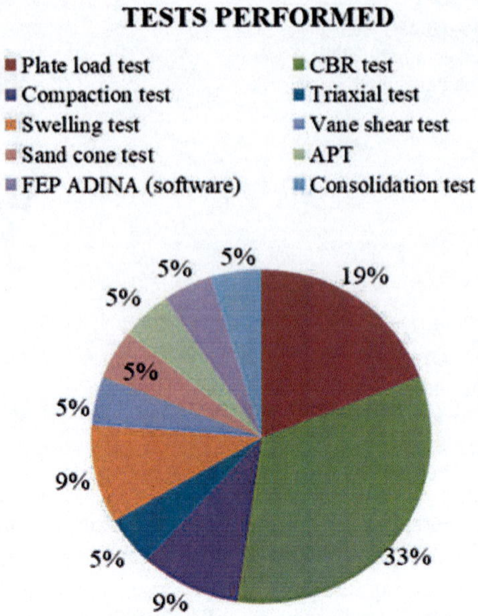

Fig. 3.3 Distribution of experimental studies on geosynthetics

literature, the percentage increase in the strength concerning CBR values while using different geosynthetic material is depicted in Fig. 3.4.

Comparatively, the glass grid incorporated in geomat materials shows an excellent increase in the CBR value [9]. The incremental percentage can be influenced by the type and fashion of installing the reinforcement material. In addition, the influence of encapsulation material and the installation fashion has a tremendous effect on the strengthening behavior of stone columns encased with geosynthetic materials [10].

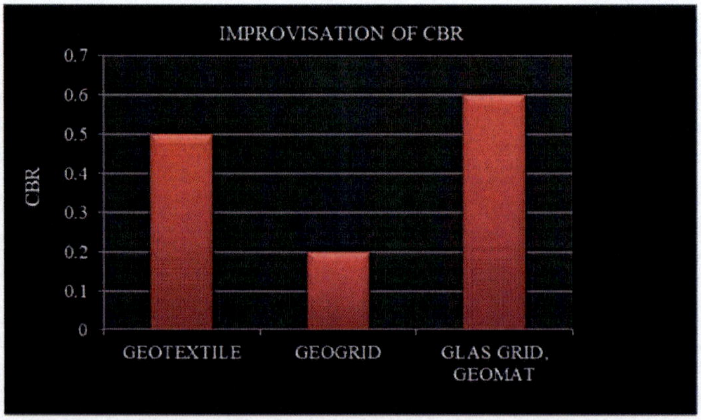

Fig. 3.4 Percentage increase in the CBR value using different geosynthetic materials

Experimental investigations are carried out to study the suitability of geosynthetics, the performance of pavement layers (subgrade, base coarse layer) after and before installation of geosynthetics, optimum depth, or position of geosynthetics. Also, various loading conditions are applied to evaluate pavement traffic interactions. Experimental studies illustrated that the purpose of geosynthetics includes the enhancement of pavement service life with minimizing the differential and total settlement, decrement in the surface and subgrade rutting behavior, prevention of reflective cracks, improvisation in the frost heave resistance, and ultimately the strength improvisation with respect to the bearing capacity.

Conclusions

- Geosynthetics have proven to acts as a cost-effective and environmentally friendly remedy for critical engineering problems. While geosynthetics have several applications in geotechnical and environmental activities, this paper elaborates on the recent advancements in geosynthetic products and their application on the pavement surface.
- A critical thing to be noted was even though geogrids were the most widely accepted ones, the usage of geomats and glass grids in pavements had obtained an increased value of CBR of approximately 60% as compared to a CBR value of approximately 20% for geogrids.
- Geogrids and geotextiles helped to reduce the deformation to a great extent, but surprisingly geomats reduced the deformation to as low as 70%.
- Both the CBR values and deformation reduction indicate that the geosynthetics other than geogrids and geotextiles need to be explored and contains a wide scope in the future to solve the technical difficulties and maximize the output.
- Geocells has proven to be improvising the slope stability factors by increasing the factor of safety to about 10 to 46%, depending upon the cohesion and frictional angle of soil.

References

1. Christopher BR (2014) Cost savings by using geosynthetics in the construction of civil works projects. In: 10th International conference on geosynthsis ICG 2014
2. Palmeira E, Tatsuoka F (2008) Advances in geosynthetics materials and applications for soil reinforcement and environmental protection works. Electronics Jounerals Geotech
3. Tschegg EK, Ehart RJA, Ingruber MM (1998) Fracture behavior of geosynthetic interlayers in road pavements. J Transp Eng 124:457–464. https://doi.org/10.1061/(ASCE)0733-947X(1998)124:5(457)
4. Duncan-Williams E, Attoh-Okine NO (2008) Effect of geogrid in granular base strength—An experimental investigation. Constr Build Mater 22:2180–2184. https://doi.org/10.1016/j.conbuildmat.2007.08.008

5. Palmeira EM, Antunes LGS (2010) Large scale tests on geosynthetic reinforced unpaved roads subjected to surface maintenance. Geotext Geomembr 28:547–558. https://doi.org/10.1016/j.geotexmem.2010.03.002
6. Yu Z, Woodward PK, Laghrouche O, Connolly DP (2019) True triaxial testing of geogrid for high speed railways. Transp Geotech 20. https://doi.org/10.1016/j.trgeo.2019.100247
7. Rajesh U, Sajja S, Chakravarthi VK (2016) Studies on engineering performance of geogrid reinforced soft subgrade. Transp Res Procedia 17:164–173. https://doi.org/10.1016/j.trpro.2016.11.072
8. Abu-Farsakh MY, Akond I, Chen Q (2016) Evaluating the performance of geosynthetic-reinforced unpaved roads using plate load tests. Int J Pavement Eng 17:901–912. https://doi.org/10.1080/10298436.2015.1031131
9. Singh M, Trivedi A, Shukla SK (2019) Strength enhancement of the subgrade soil of unpaved road with geosynthetic reinforcement layers. Transp Geotech 19:54–60. https://doi.org/10.1016/j.trgeo.2019.01.007
10. Rathod D, Abid MS, Vanapalli SK (2021) Performance of polypropylene textile encased stone columns. Geotext Geomembr 49:222–242. https://doi.org/10.1016/j.geotexmem.2020.10.025

Chapter 4
Modeling and Analysis of Back-To-Back Walls with Combined Wrap-Around and Full-Height Rigid Facing for High-Speed Rail Applications

K. Vamsi Krishna, S. Sasanka Mouli, and B. Umashankar

Introduction

In order to improve the soil properties, introduction of artificial elements like steel reinforcements and rubber tire chips into the soil is called mechanically stabilized earth (MSE). When the mechanically stabilized reinforced retaining walls are closed to one another, then they are called as back-to-back MSE walls. These walls are frequently used in areas where the distance between the walls is not far apart. Depending upon the requirement of the project, MSE walls can be built with either flexible facia (e.g., wrap around) or rigid facia (e.g., segmental panels, FHR facings, and modular blocks). The types of facia include (1) segmental panel facing, (2) wrapped around wall facing, (3) full-height panel facing, and (4) wrapped around wall with FHR facing. Wrap-around wall, segmental wall, and full-length panel facing systems are some of the conventional types of facing systems.

In terms of cost, materials, labor, and time, numerically modeling geosynthetic-reinforced soil (GRS) retaining walls has significant advantages over experimental modeling. Many numerical studies were available in the literature dealing with the analysis of the MSE walls. Hatami and Bathurst 2005 [1] analyzed three full-scale instrumented MSE walls with different reinforcement stiffnesses to study their performance under working stresses. Also, a robust numerical model of MSE wall was developed using FLAC software. Interfaces and upgradation of elastic modulus of backfill were incorporated at each stage of the modeling. Hatami and Bathurst et al. 2006 [2] studied the effect of surcharge loads on segmental panel facing. Results of above numerical study were compared with obtained reinforcement strain values, wall displacements, and boundary loads.

K. Vamsi Krishna (✉) · B. Umashankar
IIT Hyderabad, Kandi, Sangareddy 502284, India
e-mail: kadagana.vamsi@gmail.com

S. S. Mouli
VNR VJIET, Hyderabad 500090, India

K. Muthukkumaran et al. (eds.), *Transportation and Environmental Geotechnics*,
Lecture Notes in Civil Engineering 298, https://doi.org/10.1007/978-981-19-6774-0_4

Type of facing elements in the mechanically stabilized earth walls shows significant bearing on the behavior of the MSE walls. Tatsuoka et al. 1992 [3] carried out a comparative study on MSE walls with various facia. Full-scale MSE walls with various reinforcement lengths and various types of wall facia were developed and tested. The study concluded that full-height rigid (FHR) facing mobilizes higher lateral pressures than those of the other facia walls. Tatsuoka et al. 2014 [4] discussed briefly about geosynthetic-reinforced soil structure application in Japanese railways. They concluded that the coastal dikes that were developed by geosynthetic-reinforced soil with the FHR facing can perform superior to conventional facia type. The GRS structures perform efficiently under extreme conditions such as high earthquake loads and deep over turning due to tsunami current, and the same was demonstrated in numerous case histories [4].

Han and Leshchinsky 2010 [5] performed a parametric study in BBMSE walls using FLAC software. Parameters such as ratio of distance between the walls to wall height ratio (*W/H*) and backfill quality were varied to study the interaction between the walls at ultimate limit-state condition. The study concluded that the critical slip surface location may be influenced due to the interaction between the walls. Sravanam et al. 2019 [6] developed a robust model by using FLAC to analyze the behavior of back-to-back walls by giving surcharge loads and compaction stresses. Design charts for reinforcement tensile forces and lateral pressures were proposed. Sridhar 2020 [7] analyzed the behavior of closely spaced back-to-back MSE walls at ultimate limit-state by using FLAC 2D. Modified lateral earth pressure coefficient (K_a modified) calculated in the study was more than modified lateral pressure coefficient (K_a modified) given by FHWA guidelines for BBMSE walls. Sravanam et al. 2020[8] studied back-to-back and single MSE walls with full-height panel facia using FLAC software. They found that lateral earth pressures developed in the back-to-back MSE walls were greater than single MSE walls.

A detailed study of the literature on the topic indicates that research on influence of surcharge loading on behavior of back-to-back MSE walls with wrap-around wall facing is very limited and of wrap-around wall with FHR facing are not available. The main objective of this research is to study the behavior of back-to-back wrap-around walls with full-height rigid facing under working stresses for different width to height (*W/H*) ratios and reinforcement stiffnesses (*J*). The variations in lateral deformations and lateral earth pressures are studied and compared with conventional facing type system, i.e., segmental panel facing under surcharge loading.

GRS Retaining Wall Application in MAHSR Project

Wrap-around wall with full-height rigid facing (FHR) is a new construction technique. It has been the standard code of practice in Japan for rebuilding conventional type retaining walls and embankments that have been damaged due to heavy storms, tsunamis, floods, and earthquakes by using this type of GRS retaining walls [4].

These GRS structures have always been widely used in the constructing of high-speed rail lines, which are one of the Japan's important and critical infrastructures. Mumbai to Ahmedabad high-speed rail (MAHSR) line project is the India's first bullet train project. Most of the alignment of this project is proposed on an elevated viaduct, and rest of the project is on bridges and underground tunnels. There are eight tunnels of around 27 km long with the longest tunnel being about 21 km undersea at Thane creek. There are around 23 bridges with total length of 6.4 km. Rest of the alignment of 475 km is built on the viaduct. In this particular project, there are many places where they adopted back-to-back GRS retaining walls. Particularly, they have adopted GRS retaining wall with full-height rigid (FHR) facing which exhibited superior performance to conventional retaining wall system, based on experimentation.

Validation of the Model

Segmental Panel Facing Model Validation

BBMSE wall of the segmental panel facing model was validated with previous study by Sridhar 2020 [7]. The model was simulated using FLAC. The reinforcement stiffness values were varied from 500, 2,000, and 50,000 kN/m [9]. The ultimate limit-state condition in the model was achieved by giving rotation at the toe of the wall. A similar model was developed in the FLAC by using the same model parameters used in the past study. Table 6 gives the soil properties adopted in the present study.

In Fig. 4.1a, lateral earth pressures were compared at the end of reinforcement zone. Results of this present study agreed well with Sridhar 2020 [7], and deviation is within about 1% or less.

Back-To-Back MSE Wall Behavior Considering Two Facing Types

Geometry of the Models

A robust numerical model was developed using FLAC 2D, 8.00 (Itasca, 2015) [10] to investigate the behavior of back-to-back walls under influence of surcharge loads by considering different reinforcement stiffness values and different width to wall height ratio (*W/H*).

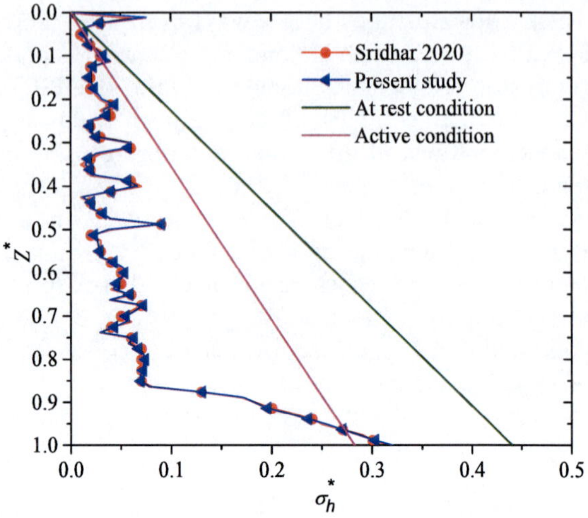

Fig. 4.1 Comparison of lateral earth pressures at end of the reinforced zone from the present study with Sridhar 2020 [7] corresponding to *W/H* = 1.4 and *J* = 500 kN/m

Height of the wall was kept constant as 8 m, and width of the wall was varied as 11.4, 13.6, and 16 m targeting the *W/H* ratios as 1.4, 1.7, and 2.0 (Tables 4.1 and 4.2).

Reinforcement Modeling

In the present study, all the MSE walls were analyzed for the connected condition. In connected walls, the reinforcement extends from one facing of the wall to the facing of the other wall. Hence, the length of the reinforcement is equal to width of the wall. Reinforcement was modeled using structural element—cable element. Different reinforcement stiffnesses were adopted which include extensible, intermediate, and inextensible reinforcements; and they are 2000, 20,000, and 50,000 kN/m [9] which were adopted in this present investigation. In FLAC, interface between the backfill and reinforcement can be introduced in the properties of the reinforcement only (refer to Table 4.3).

Backfill Properties

Backfill properties were adopted from Sravanam et al. 2019 [6] (Table 4.2). As the confining stress dependent up on the deformation modulus of backfill, a nonlinear stress-dependent stiffness model was more appropriate (as the backfill is assumed to be non-cohesive soil). Hence, back fill properties tangent elastic modulus (E_t) and bulk modulus (B) were updated at each stage of construction of walls using the

Table 4.1 Specifications for the walls considered in the study

Parameter	Value
Height of wall, H	8 m
Vertical spacing between the reinforcement, S_v	0.725 m
Distance between the walls, W	11.4, 13.6, 16 m
FHR and segmental panel thickness	0.3 m
W/H ratios	1.4, 1.7, 2.0

equations proposed by Duncan et al. 1980 [11] (Eqs. 4.1 and 4.2).

$$E_t = \left[1 - \frac{R_f(1 - sin\emptyset)(\sigma_1 - \sigma_3)}{2ccos\emptyset + 2\sigma_3 sin\emptyset}\right]^2 K_e P_a \left(\frac{\sigma_3}{p_a}\right)^n \quad (4.1)$$

$$B = K_b P_a(\sigma_3/p_a)^m \quad (4.2)$$

where K_e = elastic modulus number; P_a = atmospheric pressure; $\emptyset$ = backfill angle of internal friction; m = bulk modulus exponent; K_b = bulk modulus number;

Table 4.2 Backfill properties used for modeling in FLAC

Property	Reinforced and retained backfill	Foundation soil	Concrete panels and FHR
Constitutive Model	Mohr–Coulomb	Mohr–Coulomb	Elastic
Angle of shearing resistance, $\emptyset$	34°	50°	
Density, γ kg/m^3	2002	2002	2446
Initial deformation Modulus, E MPa	50	200	3.19e4
Poisson's ratio, μ	0.25	0.3	0.15

Table 4.3 Interface properties used for modeling in FLAC

Property	Backfill and Wall	Backfill and Foundation	Panel and Panel
Shear Stiffness, K_s, P_a/m	1.87e7	1e7	1e7
Normal Stiffness, K_n, P_a/m	3.73e7	1.095e8	1.095e8
Cohesion, c, kP_a	–	–	20°
Friction angle, δ	28°	28°	55°

c = cohesion of the backfill; σ_3 = minor principal stress; and σ_1 = major principal stress; n = elastic modulus exponent; R_f = failure ratio.

Interface Properties

Table 4.3 gives the interface properties used for modeling in FLAC. These values were adopted from Sravanam et al. 2019 [6]. Interfaces were incorporated between foundation soil and backfill, facing panels and backfill, FHR and reinforcement and between facing panels to incorporate the interaction between two different materials.

Modeling of Back-To-Back Wrap-Around Walls with FHR Facing

In the modeling of these walls, initially wrap-around walls were modeled and then the FHR facing was added. After preparing the foundation, geosynthetic layer placed on the foundation bed and soil was backfilled and compacted to 0.725 m lift thickness. At face of the wall, reinforcement was wrapped to an overlap length of 1.5 m. Similar procedure was adopted till the required height was reached. After the major residual deformations had taken place in backfill and foundation soil because of geosynthetic-reinforced soil construction, FHR facing was constructed. This facing was rigidly attached to the adjacent reinforcement layers during its service. The connection of full-height rigid facing with wrap around in a staged manner after the deformations in backfill have occurred ensures that the structure does not get damaged due to any differential settlement between them. Furthermore, before construction of full-height rigid facing, the backfill adjacent to wall face can be compacted well. In field, on top of the backfill surface, pavement and vehicular loadings are applied. In order to analyze the behavior of walls under actual field conditions, a surcharge of 30 kPa was applied on the backfill top surface. In Fig. 4.2, numerical model of wrapped around wall with FHR facing developed in the FLAC was presented.

Results and Discussions

The results in the present study were normalized with respect to total height of the wall to generalize the results for various heights of the walls. Normalized equations are presented below.

$$z^* = \frac{Z}{H} \tag{4.3}$$

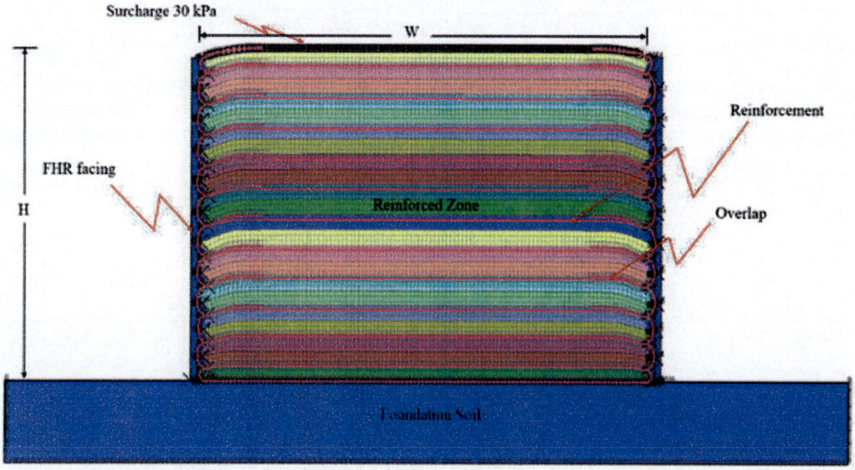

Fig. 4.2 Numerical model of wrap-around wall with FHR facing

$$\sigma_h^* = \frac{\sigma}{\gamma H} \tag{4.4}$$

where $\sigma_h{}^*$ is the normalized lateral pressure, H is the total height of the wall, σ_h is the lateral pressure at face of the wall, γ is the unit weight of backfill soil, Z^* is the normalized depth of wall, and Z is depth of each soil zone.

In Fig. 4.3a, lateral earth pressure distribution at face of the wall for various *W/H* ratios and constant reinforcement stiffness $J = 2{,}000$ kN/m was presented. The lateral earth pressure distribution for different *W/H* ratios is almost similar. The width to height ratio was found to play an insignificant role on the lateral earth pressures. In Fig. 4.3b, lateral earth pressure distribution for varying reinforcement stiffness and constant $W/H = 1.4$ is presented. In this case, lateral earth pressure distributions for stiff reinforcements (i.e., $J = 20{,}000$ and 50,000 kN/m) were similar and the values were less than reinforcement corresponding to $J = 2{,}000$ kN/m.

Behavior of Back-To-Back Walls with Segmental Panel Facing

A wall of 8 m height and different wall widths were considered. Segmental panels of height 0.725 m and thickness 0.3 m were adopted. A staged construction technique was considered in the simulation to reflect the realistic site conditions. The model was simulated in lifts of thickness 0.725 m till the required height was reached. During the construction in site, facing panels were supported by form work, and to reflect the same in the model, facing panels were restrained to move horizontally until the completion of constructing the entire wall. Foundation soil was considered

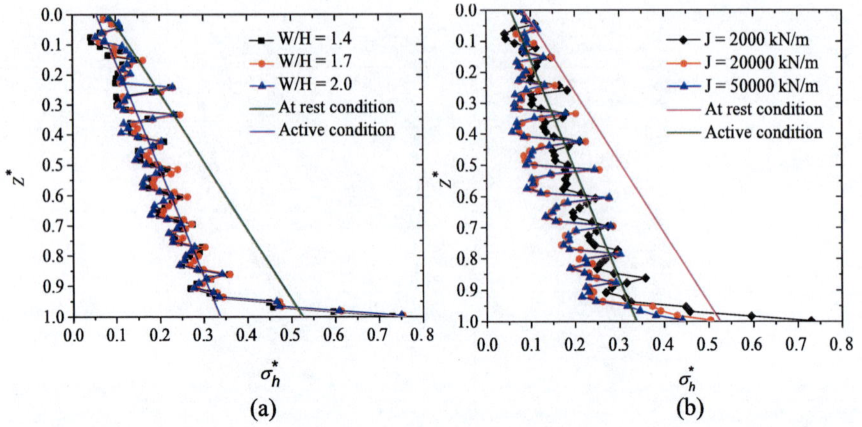

Fig. 4.3 Normalized lateral earth pressure versus normalized depth at face of the wall corresponding to **a** constant $J = 2000$ kN/m and different *W/H* ratios = 1.4, 1.7, 2.0 and **b** different $J =$ 2,000, $J = 20{,}000$ and $J = 50{,}000$ kN/m at a constant $W/H = 1.4$

as rigid. All the interface properties were taken same as those of wrap-around wall with FHR. After the entire was modeled, a surcharge 30 kPa was applied on the backfill top surface. Figure 4.4 shows numerical model of BBMSE walls developed in the FLAC.

In Fig. 4.5a, lateral earth pressure distribution for different *W/H* ratios at wall facing was presented. From the figure, it can be observed that except at the bottom of the wall, the lateral earth pressures at the wall facing were slightly lower than those at active condition. At the bottom of the wall, the lateral pressures were suddenly

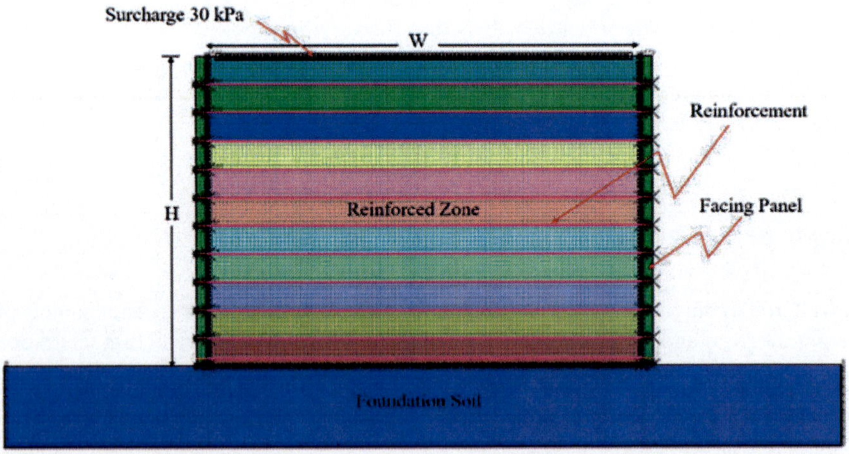

Fig. 4.4 Numerical model of BBMSE wall with surcharge

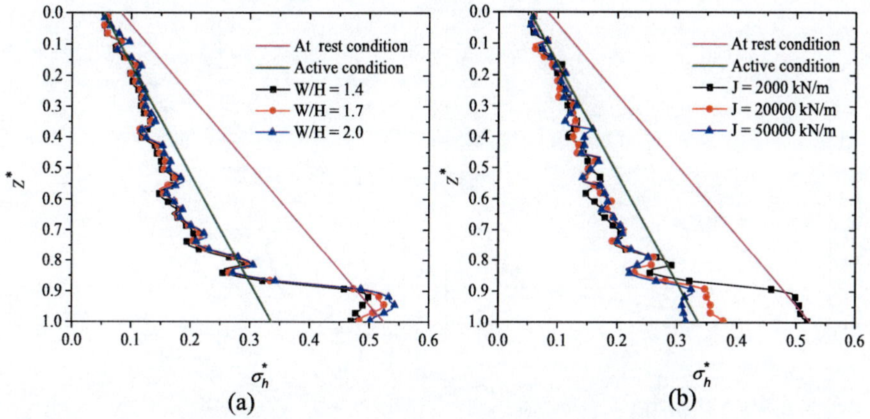

Fig. 4.5 Normalized lateral earth pressures versus normalized depth corresponding to **a** different *W/H* = 1.4, 1.7, 2.0 and a constant *J* = 2000 kN/m, and **b** different *J* = 2000, *J* = 20,000 and *J* = 50,000 kN/m and a constant *W/H* = 1.4

increasing because of the lateral restraint at the toe. It can also be concluded that the lateral pressures at the face of the wall were independent of the width to height ratio. In Fig. 4.5b, lateral earth pressures for varying reinforcement stiffness and for a given *W/H* ratio were presented. In this case, lateral earth pressures for different reinforcement stiffness were almost similar. Hence, it can be concluded change in stiffness of the reinforcement will not affect lateral earth pressures. However, in both the cases, i.e., Figs. 4.5a and b, lateral earth distribution was less than those corresponding to both active and at-rest earth pressures.

Comparison of Displacements Between Segmental Panel and Wrap-Around Wall with FHR Facing

In Fig. 4.6a and b, normalized lateral deformations versus normalized height of the wall for different width to height ratios and constant reinforcement stiffness *J* = 2000 kN/m were presented. From these plots, it can be concluded that in wrap-around wall with FHR facing, lateral displacement was much less than those of segmental panel facing. It can be concluded that lateral displacements in the FHR-facing wall were much lower (about 100 folds) than those of the segmental facing walls. Hence, these walls can be applied at the sites where lateral displacements and surface settlements are to be controlled.

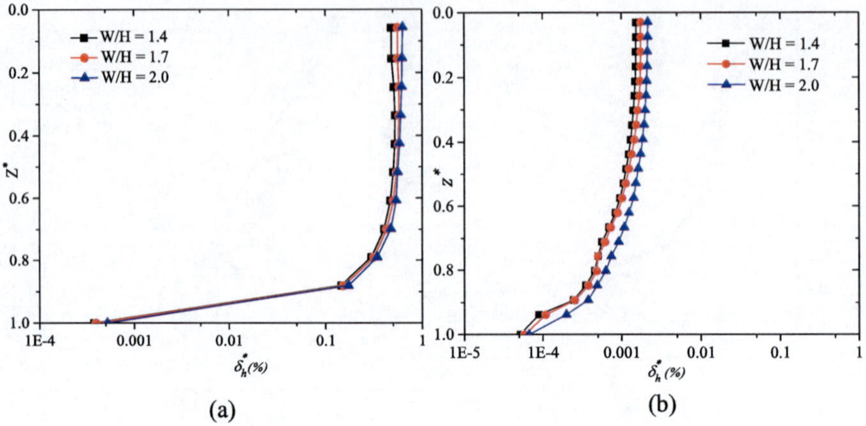

Fig. 4.6 Normalized lateral deformations *vs.* normalized height of the wall corresponding to **a** segmental panel facing and **b** wrap-around wall with FHR facing

Conclusions

In present study, behavior of wrapped around walls with FHR facing under surcharge load of 30 kPa was studied and compared with the segmental panel facing. Both the models were developed numerically by using FLAC 2D. The below conclusions were drawn from the study.

(1) Reinforcement stiffness has significant effect on lateral earth pressure distribution in wrapped around wall with FHR facing when it is subjected to surcharge loading. However, for segmental panel facing it is negligible.
(2) Reinforcement stiffness plays a major role in mobilizing the lateral displacements in wrap with FHR-facing walls. Lateral displacements are high for reinforcement with less stiffness. The same behavior was observed for the segmental panel as well.
(3) The width to height ratio has a significant effect on lateral displacements both in wrap with FHR and segmental panel facing. It was observed that wall with higher width to height ratio results in higher lateral displacement.
(4) The lateral displacements in the case of wrap around with FHR-facing walls were negligible compared to those of segmental panel facing walls. Hence, wrap-around wall with FHR-facing walls is very useful in the places where high precision structures need to be built. However, it is economical to apply segmental panel facing where flexibility in the lateral displacements is available.

References

1. Hatami K, Bathurst RJ (2005) Development and verification of a numerical model for the analysis of geosynthethic-reinforced soil segmental walls under working stress conditions. Can Geotech J 42(4):1066–1085. https://doi.org/10.1139/t05-040
2. Hatami K, Bathurst RJ (2006) Numerical model for reinforced soil segmental walls under surcharge loading. J Geotech Geoenviron Eng 132(6):673–684. https://doi.org/10.1061/(asce)1090-0241(2006)132:6(673)
3. Tatsuoka F, Murata O, Tateyama M (1992) Permanent geosynthetic-reinforced soil retaining walls used for railway embankments in Japan. In: Proceedings of the international symposium on geosynthetic-reinforced soil retaining walls, pp 101–130
4. Tatsuoka F, Tateyama M, Koseki J, Yonezawa T (2014) Geosynthetic-Reinforced Soil Structures for Railways in Japan. Transp Infrastructure Geotechnol 1(1)
5. Han J, Leshchinsky D (2010) Analysis of back-to-back mechanically stabilized earth walls. Geotext Geomembr 28(3):262–267. https://doi.org/10.1016/j.geotexmem.2009.09.012
6. Sravanam SM, Umashankar B, Madhav MR (2019) Behavior and design of back-to-back walls considering compaction and surcharge loads. Int J Geosynth Gr Eng 5(4):1–17. https://doi.org/10.1007/s40891-019-0180-z
7. Sridhar R (2020) Behavior of closely spaced back-to-back MSE walls at ultimate limit state. M. Tech dissertation, Indian Institute of Technology Hyderabad
8. Sravanam SM, Umashankar B, Madhav MR (2020) Analysis of single and back-to-back reinforced retaining walls with full-length panel Facia. Geotech Geol Eng 38(6):6281–6293. https://doi.org/10.1007/s10706-020-01435-x
9. Berg R, Christopher B, Samtani N (2009) Design and construction of mechanically stabilized earth walls and reinforced soil slopes–Volume I. Fed High W Adm I(November):2009. FHWA-NHI-10-024;FHWA-NHI-10-025
10. Itasca Consulting Group (2012) Fish reference, FLAC Man, pp. 1–84
11. Wong K, Mabry P (1980) Strength, stress-strain and bulk modulus parameters for finite element analyses of stresses and movements in soil masses. Geotech Eng

Chapter 5
Influence of Conductive Jute Geotextile-Encased Stone Column in Soft Clay

B. K. Pandey, S. Rajesh, and S. Chandra

Introduction

The improvement of soft clay is an essential construction prerequisite for any infrastructure development over it. Among various soil modification methods, stone column installation is extensively used in the engineering practice. Although the stone column has countless advantages [1], if it is installed in soft soils, they are anticipated to fail by extreme bulging [2, 3]. Encasing the stone columns with geotextile considerably decreases the extreme bulging [2–7]. Numerous studies were conducted to evaluate the behavior of soft soil improved with inclusion of encased stone columns under varied loading conditions [8–10]. Additionally, majority of the studies on the encased stone column (ESC) used geotextile, steel mesh, geogrid, and other materials [5, 11–13] as encasement. The use of natural jute geotextile as encasement of stone column has not been attempted till now. The natural jute geotextiles are commonly used for various geotechnical application as it has beneficial potentials like biodegradable, ecofriendly, economical, and having good strength, thermal properties, and toughness [7]. The short-term solutions to most geoengineering problems such as separation, drainage, filtration, and initial reinforcement make it reasonable to use natural jute geotextiles, despite their short lifespan.

As mentioned earlier, the encasement of the stone column helps in reducing the bulging, thereby helping in preventing the excessive settlement and mixing of column materials to neighboring soil. Further, due to stress concentration, the encased stone column carries a greater portion of load; subsequently, a smaller portion of the surcharge is transmitted to adjoining soil. The comparatively lower part of surcharge is transmitted to the adjacent soil, which reduces the excess pore water pressure

B. K. Pandey (✉) · S. Rajesh · S. Chandra
Indian Institute of Technology Kanpur, Kanpur, Uttar Pradesh 208016, India
e-mail: balbirpandey123@gmail.com

B. K. Pandey
Shri Mata Vaishno Devi University, Kakryal, Katra, Jammu and Kashmir 182320, India

K. Muthukkumaran et al. (eds.), *Transportation and Environmental Geotechnics*,
Lecture Notes in Civil Engineering 298, https://doi.org/10.1007/978-981-19-6774-0_5

generated and accelerates the consolidation process compared to stone columns [4, 14]. However, consolidation performance is greatly reliant on the soft soil permeability. Therefore, less permeable soil leads to uniform consolidation stage [8]. Thus, to accelerate consolidation stage, the electrokinetics may be applied together with the ESC.

The electrokinetics is introduced by using a small voltage gradient through electrodes located in the soft soil [7, 15]. The voltage gradient causes the flow of electric current across the bulk pore water of the soft soil causing enhanced flow of pore water [16]. Esrig [17] studied the influence of electrokinetics on the consolidation of soil and reported a rapid consolidation settlement after using voltage gradient. Rittirong et al. [18] investigated electrical vertical drain (EVD) behavior using a bench-scale laboratory and a field test and observed noticeable increment in the undrained shear strength within a short period of 5 days in the field. In the electrokinetic treatment, the majority of the cost (~75%) incurred will be fabrication and installation cost of electrode and cost of electrical supply [15]. Wu and Hu [19] conducted laboratory experiments on sodium bentonite using graphite, copper, and iron electrodes to investigate the alteration in microfabric, chemical composition, and geotechnical properties of the bentonite. Xue et al. [20] performed laboratory experiments to discover the impact of temperature and voltage on electro-osmotic properties of marine clay and observed that temperature increases with increase in the voltage increases, which further improves the overall performance. Wang et al. [21] analytically developed solution for consolidation of vertical drain combined with electroosmosis and surcharge preloading.

As obvious from the preceding argument, majority of the reported literature addressed the performance of stone column using conventional encasement materials. In addition, the PVDs or vacuum preloading are often coupled with voltage gradient. However, the encased stone column coupled with electrokinetics in enhancing the performance of soft clay has not been deliberated by the researchers. Hence, in this paper the influence of conductive geotextile as a encasement material for stone column in improving the physiognomies of soft soil is investigated. The encasement used in present study also acts as cathode. The anode rods used in the experiment were made of mild steel with circular cross-section. In order to start the electrokinetic process, voltage gradient was used among the electrodes. Number of tests were performed unreinforced, reinforced soft clay and ESC coupled with electrokinetics (*e*-ESC) in a customized test setup. In the course of the experiments, the settlement and cumulative discharge were monitored uninterruptedly. Further, at completion of the test, alteration in water content and pH of the soft soil were examined in detail.

Material and Methodology

Figure 5.1 illustrates the actual picture of kaolin clay (refer to Fig. 5.1a), stone aggregate (refer to Fig. 5.1b), and standard sand (refer to Fig. 5.1c) employed to prepare the sample during the experiments. The properties of the powder kaolin

Fig. 5.1 Actual picture of **a** kaolin, **b** stone aggregate, **c** standard sand used for the sample preparation

which was used for the preparation of the soft clay contain 70.25% silt and 29.75% clay fractions. The kaolin clay was having liquid and plastic limit of 50 and 26%, respectively. The coefficient of permeability of the kaolin clay was 5.09×10^{-9} m/s. Further, the kaolin clay was having pH of 8.53 and compressibility index of 0.14. The stone aggregate picked for the column construction was categorized as GP, according to USCS soil classification system. The uniformity coefficient (C_u) and coefficient of curvature (C_c) of stone aggregate were 1.83 and 1.12, respectively. The minimum and maximum unit weight of stone aggregate were 13.15 and 13.54 kN/m^3, respectively. The standard Ennore sand was used to construct the sand platform. The coefficient of curvature and uniformity of the Ennore sand was 0.97 and 1.90, respectively. The maximum and minimum unit weight of Ennore sand were 13.73 and 12.65 kN/m^3, respectively.

The encasement was constructed using a conductive jute geotextile. Figure 5.2a depicts the picture of natural jute geotextile. To make the natural jute geotextile conductive a copper thread of 23-gauge thickness was intertwined at a spacing of 29 mm. Figure 5.2b shows the encasement prepared from conductive jute geotextile. The encasement properties were evaluated using pertinent standards for geosynthetics. The permittivity, thickness, and mass per unit area of the jute geotextile were 0.19 s^{-1}, 0.96 mm, and 358 g/m^2, respectively. The secant stiffness and tensile strength of geotextile were 504.48 and 24.33 kN/m, respectively. The durability aspects of the jute geotextile were not studied in the present study. However, the natural jute geotextiles are biodegradable in nature. It blends with soft soil after their design life and is ecofriendly in nature. Further, the solution for majority of geotechnical issues is short term, and the application of natural geotextiles is advantageous despite a shorter design life. Moreover, chemical treatment can be employed to enhance the durability and design life. The circular mild steel bar having a diameter of 8 mm was chosen as an anode.

Figure 5.3 displays the schematic depiction of the test apparatus along with all the technical details used in the experiment. The test setup has inside diameter of 256 mm and thickness of 9.5 mm. It has two chambers of 200- and 300-mm height as shown in Fig. 5.3. For ease of sampling, the body of the apparatus has vertical split. Number of channels was engraved in the bottom plate and connected to a channel to collect

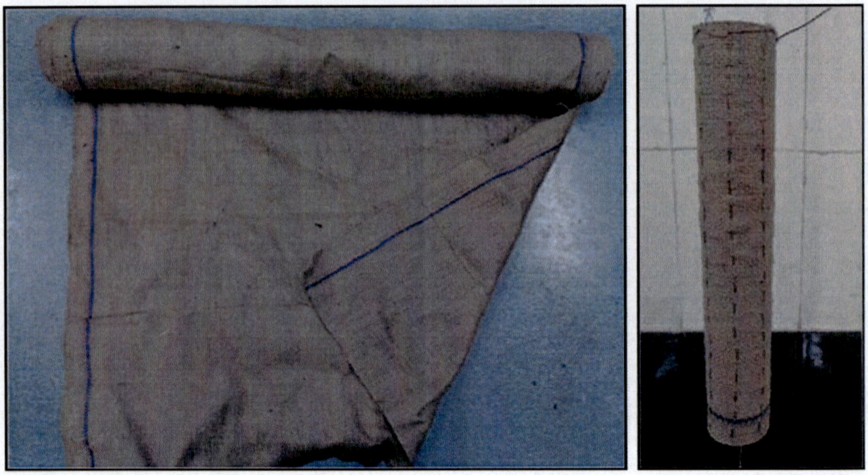

(a) (b)

Fig. 5.2 Actual picture of **a** natural jute geotextile and **b** encasement prepared from conductive jute geotextile

the water removed during the test. The column used in the experiment was having diameter of 65 mm and length of 400 mm. In order to prepare the soft clay bed, the powder kaolin was blended with an equivalent quantity of type 1 reagent water, so that no chunks were left in it. After exhaustive blending, the intermingled soil was preserved in sealed plastic bags for twenty-four hours for appropriate curing [22]. Before preparation of soft clay bed, a hollow cylindrical tube was placed at center of the setup; afterward, the filling of the soft clay was done in eight layers. Each layer was thoroughly compacted with the hand, and undrained shear strength was measured using Pocket Vane Tester. After soft clay bed reached the desired height of 400 mm, the construction of column begins. The stone aggregate used for the column construction was placed in eight layers, and each layer was well rammed, and before placing subsequent layer of stone aggregate, the cylindrical pipe was withdrawn in such a way that some part remains inside the previously laid layer. Further, after the column construction is complete sand platform was prepared as shown in Fig. 5.3.

An 8-mm-diameter mild steel rod of 600 mm length of 600 mm was placed at circumferential distance of 105 mm as anode. The surcharge was applied in four steps using the lever arm mechanism. Each increment load yields an intensity of 9, 30, 55, and 100 kPa, and each load remained for 24 h. However, in *e*-ESC test, a voltage gradient of 1 V/cm was used between the electrodes. The LVDT was used to monitor the vertical settlement. At culmination of the test, samples were mined to evaluate the alteration in water content, pH, and microfabric of soft clay.

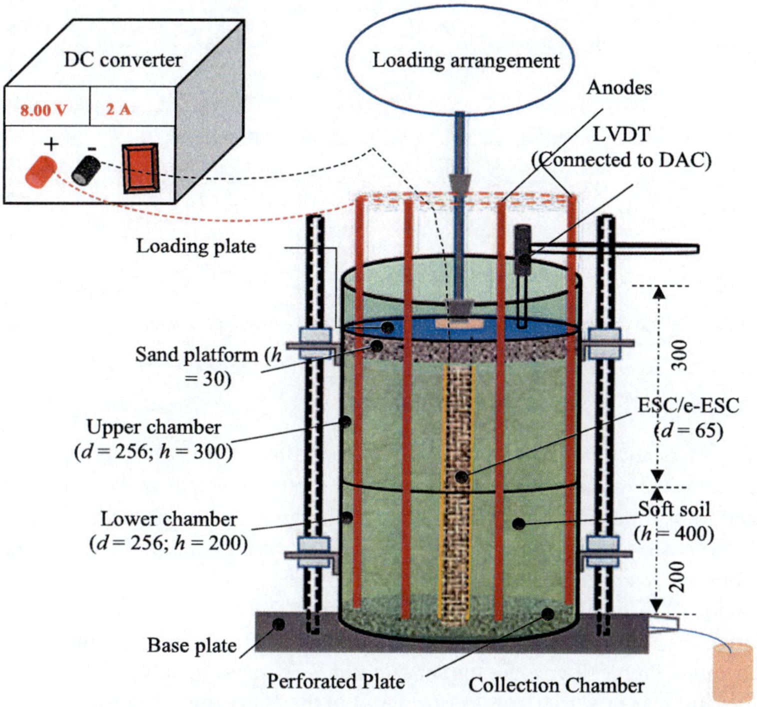

Fig. 5.3 Schematic of model test setup showing all technical details (all dimension is in mm)

Results and Discussion

Effect on Discharge

It is widely recognized that the loading generates excess pore water pressure in soft clay. This gets dissipated with time, causing pore water expulsion from soft clay. Figure 5.4 illustrates the variation of cumulative discharge during the test. For same loading, the volume of collected water in soft clay reinforced with SC was more compared to unreinforced soft clay. Further improvement was observed for ESC case. At culmination of test, the quantity of discharge in ESC case was considerably greater compared to soft clay reinforced with SC.

The discharge increases with the elapsed time. It can be further observed that the majority of the discharge for soft clay reinforced with *e*-ESC occurred during initial 8.5 h itself. However, as the test proceeds, the discharge ceded to almost zero after 18 h. When the pattern of water discharge was carefully observed during the different encasement methods, it clearly indicated that the time taken to remove half of the total collected water from the soft clay, SC, and ESC case was found to be 50–60 h. However, in case of *e*-ESC, this process was very rapid, i.e., 3.5 h only.

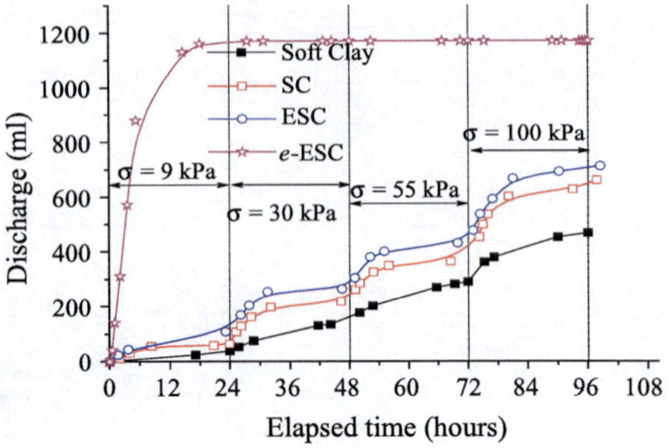

Fig. 5.4 Variation in cumulative discharge with time

Further, 99% of the total discharge was removed after 18 h itself in *e*-ESC, whereas the total removal in case of other encasement methods was merely 5–13% for the same period of time. The total cumulative discharge at a voltage gradient of 1 V/cm (i.e., *e*-ESC case) was approximately 12.23 times as compared to the ESC case after 18 h. The decrease in the discharge with time may also be attributed to reduction in diffuse double layer thickness and decrease in the zeta potential. However, the cumulative discharge is dependent on the voltage applied, electrode material, water content, and soil type. Thus, the higher cumulative discharge aids in accelerating the consolidation rate of settlement as discussed in the following section.

Variation of Settlement

In order to study the vertical deformation of the reinforced system, the settlement measured at the surface is plotted against time for the applied incremental pressure. Figure 5.5 illustrates the variation of the settlement with time and stage loading applied over unreinforced and reinforced soft clay. From Fig. 5.5, it can be seen that the rigid inclusion like SC and ESC considerably reduces the settlement of soft clay than unreinforced soft clay. In particular, the reduction in the total settlement using OSC and ESC was approximately found to be 29.28% and 40.34%, respectively, as compared to the soft clay. Further, a substantial settlement reduction was observed after the application of the final pressure (i.e., 100 kPa) in case of ESC (40.23%) as compared to OSC (13.65%). The reduction in the settlement for OSC was mainly due to the stiffening effect stone column. This effect was mobilized due to the interlocking of the soil particles achieved during the compaction of the column material. However, after the application of final incremental pressure the confinement offered by the adjacent soil was not adequate to withhold the column materials intact. This eventually resulted in the bulging of the column material resulting in the reduction of

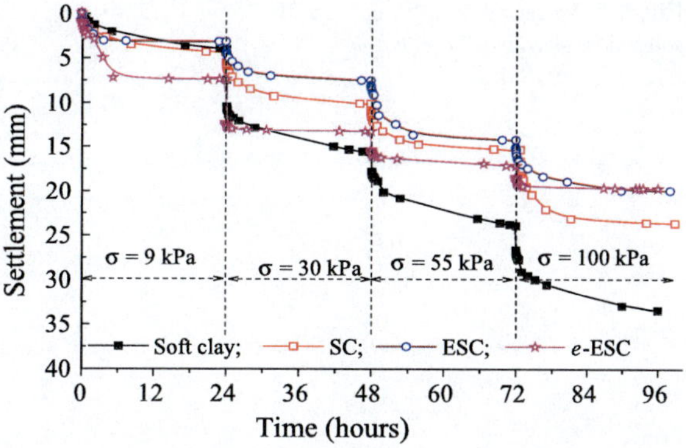

Fig. 5.5 Variation in cumulative settlement with time

stiffness and strength. Therefore, further reduction in settlement was observed even after encasing the stone column.

Subsequently, the comparison of the settlement of ESC and *e*-ESC case gives an interesting outcome. It was observed that the rate of the consolidation settlement was quicker in the e-ESC case as the voltage gradient was applied. A notable sum of settlement (~55%) in the case of OSC and ESC was after the application of the third and the fourth stage of increment in the applied pressure. However, in the case of e-ESC approximately 68% of the settlement was observed before the application of third incremental pressure as a result, it is reasonable to conclude that the use of such a voltage gradient increases the settlement of the soil during the initial phases of loading. However, the difference in the total settlement observed in case of *e*-ESC and ESC was almost insignificant (<1%).

Reduction in Water Content

The efficiency of the treatment after coupling with the voltage gradient in decreasing the water content is presented in terms of percentage water content reduction (PWR) [15]. Therefore, to understand the PWR variation of the soft clay at the culmination of the test, samples were obtained from the mid of the clay layer at different locations between the electrodes. The outcome of the study was stated in a non-dimensional form (*r*/*R*). Figure 5.6 shows the PWR variation of the soft clay at the termination of the test. As noticed from Fig. 5.6, there is significant reduction in the average water content of soft clay in the ESC case. The maximum enhancement of approximately 1.81 times was witnessed for the *e*-ESC case than ESC case in the neighborhood of the anode. Additionally, it is fascinating to note that for *e*-ESC case, a reasonably minor decrease in the water content nearby the granular column (i.e., cathode) and comparatively smaller water content close to anode was noticed. This might be due to the buildup of water neighboring the column due to the migration of water toward cathode from anode.

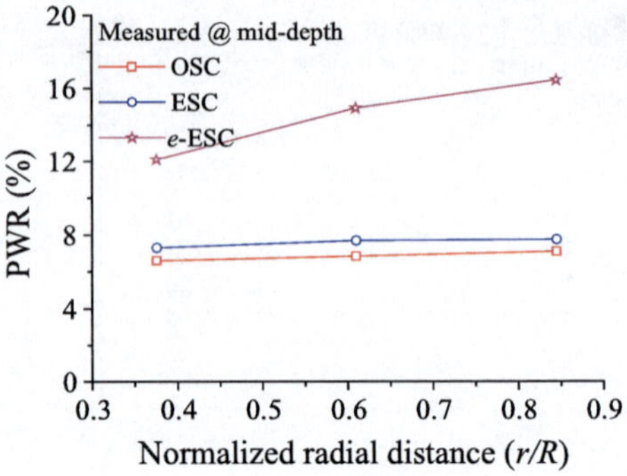

Fig. 5.6 Variation in reduction of water content between electrodes

PH Alteration

The variation of pH between electrodes is shown in Fig. 5.7. As anticipated, the pH of soft clay before test was uniform (i.e., 8.53) between the electrodes. For the soft clay reinforced with ESC, sight reduction in the pH was observed. The reduction is fairly uniform irrespective of depth. Remarkably, for the selected anode material and applied voltage gradient, an extreme deviation in the pH of soft clay was observed across the electrodes. The soft clay in the vicinity of the cathode has greater pH, while close to anode, the pH is lesser than initial pH at the start of the test. The decrease in the pH in the vicinity of the anode is probably due to the development of an extra acidic front. The acidic front is due to production of H^+ ions, which is result of the oxidation of the water [7, 15]. The reactions occurring in the vicinity of anode lead to increase in the temperature of the soft clay resulting in reduction of pH. The rise in value of pH of soft clay neighboring the cathode may be due to the development of Hydroxyl ion at the time of water removal.

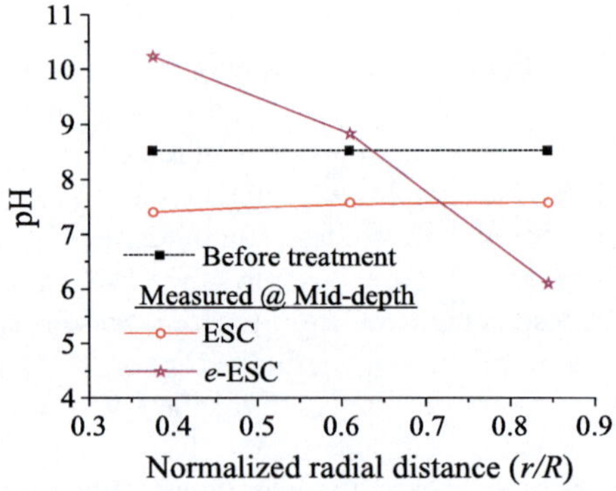

Fig. 5.7 Alteration in pH between the electrodes

Conclusion

- The coupling of electrokinetics with ESC helps in the quick pore water removal from the soil. The quantity of removed water after 18 h in *e*-ESC case is roughly 99% of the final discharge, which is unlikely in other cases where only 5–13% discharge is witnessed during this period.
- Encasing the stone column using conductive geotextile helps in reducing the settlement as compared to OSC and soft clay. Further, the *e*-ESC case speeds up the settlement rate by nearly 1.8 times than ESC case.
- The *e*-ESC case is highly effective in reducing the water content compared to other cases considered in the study.
- The pH value for *e*-ESC case decreases in anode vicinity and increases in cathode area, unlike in the ESC case where the uniform reduction of pH is observed between the electrodes.

References

1. Gniel J, Bouazza A (2009) Improvement of soft soils using geogrid encased stone columns. Geotext Geomembr 27(3):167–175
2. Pandey BK, Rajesh S, Chandra S (2020a) Numerical evaluation of geogrid-encased stone columns in soft soil under embankment loading. In: Geotechnical special publication, vol 315. ASCE, pp 543–551
3. Rajesh S (2017) Time-dependent behaviour of fully and partially penetrated geosynthetic encased stone columns. Geosynth Int 24(1):60–71
4. Castro J, Sagaseta C (2011) Deformation and consolidation around encased stone columns. Geotext Geomembr 29(3):268–276
5. Chen JF, Li YL, Xue JF, Feng SZ (2015) Failure mechanism of geosynthetic-encased stone columns in soft soils under embankment. Geotext Geomembr 43:424–431
6. Pandey BK, Rajesh S, Chandra S (2020b) 3-D finite element study of embankment resting on soft soil reinforced with encased stone column. In: Latha GM, Rao PR (eds) Problematic soils and geoenvironmental concerns. Lecture notes in civil engineering, vol 88. https://doi.org/10.1007/978-981-15-6237-2_38
7. Pandey BK, Rajesh S, Chandra S (2021) Performance enhancement of encased stone column with conductive natural geotextile under k0 stress condition. Geotext Geomembr. https://doi.org/10.1016/j.geotexmem.2021.03.004
8. Rajesh S, Jain P (2015) Influence of permeability of soft clay on the efficiency of stone columns and geosynthetic encased stone columns-a numerical study. Int J Geotech Eng 9(5):483–493
9. Singh R, Das A, Sathiyamoorthy R (2019) Efficacy of coupled solid—fluid formulation in regularizing an ill-posed finite element model. Indian Geotech J 49(4):409-420
10. Zhang L, Xu Z, Zhou S (2020) Vertical cyclic loading response of geosynthetic-encased stone column in soft clay. Geotext Geomembr 1–15. https://doi.org/10.1016/j.geotexmem.2020.07.006 (2020).
11. Dash SK, Bora MC (2013) Improved performance of soft clay foundations using stone columns and geocell-sand mattress. Geotext Geomembr 41:26–35
12. Ibrahim SF, Ahmed NG, Mohammed DE (2016) Developing a performance criterion for stone columns to improve surface pavement for weak subgrade conditions. Proc Eng 143:1309–1316

13. Yoo C, Lee D (2012) Performance of geogrid-encased stone columns in soft ground: full-scale load tests. Geosynth Int 19(6):480–490
14. Zhang Y, Chan D, Wang Y (2012) Consolidation of composite foundation improved by geosynthetic-encased stone columns. Geotext Geomembr 32:10–17
15. Pandey BK, Rajesh S (2019) Enhanced engineering characteristics of soils by electro—osmotic treatment: an overview. Geotech Geol Eng 37(6):4649–4673
16. Jones CJFP, Lamont-Black J, Glendinning S, Bergado DT, Eng T, Fourie A, Liming H, Pugh RC, Romantshuk M, Simpanen S, Yan-Feng Z (2008) Recent research and application in the use of electrokinetic geosynthetics. In: EuroGeo4 conference, pp 1–30
17. Esrig MI (1968) Pore pressure, consolidation and electrokinetics. J Soil Mech Found Div ASCE 94(4):899–921
18. Rittirong A, Douglas RS, Shang JQ, Lee EC (2008) Electrokinetic improvement of soft clay using electrical vertical drains. Geosynth Int 15(5):369–381
19. Wu H, Hu L (2014) Microfabric change of electro-osmotic stabilized bentonite. Appl Clay Sci 101:503–509
20. Xue Z, Tang X, Yang Q (2017) Influence of voltage and temperature on electro-osmosis experiments applied on marine clay. Appl Clay Sci 141:13–22
21. Wang L, Huang P, Liu S, Alonso E (2020) Analytical solution for nonlinear consolidation of combined electroosmosis-vacuum-surcharge preloading. Comput Geotech 121:103484
22. Mazumder T, Rolaniya AK, Ayothiraman R (2018) Experimental study on behaviour of encased stone column with tyre chips as aggregates. Geosynth Int 25(3):259–270

Chapter 6
Simulation of Geosynthetic Encased Stone Column Under Embankment Loading in Soft Clay

Jitendra Kumar Pradhan and Abdhesh Kumar Sinha

Introduction

Soft clay deposits owing to their low hydraulic conductivity, poor drainage characteristics, high compressibility, low stiffness, and low shear strength have always been a cause of concern for geotechnical engineers. As a result of these attributes, it becomes one of the most challenging soils to deal with and often the soft soil ground needs significant improvement before designing any construction on such ground. For the improvement of such soils, the stone column approach can be a better solution. Stone columns are used to improve the soft soil characteristics by increasing bearing capacity, reducing total and differential settlements, and accelerating the consolidation process, reduction in liquefaction potential providing better stability, and post-construction performance of the structure.

The accelerated consolidation process in stone column-supported foundations results in significant reduction in direct stresses developed on soft soils [1]. Vibro replacement technique is more appropriate for a wide range of grain sizes in loose soils, whereas vibro compaction is limited only to compact sand and gravels but can be applied in non-cohesive natural deposits [2]. The use of higher angle of internal friction and stress concentration factor in design may lead to local bearing failure because of the decrease in the shear strength of the clay which can be sorted by slope stability analysis using limiting angle of internal friction concept of the aggregates [3]. Encased columns might be an efficient alternative to conventional piles and ordinary stone columns since they can be constructed even in extremely poor soil condition. Also, the entire encased system is strain-dependent and the behavior of

J. K. Pradhan (✉)
Post Graduate Student, National Institute of Technology, Jamshedpur, Jharkhand 831014, India
e-mail: pradhanjk96@gmail.com

A. K. Sinha
Associate Professor, National Institute of Technology, Jamshedpur, Jharkhand 831014, India
e-mail: aksinha.ce@nitjsr.ac.in

K. Muthukkumaran et al. (eds.), *Transportation and Environmental Geotechnics*,
Lecture Notes in Civil Engineering 298, https://doi.org/10.1007/978-981-19-6774-0_6

the column is softer than that of the traditional piles as the columns are interactive, ductile, self-regulating system [4]. Encasing the stone column can also improve the functionality of the stone column as it acts as a single pile system. It has been reported that lesser diameter encased stone columns have superior quality than that of larger diameter stone columns because of mobilization of higher confining stresses in larger diameter columns [5]. Upon a laboratory test conducted on geogrid encased stone columns by varying the encasement depth and compared the different encasement depth with fully encased column, it has been observed that the vertical strain of the partially geosynthetic encased stone column decreased by increasing the encasement length for both the single and group of stone columns. Also, the fully encased stone columns are stiffer than all other conditions of the stone column test [6]. The load carrying capacity of the stone column can be further increased by installing larger diameter stone columns and stiffer encasement which may be helpful in enhancing the stiffness of the stone column [7]. It is reported that the depth of the radial bulging of an enclosed stone column is about four times the diameter of the column [8]. The efficiency of the encasement stiffness has been observed to be influenced by the construction condition, and the better performance of the encased stone column is due to the collective impact of decreased excess pore water pressure and high column stiffness [9].

In the present paper, parametric studies of encasement length, column diameter, friction angle of aggregates, and cohesive strength of clay have been considered for the simulation. The outcomes of the analysis are also compared with the application of geosynthetics.

Geometry and Model Description

The current research entails a significant number of stone columns that sustain an embankment of 22 m wide and 6 m high, with a 1 V:2H side slope in soft clay strata of 9.5 m. The embankment was constructed in three stages, each with a height of 2 m. The construction duration for each stage was around 10 days, followed by a 50-day rest period for consolidation. Before laying the embankment stages, a granular sand layer was intended with a construction period of 2 days followed by a 5-day rest period for the dissipation of excess pore water pressure partially from the clayey layer.

The unit cell idea was adopted for PLAXIS simulation. The depth of the stone column was considered to be approximately the same as the depth of soft clay strata, with a diameter of 0.8 m placed at 2 m c/c spacing in square pattern. The ground water table positioned at the upper surface of the clayey layers. Several ordinary stone columns (OSC) and geosynthetic-encased stone columns (GESC) with an assumed axial stiffness were modeled. The axial stiffness of geosynthetics was set at 2500 kN/m and the encasement depth of GESC was changed from 0.5 to 1.0 times that of OSC. Figure 6.1 depicts the PLAXIS model and mesh model of OSC and GESC.

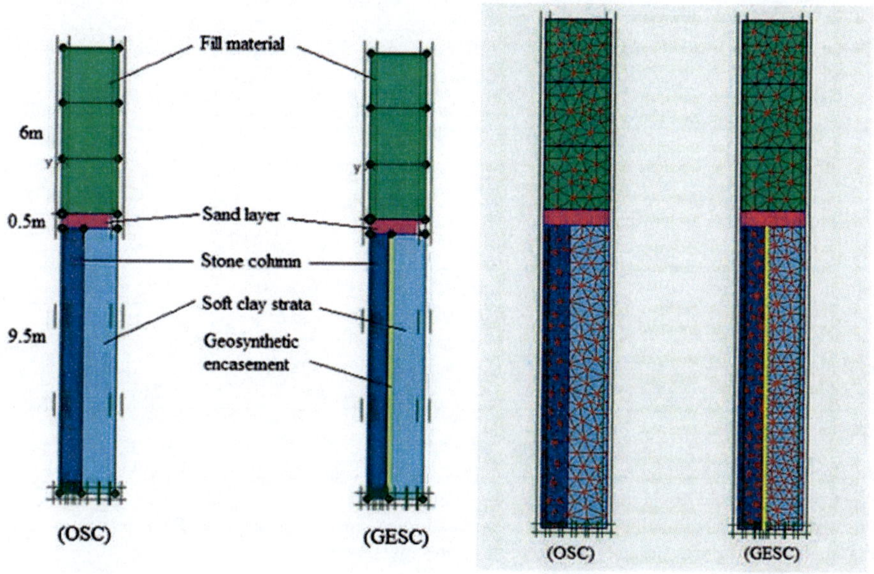

Fig. 6.1 Cross section of geometry and mesh of unit cell of OSC and GESC model using axis-symmetric condition

The displacement and hydraulic flow boundary for the model were considered accordingly. At the base of the model, a full displacement fixity and no displacement boundary condition in the perpendicular direction of the symmetry plane were considered. Similarly, for the hydraulic flow boundary, the symmetry plane was kept as a closed consolidation boundary and the bottom boundary was kept as an open consolidation boundary, assuming the presence of granular soil strata under the soft clay, followed by an impervious layer.

The stone-geosynthetic and geosynthetic-clay interfaces are both expected to have full strength which might be due to

(i) The installation of stone columns results in undulating interfaces.
(ii) The interfaces are considered as internal drainage nodes.

As a result, preferred slippage is not conceivable, and extra internal interface pieces are not essential. PLAXIS was used to assign the attributes of soil components reported in Table 6.1 that were acquired from field situations [10]. The encasement material was chosen to be an elastic material with orthotropic properties that can resist only tensile loads and no compressive loads. The embankment finite element model was simulated using 15 node triangular elements with axis-symmetric model features and finer mesh qualities.

Table 6.1 Material data used in the analysis

Parameters with unit	Soft clay (undrained)	Fill material (drained)	Sand blanket (drained)	Stone column (drained)
	Mohr–Coulomb	Mohr–Coulomb	Mohr–Coulomb	Mohr–Coulomb
γ_{unsat} (kN/m^3)	15	17	19	12
γ_{sat} (kN/m^3)	16	18	22	22
k_x (m/day)	7.36×10^{-5}	1	1	1
k_y (m/day)	3.68×10^{-5}	1	1	0.5
E_{ref} (kPa)	2×10^3	20×10^3	20×10^3	20×10^3
υ	0.3	0.33	0.33	0.33
c_{ref} (kPa)	28	0.1	0.1	0.1
ϕ (°)	1	30	35	38
ψ (°)	0	0	0	8
R_{inter}	1	–	–	1

c = Cohesion, E = Elastic modulus, ϕ = Angle of internal friction, γ_{unsat} = Unit weight (subscript unsat and sat corresponds to unsaturated and saturated, respectively), υ = Poisson's ratio, k = Permeability (subscripts x and y correspond to horizontal and vertical, respectively)

Staged Construction

The model simulation was accomplished with the incorporation of nine steps in staged construction mode. The plastic analysis was performed in the first step by activating the stone column and geosynthetic encase (if any) by substituting the previously determined soft clay in the initial phase with the column material. After the first step, the steps followed were evaluated for consolidation analysis. The sand layer above the stone column was activated in the second step, followed by a consolidation phase as third step. As the embankment was constructed in three stages, the remaining steps can be described by activating the individual layer of fill, followed by a consolidation phase. For improved computation and site safety, the ninth step was assigned as the minimum pore water pressure loading input type with a range of less than or equivalent to 0.1 kPa.

Results and Discussion

The overall calculation approach is summarized using three parameters: normalized encasement length (ζ), settlement reduction factor (β), and excess pore water pressure reduction factor (η). The normalized encasement length is the ratio of the encasing section of the stone column (L_{GESC}) to the entire length of the stone column (L_{OSC}). The settlement reduction factor is defined as the ratio of the settlement at a node after treatment (i.e., after the application of OSC or GESC) to the corresponding value before treatment (i.e., without the installation of any OSC or GESC). The excess pore water pressure reduction factor can be expressed as the ratio of the excess pore water pressure generated at a node after treatment to that of the untreated state. The settlement and excess pore water pressure data corresponding to minimum pore water pressure and at the completion of construction process, respectively, are gathered in this study. All the monitoring points mentioned are corresponding to the critical nodes found during the simulation.

Influence of Encasement Length

By varying the normalized encasement length from 0.5 to 1.0, the impacts on total settlement, excess pore water pressure, and lateral deformation were observed. For $\zeta = 0$, the model is OSC, and for any other variable, it is GESC. A substantial drop in the settlement reduction factor was found as encasement length increased, explaining a reduction of about 55.5% for an encasement length of 0.7 L_{OSC} as shown in Fig. 6.2. Encasement causes the column materials to acquire better lateral support, resulting in a shorter course for aggregates to settle upon load application than that of OSC. Figure 6.3 depicts the decrease of the excess pore water pressure reduction factor with the corresponding normalized encasement length. The excess pore water pressure reduction factor was determined to be about 0.3 after an encasement length of 0.7 L_{OSC}. The encasement serves as an interface between the soil around and the aggregates, preventing the soil particles from entering the voids of the stone column. Consequently, lesser clogging is expected to occur than that of OSC, and the pore water will have simpler routes to move and disperse. Also, the encasement avoids lateral movement of aggregates, which prevents the column from deforming (shown in Fig. 6.4). At greater depths, as the encased depth increases, the lateral distortion of the stone column gradually reduces due to which the encasement keeps the stone column contents intact within it, preventing them from shifting from their own location after applying load.

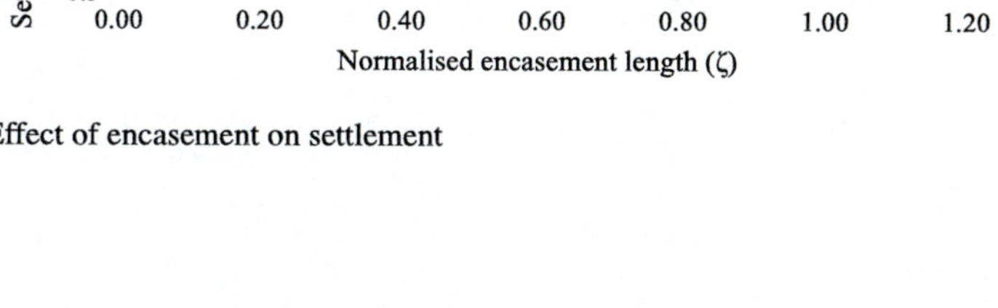

Fig. 6.2 Effect of encasement on settlement

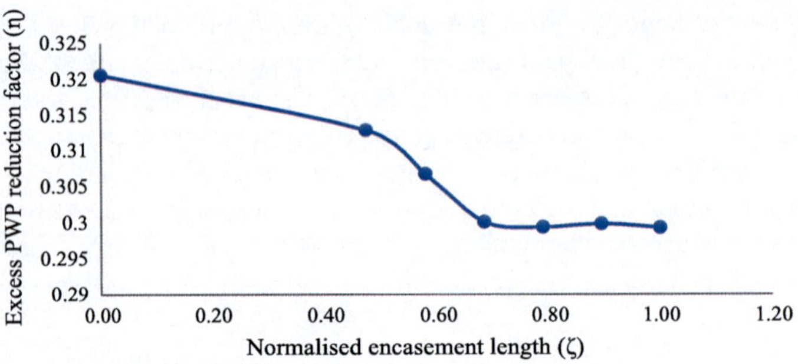

Fig. 6.3 Effect of encasement on Ex PWP

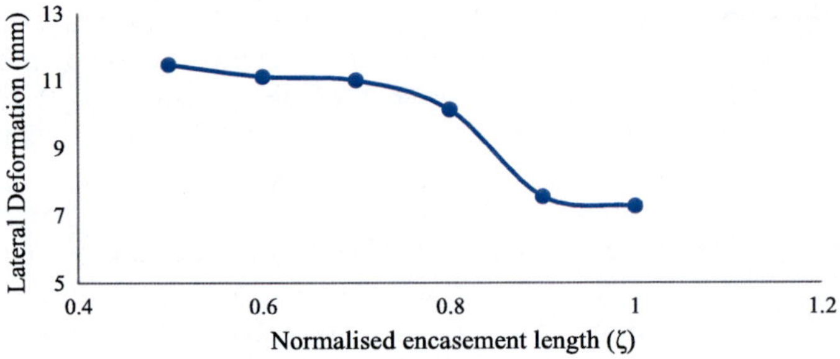

Fig. 6.4 Effect of encasement on lateral deformation

Influence of Diameter

The impact of diameter was explored by varying the diameter of the stone column from 0.4 m to 1 m, and it is observed that total stress reduces with increasing diameter of the stone column for both the case of OSC and GESC, and the variation is presented through Fig. 6.5. Further, it can also be seen that the stress mobilized in the case of GESC is greater than the corresponding value in OSC. While comparing with GESC, the rate of total stress reduction appears to be significantly less in the case of OSC for larger diameter columns. The generation of high supplementary confining stress for smaller diameter GESC is also found, making it superior to having a larger size. Numerically, the stress induced by a smaller diameter stone column is large since the diameter is in the denominator while computing total stress.

Influence of Internal Friction Angle

The internal friction angle of aggregates influences the settling reduction factor significantly. When the friction angle is increased, the settlement reduction factor also reduces in both cases of OSC and GESC, and the same is more prominent in case of GESC than that of OSC, as can be seen in Fig. 6.6. Aggregates with a high friction angle have a better shear strength, resulting in less mobilization and slippage. As a result, there will be less settlement. Furthermore, encasement limits mobility even further. Total strain was likewise shown to decrease as the angle of internal friction increased in both OSC and GESC. GESC, on the other hand, showed less strain than OSC as shown in Fig. 6.7.

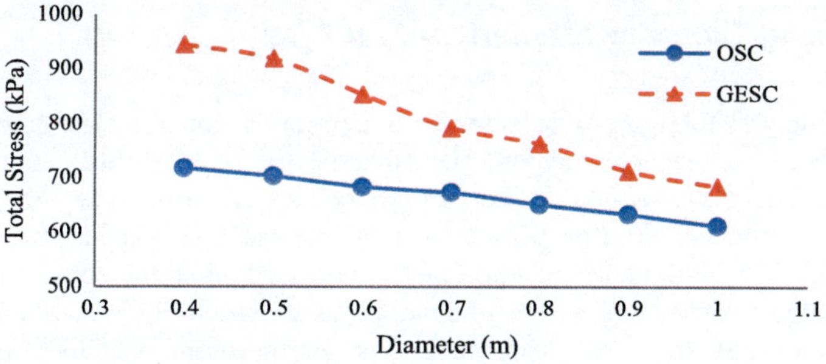

Fig. 6.5 Effect of diameter on total stress

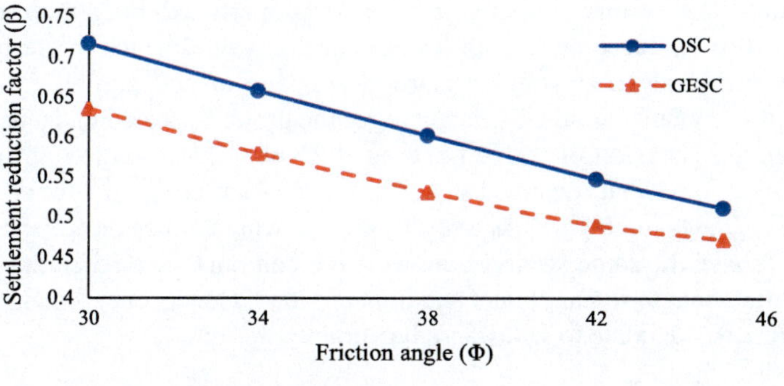

Fig. 6.6 Effect of friction angle on settlement

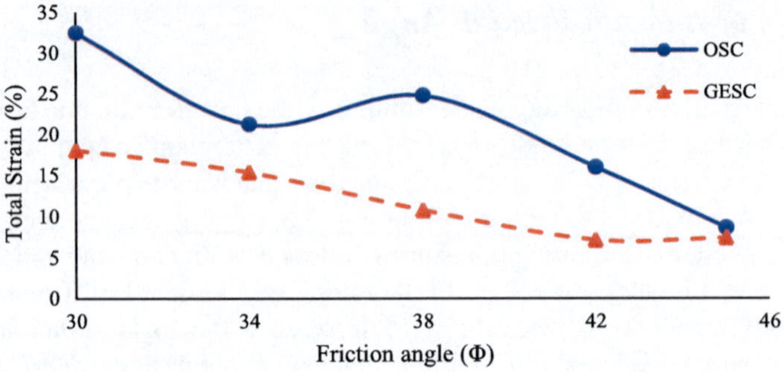

Fig. 6.7 Effect of friction angle on strain

Influence of Cohesive Strength of Soft Clay

It is extremely challenging to substitute the mentioned site soil by another type of soil. As a consequence, it was hypothesized that the soil for different locations existed with the soil dataset provided. In order to satisfy such scenario, except the cohesive strength, all other material properties were maintained constant for this, as stated in Table 6.1. The settlement reduction factor increased when the cohesion value of the associated ground increased in all contexts, as shown in Fig. 6.8, which is more in the case of OSC than the encapsulated type. Furthermore, with the influence of encasement, larger encasement depth reduces the respective settling for each type of ground cohesiveness. Since the encasement offered more lateral support (i.e., more stiffness), in case of GESC mobilization is somewhat less than that in case of OSC upon loading. The ground with greater cohesive property exhibited high settlement than the corresponding ground with lower cohesion value, which might be related to lack of sufficient shear strength causing slippage of soil particles. The strain generated was shown to gradually decrease for the grounds treated with encasement. Also, when the cohesion of the soil exceeded 25 kPa, the deviation of the rate of strain induced somewhat remained steady as can be seen in Fig. 6.9 for encasement lengths of $0.7L_{OSC}$ and higher. As a result, ground with a cohesion value more than 25 kPa will have the same strain values as if the columns were receiving adequate lateral confinement to remain intact and transmit load. Hence, encasement length of $0.7L_{OSC}$ may be adequate to withstand that strain criterion.

Conclusions

This paper represents the numerical study of several OSC and GESC models, and the following conclusions from the parametric studies can be drawn.

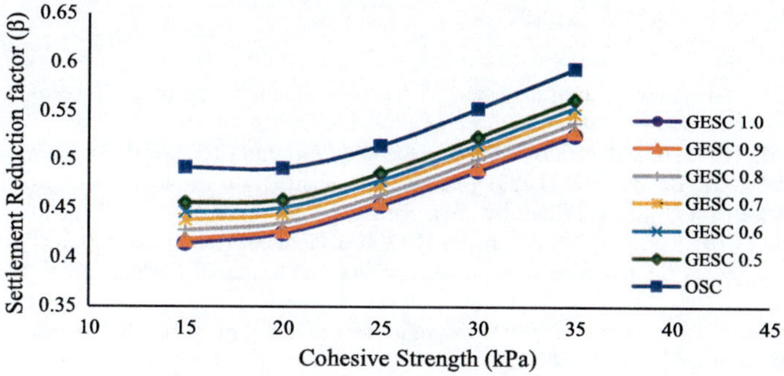

Fig. 6.8 Effect of cohesion on settlement

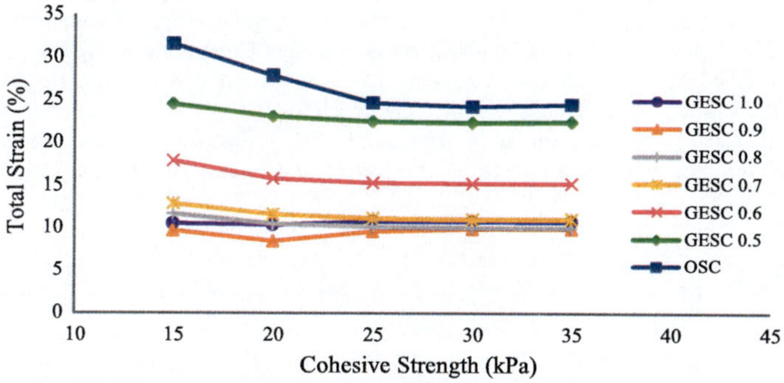

Fig. 6.9 Effect of cohesion on strain

- As encasement length increases, the settlement reduction factor and excess pore water reduction factor decrease. It also lessens lateral deformation.
- Stone columns with a smaller diameter mobilize more stress than with larger diameter. Encasement provides additional confinement, which further mobilizes greater stresses. As a result, the performance of GESC with a smaller diameter excels that of with larger size.
- Column materials having higher internal friction angle cause lesser mobilization of particles. Hence, lesser settlement is observed.
- Increasing the internal friction angle, the strain developed reduces.
- The settlement reduction factor is higher for places with high cohesion values than for areas with lower corresponding values, and it reduces progressively as the encasement length increases. The rate of change in strain induced in the column remained considerably lower for normalized encasement lengths of 0.7 or more. As a result, encasement length of $0.7L_{OSC}$ will be sufficient to moderate strain characteristics.

References

1. Han J, Ye S-L (1999) Simplified method for consolidation rate of stone column reinforced foundations. J Geotech Geoenviron Eng 127:2–5
2. Priebe HJ (1995) The design of vibro replacement. Ground Eng 28(10):31–37
3. Barksdale RD, Bachus RC (1983) Design and construction of stone columns. FHWA/RD-83/026, Federal Highway Administration, Washington
4. Alexiew D, Brokemper D, Lothspeich S (2005) Geotextile encased columns (GEC): load capacity, geotextile selection and pre-design graphs. Issues in Foundation Engineering, pp 1–14
5. Murugesan S, Rajagopal K (2006) Geosynthetic-encased stone columns: numerical evaluation. Geotext Geomembr 24(6):349–358
6. Gniel J, Bouazza A (2009) Improvement of soft soils using geogrid encased stone columns. Geotext Geomembr 27(3):167–175
7. Marto A, Moradi R, Helmi F, Latifi N, Oghabi M (2013) Performance analysis of reinforced stone columns using finite element method. Electron J Geotech Eng 18(B):315–323
8. Yoo C (2015) Settlement behavior of embankment on geosynthetic-encased stone column installed soft ground—a numerical investigation. Geotext Geomembr 43(6):484–492
9. Kong G, Zhou Y, Liu H (2018) Nonlinear model analysis of radial bulging deformation of geosynthetic-encased stone columns. Int J Geomech 18(10)
10. Gaber M, Kasa A, Abdul-Rahman N, Alsharef J (2018) Simulation of stone column ground improvement (comparison between axisymmetric and plane strain). Am J Eng Appl Sci 11(1):129–137

Chapter 7
Improvement of Railway Embankment Formation by Laying Geogrid Beneath Ballast Bed—A Case Study

Akshay Kumar Jha

Introduction

Railway Track consists of rail fastened to sleepers with elastic fastening and sleeper is supported on ballast laid on sub-ballast and stable embankment. Axle load of vehicle is transferred through rail, sleepers, and ballast to embankment. Hence, embankment is to be designed for static load and dynamic loads along with other unforeseen factors like defects of rolling stock, track, etc. On Indian Railway, currently freight stocks of axle load of about 22.9–25 Ton are running. In good old days, the embankment construction was not mechanized. However, the practice of leaving finished embankment for one rainy season for consolidation was a healthy practice being followed. Moreover, due to non-availability of granular soil (meeting the specifications) near the site of construction, in few stretches of track, soil with higher fine content has been used. Such formation does not serve the required purpose as the bearing capacity of soil is inadequate leading to uneven settlement of formation and hence ballast bed. Hence, there is need to attend track frequently and speed of train needs to be reduced to ensure safety of running traffic.

Providing Geosynthetics at ballast subgrade interface during ballast overhauling improves bearing capacity, minimizes ingress of moist fines from subgrade into ballast, and improves drainage of water from ballast layer. On Kazipet–Kondapalli section of South Central Railway, patches of track having weak formation between Madhira and Kondapalli stations (Fig. 7.1) require frequent maintenance by way of heavy on track machines by way of tamping and cleaning of ballast. To improve bearing capacity of subgrade and improve maintainability of track, Geogrid was laid at ballast subgrade interface in about 19 km length in 2016–2017. This has improved line and level of track and reduced maintenance efforts.

A. K. Jha (✉)
South Central Railway, Secunderbad 500017, India
e-mail: akshayghungru@gmail.com

K. Muthukkumaran et al. (eds.), *Transportation and Environmental Geotechnics*, Lecture Notes in Civil Engineering 298, https://doi.org/10.1007/978-981-19-6774-0_7

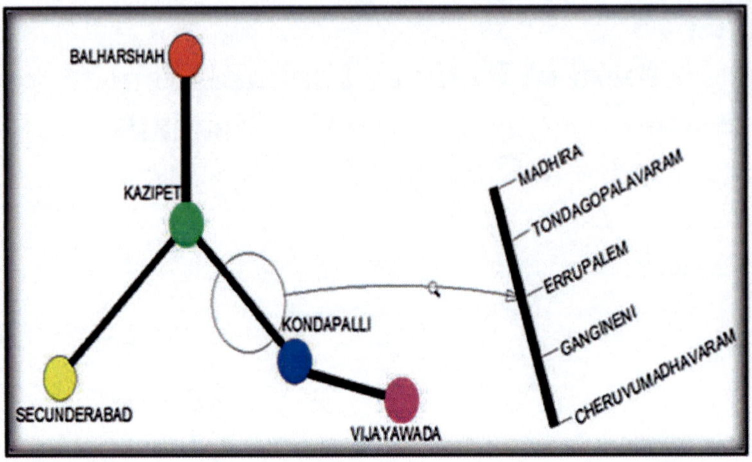

Fig. 7.1 Stretch where geogrid was laid in Madhira–Kondapalli section

Weak Formation Problem

Railway track rests on competent foundation, which is in the form of embankment, which is supposed to be non-deformable. In case, proper material is not used for construction of embankment, and quality control measures like compaction, Optimum Moisture content (OMC), and achieving maximum dry density (MDD) are not adhered to; in such cases, formation can deform/settle. Wheel loads being transmitted more than design load can also cause deformation of formation layer.

Properties of Embankment Soil

The soil sample for testing soil parameters were collected from km 534.22 to 534.24 between Madhira and Tondalagopavaram block section. Soil sample from the above location was collected and tested for Grain size analysis, consistency limits, OMC, MDD, Coefficient of uniformity (Cu), Coefficient of curvature (Cc), Differential Free swell Index and Shrinkage Limit were tested at Bureau Veritas Pvt. Ltd labs, Hyderabad. The summarized properties of soil are tabulated in Table 7.1.

The differential free swell index and shrinkage limit of soil sample collected from km 534/22–24 between Madhira and Tondalagopavaram stations were tested at Bureau Veritas Pvt. Ltd labs, Hyderabad. Liquid limit is 32 which is between 30 and 35, Plasticity Index is 8 which is < 12, Shrinkage Index is 12 which is less than 15 and Free swell percent is 30 which is less than 50. As per Para 3.9 of IS:1498-1970, Reaffirmed in 2002 read along with Table 8 of same code indicates that the soil is having Low Degree of expansion and Degree of Severity is non-critical.

Table 7.1 Salient geotechnical parameters of typical soil samples picked up from MDR–KI section (sample location is DN line @ 534/22–24 between Madhira and Tondalagopavaram)

Testing parameter	Value
Liquid limit	32.02
Plastic limit	23.97
Plasticity index	8.05
OMC (in %)	21
MDD (in gm/cc)	2.37
C_u	45
C_c	0.45
Differential free swell index (in %)	30
Shrinkage limit (in %)	12

The soil based on laboratory test result is categorized as fine-grained soil with clay of low plasticity as per IS Classification.

Rehabilitation of Weak Formation by Laying Geogrid at Formation Level Using Ballast Cleaning Machine (BCM)

Soil has very good compressive strength but is weak in tension. Biaxial geogrid has been used for rehabilitation of formation to improve track health. The schematic diagram showing the location of laying of Geogrid is shown in Fig. 7.2.

Laying of Geogrid over Geotextile at ballast subinterface is done during overhauling of track ballast by Ballast Cleaning Machine. These machines take out the ballast using cutter bar and conveyor, screen ballast through different screen sizes, and put the clean ballast back by throwing the muck away. While ballast is being scooped out screened and put back, Geogrid is laid simultaneously. The output with

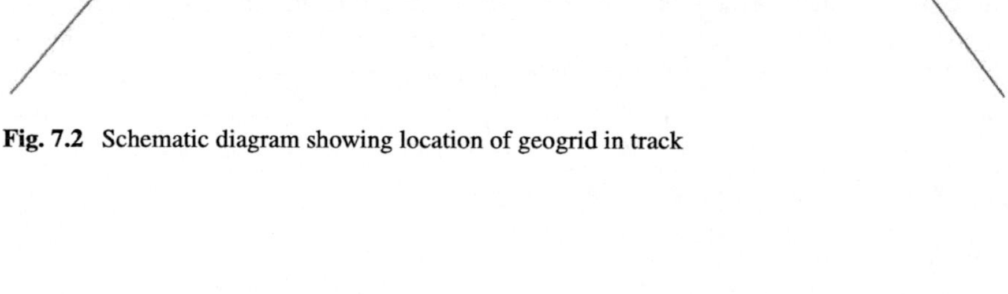

Fig. 7.2 Schematic diagram showing location of geogrid in track

Fig. 7.3 Photographs of geogrid at site

this method is about 75 m per hour of traffic block, which includes the process of insertion of geogrid roll and laying of same on ballast subgrade interface.

Properties of Geogrid Used for Rehabilitation of Formation

Geogrid of aperture size 38 mm × 38 mm made up of polypropylene proposed to be used for laying was tested at IIT, Madras (Fig. 7.3).

Geogrid tensile strength is 30 kN/m in both machine and transverse direction at 13% elongation, load at 5% strain is 23.5 kN/m in machine direction and 23 kN/m in transverse direction, and mass per unit area for Geogrid is 510 gm/m^2.

Locations of Laying of Geogrid

On Madhira–Kondapalli section, Geogrid was in patches at weak formation location in km 533–536, 547–552 on upline totaling to 10 km and in 9 m stretch on down line in km 552–553, 556–560, and 562–563.

Method of Laying and Traffic Block Requirement

The Geogrid is laid at the interface of Formation and Ballast. The process of laying involves temporary suspension of traffic, movement of Heavy on track machines like BCM, tampers, etc., to work site in a traffic block, Insertion of Cutter bar of BCM into the track, tying up rolls of Geogrid to Ballast collecting chutes. These rolls of Geogrid get laid at the formation level as Ballast cleaning machine moves.

The average output is 175–185 m per effective block hour as compared to 200 m per effective block hour without insertion of Geogrid.

Impact of Laying Geogrid—Performance Evaluation

Five parameters have been studied for comparing the performance of railway track before and after laying Geogrid. The effect of improvement in formation by way of laying of Geogrid is reflected in the performance/maintenance requirement of track. The maintenance requirement is covered by way of no. of attentions given to track both manual and machine, and performance is captured by no. of acceleration peak exceeding 0.15 g, Ride Index, and Unevenness Index. Unevenness Index is a measure of track deterioration due to traffic and improper formation. In a sense, this is an indirect measurement of effect of laying Geogrid like that of acceleration peaks. The actual performance can be compared by putting strain gauges at PSC sleeper bottom and measuring vertical deformation of ballast bed. However, such instrumentation could not be done. Hence, indirect evaluation method is adopted.

The parameters recorded pre- and post-laying of geogrid have been averaged out for a km and compared to evaluate the impact of laying Geogrid. Keeping in view the large no. of data points over a period, the averaging of data for a particular parameter pre-laying and post-laying has been done so that effect of laying could be compared.

Performance Improvement in Terms of Unevenness Index (UI)

UI is a parameter calculated based on Unevenness measured on a chord of 9.6 m on Left and Right rails at an interval of 0.25 m and calculated by following formula.

$$\text{Unevenness Index(UI)} = 100e^{-\left[\frac{\text{Measured Value of SD}-\text{SD for New Track}}{\text{SD of Urgent maintenance}-\text{SD for New Track}}\right]} \quad (7.1)$$

where SD for New track = 2.5 and SD for Urgent maintenance = 6.2. These values are specified in Refs. [1, 2].

The Track Recording Car (TRC)-recorded data provides Standard deviation of block of 200 m. The UI is calculated block wise and averaged over a km by combining five such blocks.

On UP line out of 10 km stretch UI has improved in a length of 8 km. This improvement is based on average value of UI measured by TRC between 2014 and 2016 totaling to 7 runs in pre-geogrid installation period and runs during year 2017–2019 in post-laying period. Improvement of UI took place in about 80% stretch of track.

Similarly, in DN line geogrids were laid in 9 km and improvement took place in 4 km stretch based on UI. Thus, 44% has improved based on UI. Considering overall length of UP and DN line based on UI improvement has taken place in 12 km out of a stretch of 19 km. The percentage of improved length is 63%. The changes in UI are depicted in Fig. 7.4.

However, in terms of percentage improvement in calculated value of UI the averaged over entire length of 19 km over a period of two years if compared with two-year period of post-laying parameter, improvement is about 4%.

Performance Based on Number of Vertical and Lateral Acceleration Peaks Exceeding 0.15 g

Out of 19 km stretch where Geogrid has been laid beneath formation level during deep screening of ballast by BCM, number of km where vertical and lateral acceleration peaks have gone down is 14, thus leading to improvement in 74% length of the track. The kilometer wise improvement/deterioration in peaks recorded is shown in Fig. 7.5.

In order to assess the pre- and post-laying performance of track, OMS peaks of 10 km on UP line 9 km on DN have been averaged for each year for up and down line per km. The comparison of OMS peaks of pre- and post-laying is shown in Fig. 7.6.

The number of peaks in reference to the last recorded value of pre-laying in UP line shows reduction in number of peaks of acceleration indicating that there is a significant improvement in track. Also, in DN line even though there is an increase in number of peaks to 16 in the year 2018 which can be taken as an aberration but compared to pre-laying data of 2015 there is an improvement as number of peaks has decreased from 9 to 6.

Performance Improvement Based on Spot Manual Attention Given to Track

Although for planned maintenance of track heavy on track tampers are used for tamping on limited stretch of track but spot manual attention of special features of track/bad spots become a necessity as heavy on track machines cannot be moved for very small patch of track requiring attention. The comparison between averaged number of Fig. 7.7: km wise Variation of No. of spot attentions per year for pre- and post-laying indicates that number of manual attentions given to track has reduced post-laying of Geogrid for all the 19 km where formation has been treated by laying Geogrid. The km wise variations in no. of spot attentions of pre- and post-laying of Geogrid are shown in Fig. 7.7.

In order to assess the pre- and post-laying performance of track spot manual attention of 10 km on UP line, 9 km on DN line have been averaged for each year off pre-laying and post-laying using Eq. (7.2). The pre-laying years are 2013, 2014,

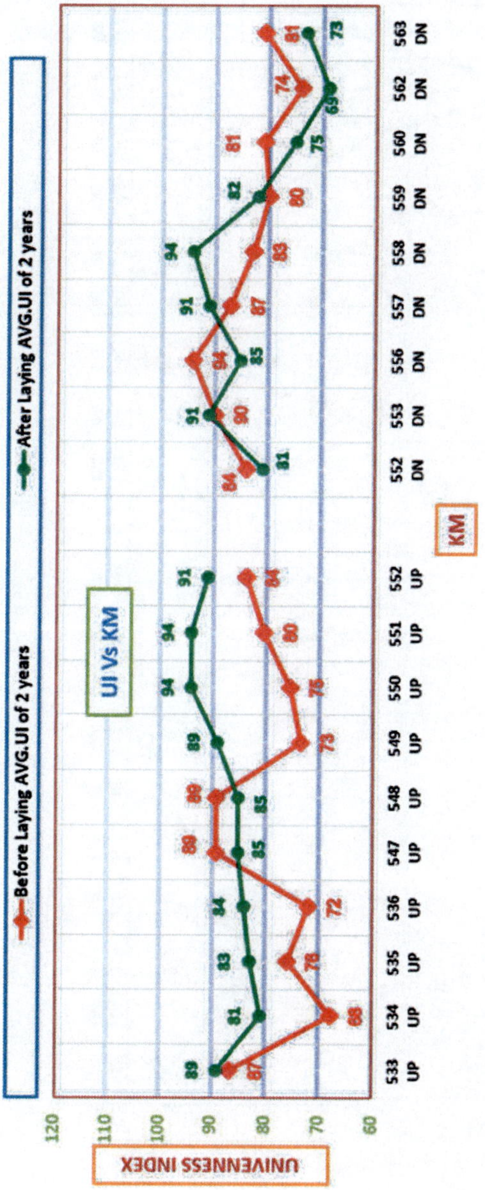

Fig. 7.4 Change in unevenness index with km

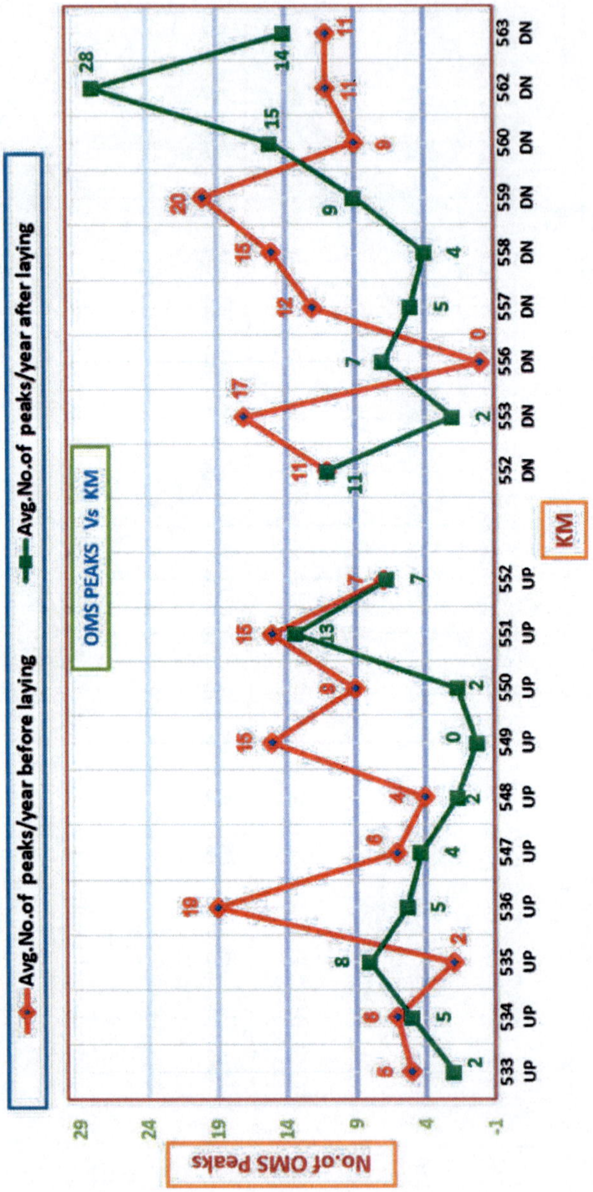

Fig. 7.5 km wise variation of oscillation peaks exceeding 0.15 g

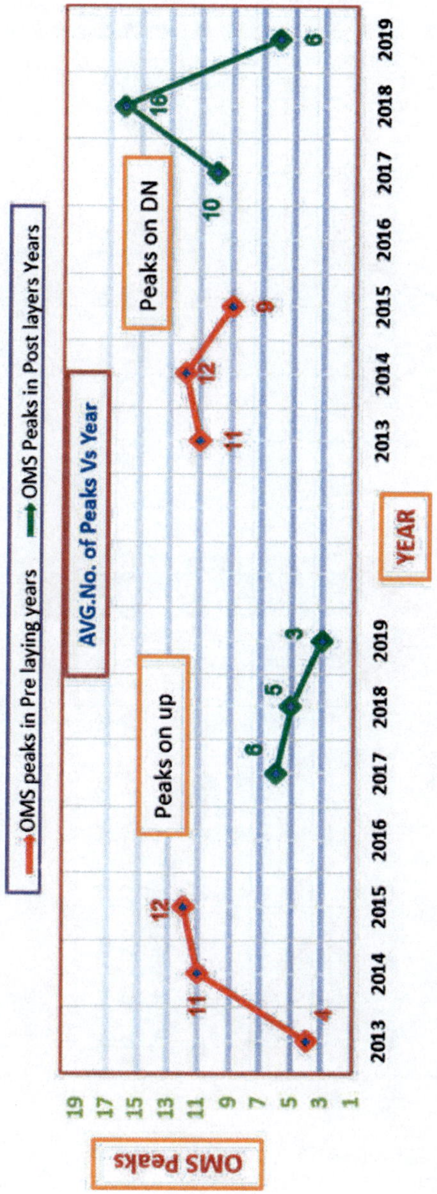

Fig. 7.6 Year wise variation of OMS peaks exceeding 0.15 g

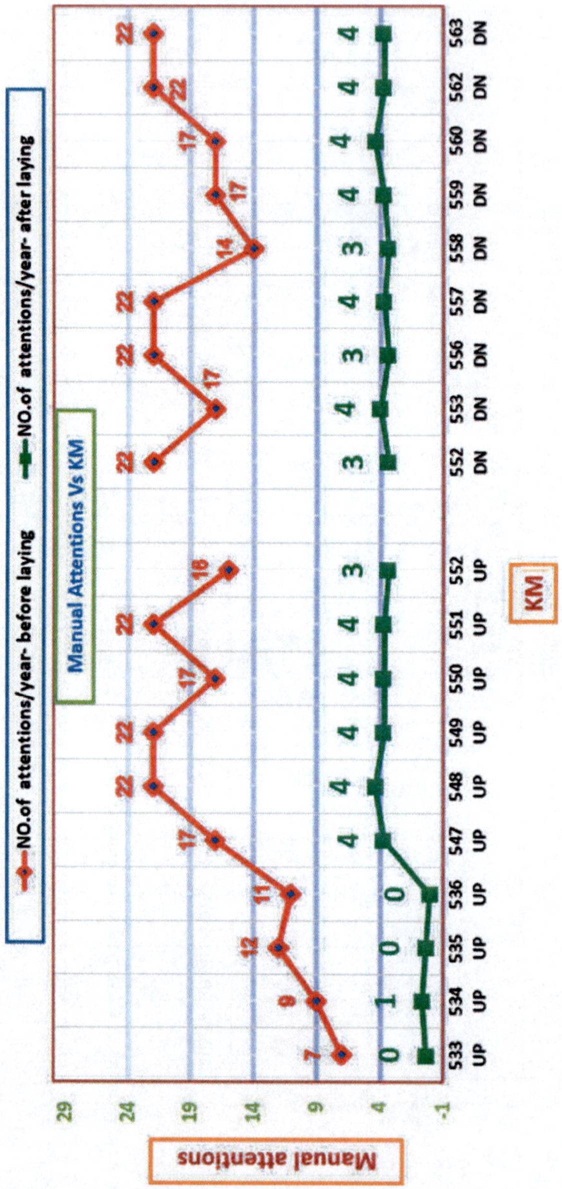

Fig. 7.7 km wise variation of No. of spot attentions

Fig. 7.8 Comparison of year wise manual spot attentions pre- and post-laying

and 2015 and post-laying years are 2017, 2018, and 2019. The comparison of pre- and post-manual attentions is shown in Fig. 7.8.

From Fig. 7.8, it is evident that there is an improvement on both UP and DN lines as number of manual attentions has significantly come down from last recorded value of 14 during pre-laying period of 2015 to an average value of 2 in post-laying period which is about 82% improvement.

Performance Improvement Based on Number of Machine Tamping

In view of heavy track structure consisting of PSC sleepers, elastic fastening, and heavy rails, the maintenance has to be mainly carried out by heavy on track tampers. All the 19 km where Geogrid has been laid to improve formation have shown improvement post-laying. The km wise variation in improvement is depicted in Fig. 7.9.

By way of inclusion of Geogrid at formation level, the bearing capacity of formation is supposed to improve, which is evident from Fig. 7.9. However, whether this improvement is dependent on time elapsed after laying of Geogrid needs to be ascertained. One of the indirect ways is to study the variation of average no of machine attentions per km pre- and post-laying. The data in Fig. 7.9 indicates that machine attention numbers have come down from three to one and are holding with respect to time as well as number of attentions is either constant or decreasing with increasing time from date of laying.

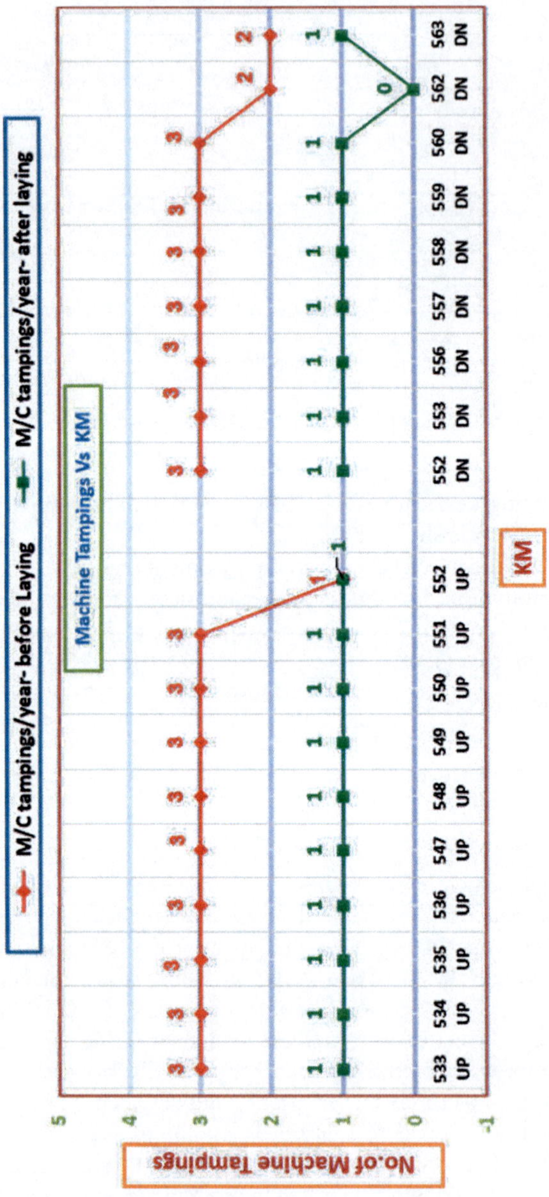

Fig. 7.9 km wise variation of No. of machine tamping

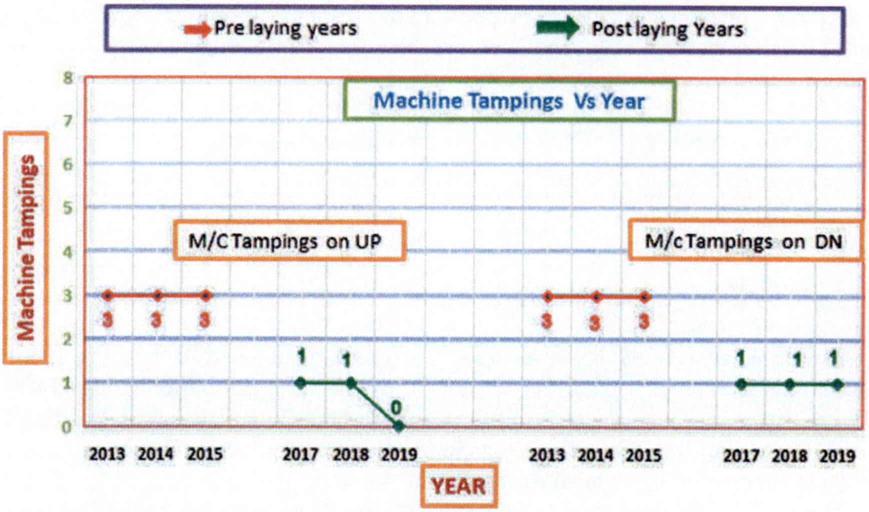

Fig. 7.10 No. of machine tamping pre- and post-laying

From Fig. 7.10, it is evident that there is an improvement on both UP and DN lines as number of machine tamping has significantly come down from last recorded value of 3 during pre-laying period of 2015 to an average value of 1 in post-laying period which shows a reduction of 66% in machine tamping.

Performance Improvement in Terms of Ride Index (RI)

Oscillation monitoring of track captures vertical and lateral acceleration every 10 ms (100 values per second), and these values are used for calculation of Sperling Ride index (7.2) normally called as Ride Index (RI). Ride Index is calculated by considering all the acceleration peaks irrespective of its value/threshold. The RI is defined as

$$\mathrm{RI} = 0.896\sqrt[10]{\left\{\left[\sum_{i=1}^{n} \frac{b_i^3}{f_i} * F(f_i)\right] * \frac{1}{n}\right\}} \tag{7.2}$$

where $n =$ no. of completed half waves (cycles)

$b_i =$ Acceleration Peak (absolute) value (cm/s^2) calculated for each of the half wave.

$f_i =$ Frequency of the ith half wave (Hz) $=$ s/2 N, where s $=$ sampling frequency of the sensors, $N =$ number of samples of ith wave half wave.

$F(f_i)$ = Function of frequency to reflect human response to various frequencies. Its value is different for vertical and lateral acceleration. Correction factors for various frequency values are as under,

For vertical acceleration	For lateral acceleration
0 for $f < 0.5$ Hz	0 for $f < 0.5$ Hz
$0.325 \times f^2$ for $0.5 < f < 5.4$ Hz	$0.8 \times f^2$ for $0.5 < f < 5.4$ Hz
$400/f^2$ for $5.4 < f \leq 20.0$ Hz	$650/f^2$ for $5.4 < f \leq 20.0$ Hz
1 for $f > 20$ Hz	1 for $f > 20$ Hz

Lower the value of RI, better the track quality. Comparison of RI for entire 19 km stretch has been done, and a lower value of RI has been recorded after laying. Out of 19 km stretch between Madhira and Kondapalli where Geogrid has been laid based on RI of pre- and post-laying there is an improvement in all 19 km. The improvement in RI is depicted in a graphical form in Fig. 7.11.

In Fig. 7.11, RI shown against a km is average of RI recorded during last three previous years. These are kilometer wise values averaged over three years period. Similarly post-laying values of RI are kilometer specific averaged over three years pre- and post-laying. The reduction in RI value indicates that performance of track has improved.

To assess the pre- and post-laying performance of track RI of 10 km on UP line have been averaged for each year off pre-laying and post-laying. The pre-laying years are 2013, 2014, and 2015, and post-laying years are 2017, 2018, and 2019. The comparison of pre- and post-RI vertical and lateral is shown in Fig. 7.12.

This improvement is to be seen in light of the fact that machine tamping and manual attentions have reduced post-laying by 66% and 82%, respectively. That means with reduced maintenance efforts RI has improved. This indirectly indicates that laying of Geogrid has improved performance of formation.

Cost of Laying and Reduction in Cost of Maintenance

The usual method involves procurement of Geogrid rolls from the manufacturer. The same is transported by road to site of laying. The item in schedule of contract is supply of Geogrids and laying of Geogrid. The Cost of supply of Geogrid along with laying per km is about Rs. 259,660. There is reduction in maintenance cost as is evident from reduction in manual and machine maintenance efforts. However, it is difficult to quantify the savings in maintenance cost as of track maintenance has so many components. However, there is reduction in spot attention and machine attention as discussed in "Performance Improvement Based on Spot Manual Attention Given to Track" and "Performance Improvement Based on Number of Machine Tamping" above.

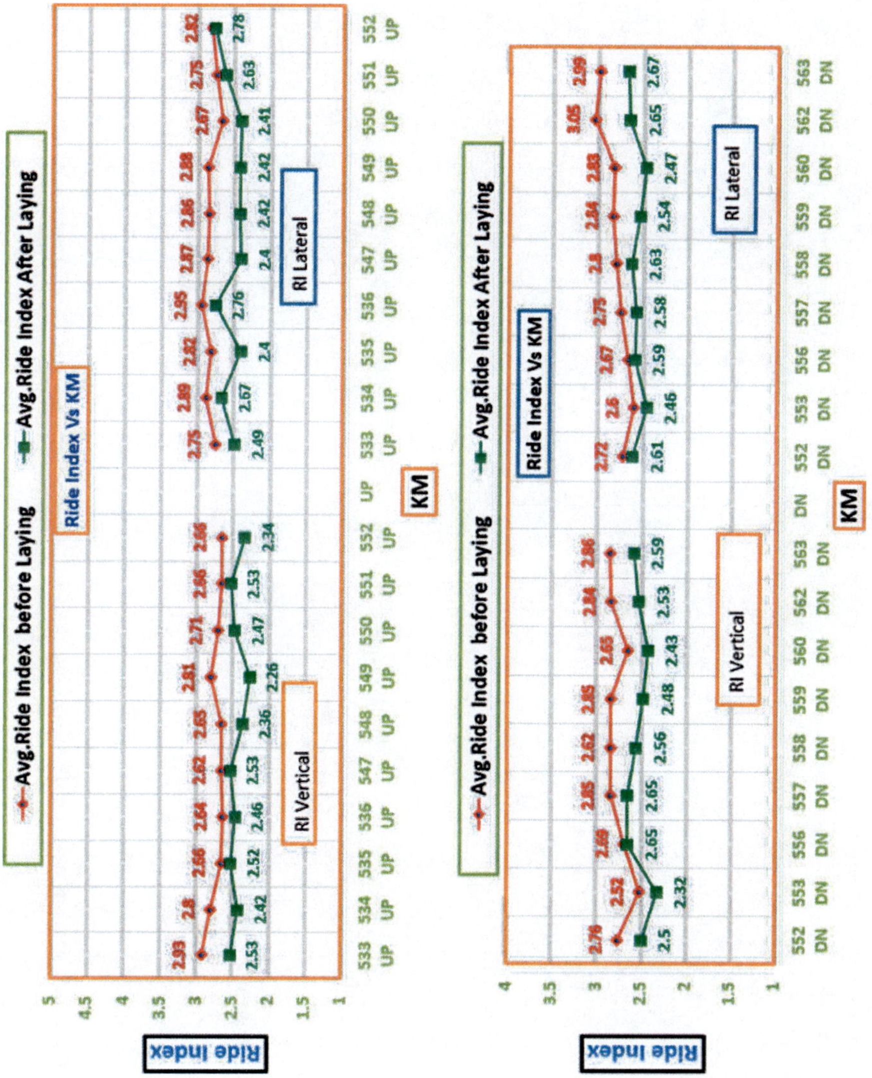

Fig. 7.11 km wise variation with ride index

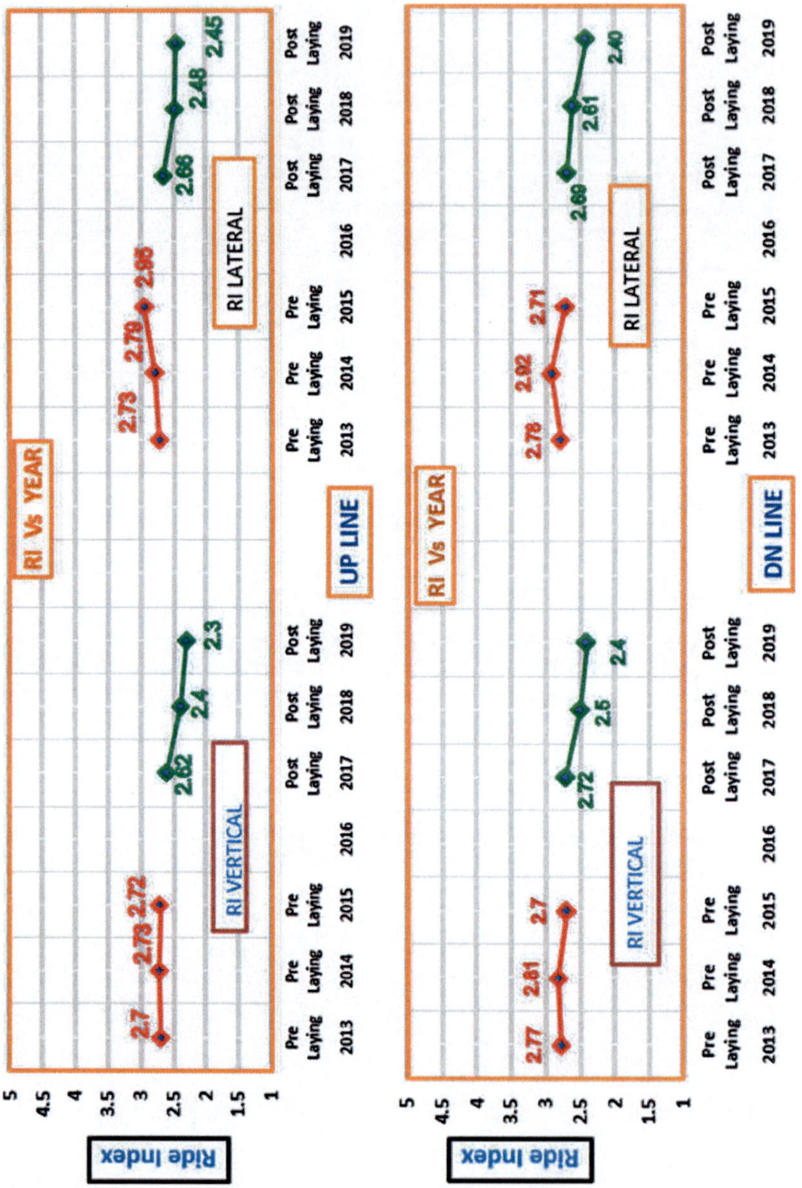

Fig. 7.12 Year wise variation with ride index for UP and DN line

Summary and Conclusion

On overall basis in a stretch of 19 km, Geogrid has been laid at ballast formation interface. The performance has been evaluated based on five parameters namely Unevenness Index, OMS acceleration peaks, Manual/Spot attentions, Machine tamping, and Ride Index. It has been observed that except Unevenness Index all other parameters where Geogrid has been laid have shown improvement barring few aberrations which could be due to specific site feature/event. Unevenness Index has improved in 12 km and balance 7 km variation is to be investigated further before concluding that it has deteriorated. Unevenness Index value improved in first year of post-laying and deteriorated in second and third years. Proper Instrumentation of sleeper bottom for vertical settlement for strain measurement needs to be done for accurate and detailed evaluation of effect of Geogrid, which is difficult. But this will lead to real-time monitoring of settlement. On overall basis, this is an effective method of formation treatment which can be done using opportunity of ballast overhauling cycle by ballast cleaning machine.

Acknowledgements Author is thankful to Sri Bhargava Krishna Sajja, Senior Divisional Engineer/South/Secunderabad Division, Sri Sudhir Kumar Verma Executive Engineer/Track Machine, Sri P. Prabhudasu, SSE/D/HQ, Ms. B. Geeta Sahithi SSE (Pway)/HQ, and B. G. Malleshwara Rao (SSE/P.way/MDR), for their valuable contribution by way of providing data, valuable suggestions, technical discussions, etc., without which this paper would not have come to present shape.

References

1. Track monitoring. IRICEN Publication (2013)
2. Track geometry management on Indian railways for better ride quality, 1st edn. IRICEN Publication (2019)

Chapter 8
Review on the Application of Geosynthetic for Ground Improvement

Kaustav Das, Subhojit Chattaraj, and Kaushik Bandyopadhyay

Introduction

Due to globalization, amount of suitable land and component land is decreasing day by day. As a result of this, some advanced techniques are required that could convert the weak soil into good, engineered soil. Among various soil improvement techniques, hydraulic conductivity modification, reinforcement, and confinement have been used for ground improvements. Among all these soil improvement techniques, incorporation of reinforcement is mostly used due to its cost effectiveness and simplicity in application [1]. Soil does not possess good tensile strength due to its environmental condition. The tensile strength can be improved by providing reinforcement along the direction of the tensile stress [2]. Geosynthetic is a product made from polymer material and applied to soil to enhance its engineering properties. Prefix "geo" suggests that it is applied with soil, rocks, or earth. Reinforcement, separation, filtration, drainage, containment, barrier, surface erosion control, and protection are some basic functions of geosynthetic. These synthetic materials have been used in airport runways, railways, roadways, retaining structures, embankments, and landfills. Geogrids, geocells, geotextile, geomembrane, and geocomposites are various products of geosynthetics used to improve bearing capacity of soil [3]. Using geosynthetic reinforcement, shear failure prevention and load settlement behavior of footing can be enhanced [4]. In this context, researchers did several experiments to find out preferable reinforcement configuration for soils to improve its engineering properties [5]. Figure 8.1 exhibits the framework of strengthened soil.

K. Das (✉) · K. Bandyopadhyay
Department of Construction Engineering, Jadavpur University, Kolkata 700 098, India
e-mail: kd3853@gmail.com

S. Chattaraj
GKCEM, Baruipur, West Bengal 743387, India

K. Muthukkumaran et al. (eds.), *Transportation and Environmental Geotechnics*,
Lecture Notes in Civil Engineering 298, https://doi.org/10.1007/978-981-19-6774-0_8

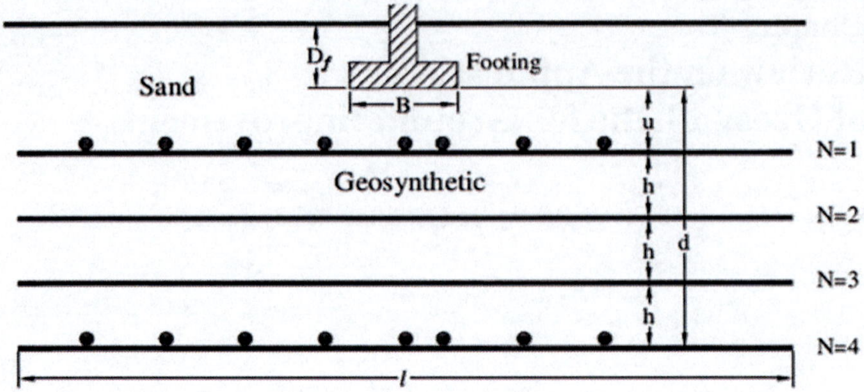

Fig. 8.1 Various parameter of a soil which is reinforced [6]

Various Types of Geosynthetics

Geosynthetic is a planar polymeric product used with soil, earth, and rock or with any other geotechnical related material. Geotextiles, geonets, geogrids, and geocells are some of the mostly used geosynthetics for ground improvement [7]. Geotextile is a material that is made of synthetic fibers rather than the natural one weaved in random manner and porous in nature. These are used for Separation, Reinforcement, Filtration, Drainage, and Liquid barrier (when impregnated) [8]. A geonet is a geosynthetic material having a similar structure as geogrid. It consists of parallel ribs which are internally connected for drainage of liquid and gases. Sometimes, geonets are laminated with geotextiles on both sides or maybe on one side for drainage use. For improving permeability of soils like clay, geonets are used but they do not act as reinforcing material [9]. Geogrid is another type of geosynthetics which are used to reinforce soils. They are basically used to reinforce retaining walls, sub-bases, and subsoils. As we know soil is weak in tension, these geogrids transfer forces to a larger area of soil as they are strong in tensions [10]. Geocell is a 3D geosynthetic material that is produced by interconnection of polymeric strips in a cellular fashion. The cells are filled with granular materials to distribute buildup loads over a large area due to static and cyclic loadings. Geocells are applied in retaining walls, foundations, pavements, and steep slopes to provide supports under various load. Cellular confined geocells are used in erosion control [11].

Ground Improvement Techniques

In recent days, the ground improvement techniques are popular because they can be used in wide range of soils. The reason behind adopting these techniques is stated below [12].

(a) Bearing capacity and shear strength can be enhanced.
(b) Soil distorted less under applied stresses.

(c) Settlements and deformation can be controlled.
(d) Consolidation of the soil can be improved.
(e) Prevention of physical or chemical changes in soil due to environmental effects.
(f) Void spaces below the ground surface can be filled.
(g) Liquefaction potential of a soil can be improvised.

There are various methodologies available for improving the ground conditions, and based on that, ground improvement techniques are categorized into five different types, as given below [13].

(a) Modification by mechanically;
(b) Modification by using hydraulic appliances;
(c) Modification of soil by physical effort or inducing chemical admixtures;
(d) Using external materials for strengthen the soil;
(e) Others.

The above mentioned methods are used to strengthen the soil as per the requirements. But various in-field techniques are used for the modification of soil, and they are represented by the schematic diagram (Fig. 8.2). Some of them are used for increasing the density of in-situ soil; sometimes, admixtures are used to improve the soil behavior, and some are used to control swelling and shrinkage of soil. Among all these methods, geosynthetics become popular in modern era for ground improvements. So, in the following sections importance is given to the application of geosynthetics for improving ground quality.

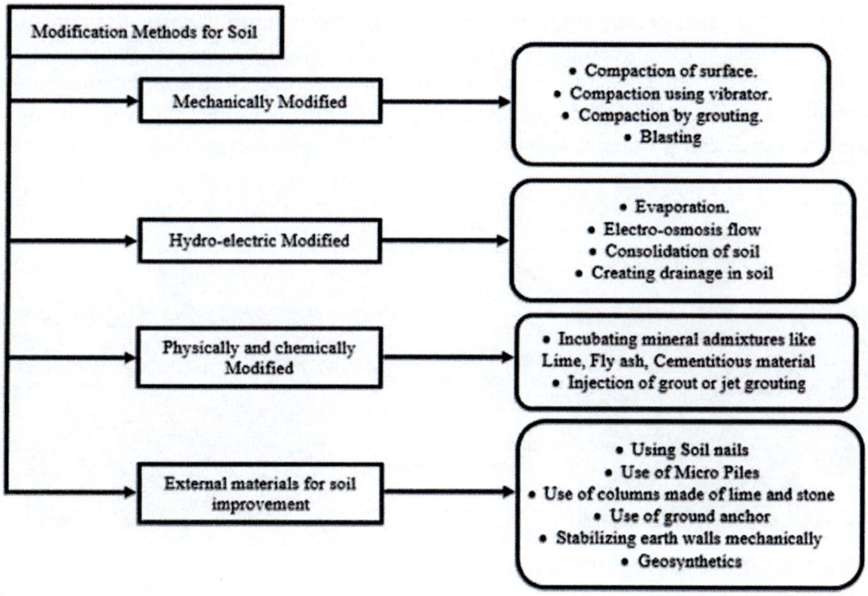

Fig. 8.2 Various soil modification techniques

Various Functions of Geosynthetics

As geosynthetics offer wide range of applications, so their uses are increasing rapidly in various geotechnical projects. But the selection of geosynthetics products depends on the requirement of the project. If the function of the project is multi-objective, sometimes combined geosynthetic products are used. This section provides a basic idea about the applications of geosynthetics which are categorized in six different types [14–19].

Separation

The process which prevents the mixing of two different layers of soil is known as separation. Geosynthetics are used to separate soils with different grain size. Geotextiles prevent the road base layer to penetrate the subgrade layer; because of that, integrity of the design and thickness of the pavement remain intact. Figure 8.3 demonstrates the separation function of geosynthetics.

Filtration

Geosynthetic materials allow the flow of water across the plane so that the soil particles can remain intact. Their working principal is like sand filter which allows the water to pass through soil without disturbing the upstream soil particles. They are used in coastal regions and in riverbanks to restrict soil erosion. Figure 8.4 shows the filter function of geosynthetics.

Drainage

Drainage is a process where water runs along the plane and improves the consolidation process of the soft soils. Where the soil is impermeable, geosynthetics act as a drain to carry water through the soil. Geotextiles are used to disperse pore water pressure at the base for roadways. Geocomposite drains are used as pavement edge drain, abutment drain, and slope seize drain. PVDs (prefabricated vertical drains)

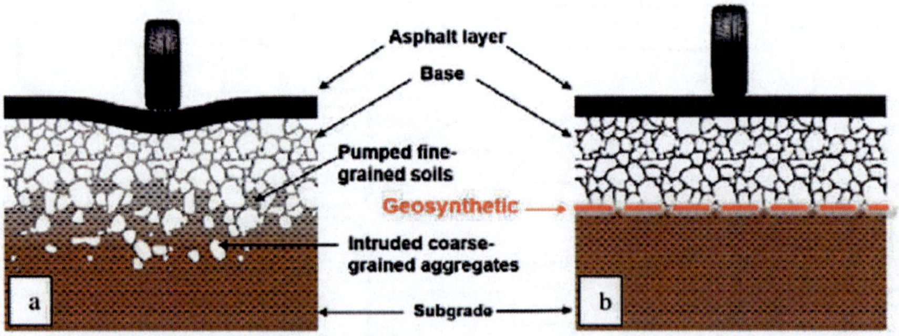

Fig. 8.3 Separation function of geosynthetics

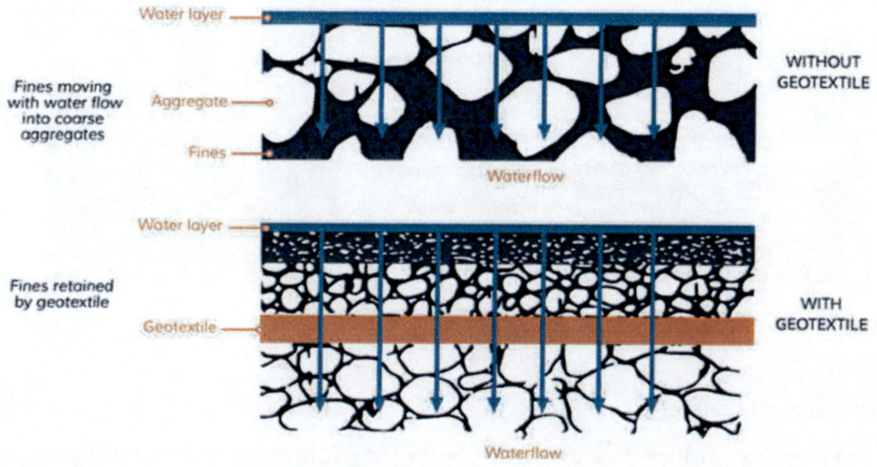

Fig. 8.4 Filtration function of geosynthetics

are used to improve the soft cohesive foundation clay under applied load. Figure 8.5 shows the drainage function of geotextiles.

Reinforcement

Reinforcements are used to improve the strength and durability of soil by imparting tensile strength to the soil. To enrich the stress distribution and to control the deformation of an unreinforced soil, geosynthetics are used as reinforcement material within the soil. In the field, geogrids are used as reinforcing material for increasing the tensile strength of soil mass. Figure 8.6 demonstrates the reinforcing function of geosynthetics.

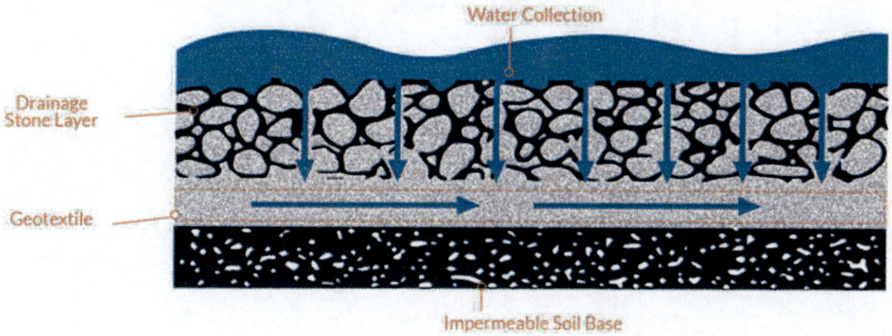

Fig. 8.5 Drainage function of geosynthetics

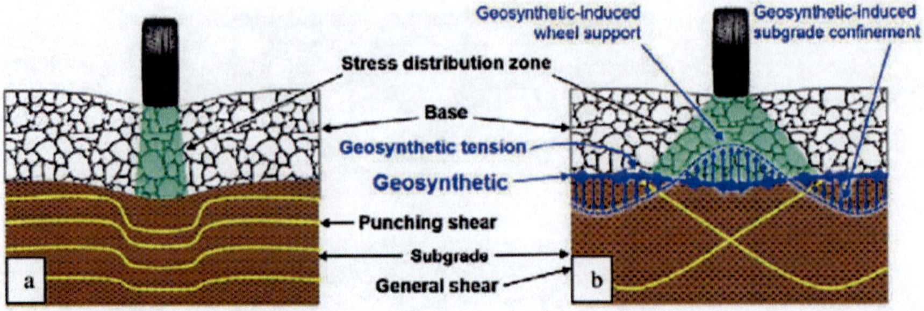

Fig. 8.6 Reinforcement function of geosynthetics

Protection (Cushion)

The way of controlling local damage of a geosynthetic material by the help of geotextile or geotextile-related material is termed as cushion. The main function of protection is to protect the primary geosynthetics from breach and abrasion. To reduce constructional damage, geotextile layers are placed over geomembrane. Figure 8.7 represents the protection function of geosynthetics.

Barrier (Waterproofing)

Geomembranes, GCLs, and geotextiles field-coated and thin-film are used as barrier material. These barriers are used where containment of soil can be made by waste materials; as a result, swelling of soils can happen. The performance of these geosynthetic barrier is more than thick soil layers, and they are economical and robust in nature. Figure 8.8 demonstrates the barrier function of geosynthetics.

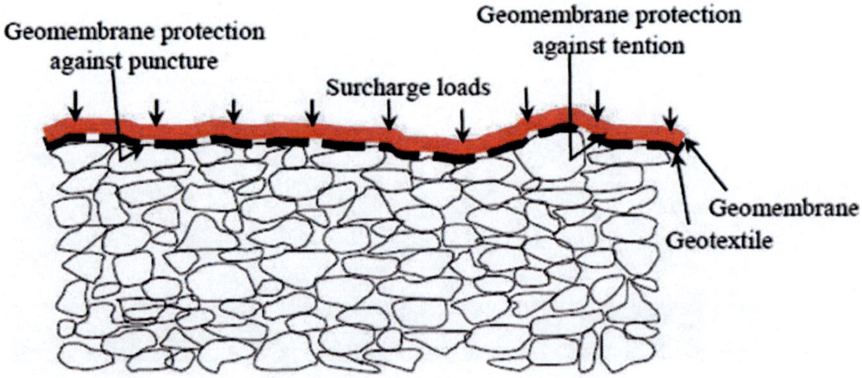

Fig. 8.7 Protective layers of geosynthetics

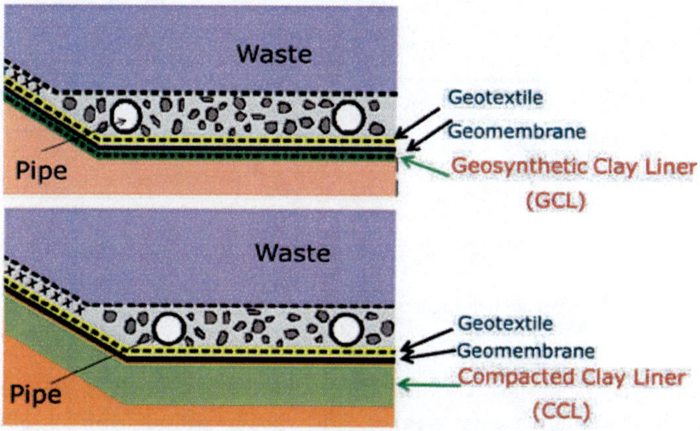

Fig. 8.8 Geosynthetics working as waterproofing agents

Previous Studies on Geosynthetic-Reinforced Soil

In this section, an effort has been made to review the applications of geosynthetics that have been used for various ground improvement projects. Geosynthetics are applied to enrich the tensile behavior of a soil medium which has strong compressive characteristics. As a result, the combination of soil and geosynthetics attain tensile and compressive strength both. The tensile forces of soil are transferred to geosynthetics by the help of induced friction. Parameters that influence the reinforcing function are tensile modulus, tensile strength, and shear strength at the interface of geosynthetic and soil medium [20]. Tensile strength helps to resist the tensile stresses that are transformed to soil. Interface shear strength is the main parameter for transferring stresses from soil to geosynthetics. Tensile modulus helps to restrict the total deformation of soil under allowable limit. The reinforcing function is divided into two parts tensile element and tensile membrane [21]. Figure 8.9a shows that a tensile member is supporting the planar load, and Fig. 8.9b shows that both planar and normal loads can be supported by tension membrane.

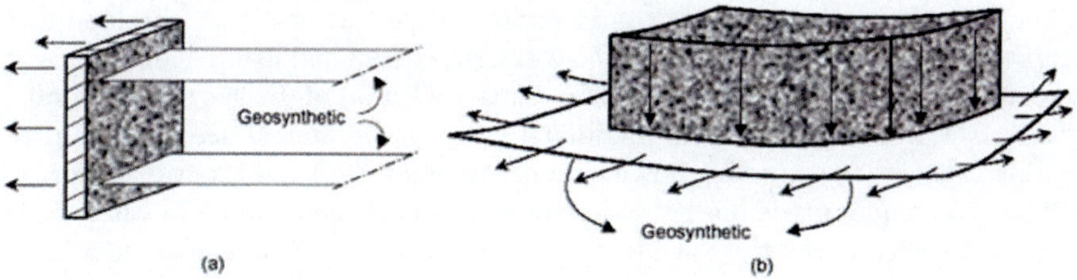

Fig. 8.9 **a** Tensile member is supporting the planar load, **b** normal and planar loads are supported by tension membrane [21]

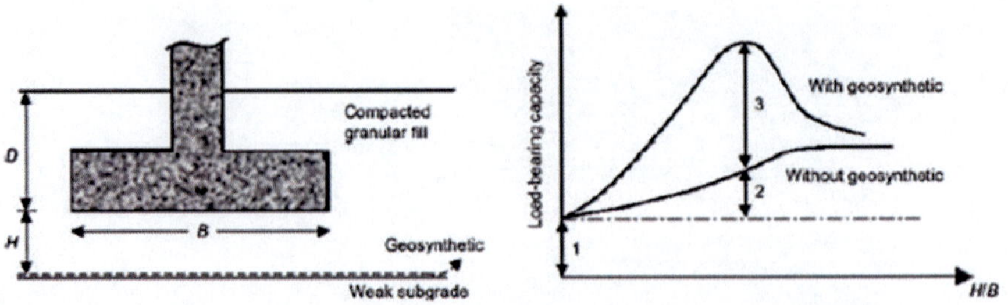

Fig. 8.10 Increment of ultimate bearing capacity by applying geosynthetics [25]

Later on, it has been observed that geosynthetics function as a tensile member due to the shear and anchorage mechanism between soil and reinforcement. Due to these, three different mechanisms are revealed—they are shear, anchorage, and tension membrane. Shear occurs due to sliding of soil particles over geosynthetics. Anchorage occurs due to pull-out action of geosynthetics, and the membrane mechanism occurs due to the deformation of geosynthetics under applied loads [22, 23]. Recently, geosynthetic-reinforced structures are used worldwide due to their high-performance index. These reinforced structures are divided into two group earth and load supported structure [24]. Load supported structures are used for supporting their own weights, and they do not have to carry external loads, e.g., foundations, pavements, and railways. Earth supporting structure must support their own weight as well as external weight, e.g., steep slope, embankments, and retaining walls. Here, more emphasis is given on the reinforcing function of geosynthetics for improving bearing capacity of soils. Figure 8.10 shows that the reinforcing behaviors of geosynthetics increase the ultimate bearing capacity of foundations [25].

A systemic research was carried out on circular footing where geogrid is incorporated with granular coarse filled layer on clay. It has been observed that for a single layer support, geogrid at 0.1D and 0.5D depth underneath the base of footing has shown better result where the granular layer has a thickness of 0.67D. The bearing capacity was increased up to 90%, whereas the settlement of soil was reduced to 53%. Inclusion of a second geogrid layer increases the bearing capacity up to 230% and vertical settlement is reduced to 60% only when the second layer is placed in between 0.15D and 0.3D under the first layer of geogrid [26]. While investigating effect of anchorage system in geogrid-reinforced soil, it has been observed that bearing capacity ratio for grid-anchorage system is 2.98 and using geogrid only it was 1.65. It clearly shows BCR (Bearing Capacity Ratio) of the geogrid anchorage system is 1.8 times greater than traditional geogrid. Traditional geogrids improve bearing capacity by restricting downward movement of soil and geogrid anchorage increase the angle of friction between soil and reinforcement [27]. In case of 3D geogrids which are triangular and rectangular in pattern, BCR was reported as 3.05 and 2.7, respectively, and load carrying capacity was also 1.85 times greater than single layered geogrid [28]. A series of direct shear tests conducted to examine shear behavior of normal and geogrid-reinforced sub-ballast interfaces. Nominal shear stress and rate of shearing varied from 20 to 100 kN/m^2 and from 2.5 mm/min to 10 mm/min, respectively. It was observed that use of geogrid reduced the breakage from 2.84 to 2.08% when shear stress was 20 kN/mm^2 and shearing was 2.5 mm/min [29]. From an investigation on the impact of geogrid on settlement and loadbearing capacity of a strip footing on soil, it has been observed that bearing capacity of the sand significantly improved and load bearing capacity was increased up to 20% and settlement of the soil is also decreased up to 25% [17]. FLAC-3D was used to model a square footing on geocell-reinforced soft clay. It has been observed that bearing capacity of geocell and geogrid reached up to 150 kPa, where individual bearing capacity when reinforced by geogrid was 70 kPa on use of geocell it became 110 kPa. The maximum improvement in bearing capacity was observed when the soft clay was reinforced with both geogrid and geocell [30]. Bearing capacity of a layered foundation system of clay and sand which is reinforced by planar geogrid and 3D geocell has also been experimented. When geogrid is used between sand bed and clay subgrade bearing weight decreased from 315 kPa at depth 0.63D to 203 kPa at depth 2.19D. But when geocells are used, bearing capacity was noted as 412, 510, 538, and 559 kPa at the depth of 063D, 1.15D, 1.67D, and 2.19D. When geocell-reinforced sand bed was used as base and geogrid overlayed clay was used, bearing capacity was observed as 720 kPa at depth 1.15D [31]. A strip footing rested on sand with biaxial geocell mattress have been examined, and the results showed that biaxial grid is better due to its high stiffness. Bearing capacity was 8 times of the unreinforced soil when geocells were placed at a depth of 0.1B from bottom of footing and formation of these geocells was V-stripe in fashion [32]. While examining bearing capacity of geotextile-reinforced sand, cement treatment had been used at the interface of geotextile and sand. The results shown that for a single layer of geotextile bearing capacity increased up to 1.46–2.22 times of unreinforced bearing capacity and when cement treatment is used bearing capacity increased up to 1.71–2.34 times [33]. But when geogrid and geocells are used, bearing capacity was 1.1–1.15 times of geotextile-reinforced footing [34]. A model test was conducted on geotextile-reinforced sand where number of layers varies from 1 to 4. Peak stress increased from 140 to 1060 kN/m^2 when number of layers increased from 1 to 4 and BCR also improved with geotextile layers, but rate of increment reduced when number of layers was increased in addition to reduction of settlement [35].

Conclusion

This is mainly a review paper which shows the various applications of geosynthetics and its implication as geosynthetic-reinforced soil systems. Also, an effort has made to study its effect on bearing capacity of soil and load displacement curves of that soil system. After a thorough review of literature on geosynthetic-reinforced soil, the following conclusions are drawn.

- Bearing capacity increases when 3–4 layers of reinforcement have been used beyond that no such significant result has been found.
- V strip geocells are more effective than diamond pattern.
- BCR of a grid-anchorage network is 1.8 times of normal geogrid reinforcement.
- Influence of depth on the placement of geogrids should be within the limit of 0.8–1.6B for better results.
- Three dimensional triangular geogrids show more efficiency than the conventional ones.
- Apart from these, use of geosynthetics in subgrade soil shows better performance for roads and railways.

References

1. Biswas A, Krishna AM (2017) Geocell-reinforced foundation systems: a critical review. Int J Geosynth Ground Eng 3:17. https://doi.org/10.1007/s40891-017-0093-7
2. Sarant S (2005) Engineering aspects of reinforced soil. Ind Geotech J 35(1)
3. Binquet J, Lee KL (1975) Bearing capacity tests on reinforced earth slabs. J Geotech Eng Div 101(12):1241–1255
4. Mandal JN, Manjunath VR (1995) Bearing capacity of strip footing resting on reinforced sand subgrades. Constr Build Mater 9(1):35–38
5. Al-Sinaidi AR, Ali AH (2006) Improvement in bearing capacity of soil by geogrid-an experimental approach. In: IAEG 2006, United Kingdom
6. Abu-Farsakh M, Chen Q, Sharma R (2013) An experimental evaluation of the behavior of footings on geosynthetic-reinforced sand. Soils Found 53(2):335–348
7. Jawad ZH, Shakir RR (2021) Behavior of foundation rested on geogrid-reinforced soil: a review. IOP Conf Ser Mater Sci Eng 1094(1)
8. Abdel-Rahman MM (2021) Review of soil improvement techniques. In: Advancements in geotechnical engineering. Springer, Cham, pp 175–199
9. Oleiwi GH et al (2021) A critical review of subgrade soil reinforced with geo-synthetics. IOP Conf Ser Mater Sci Eng 1076(1)
10. Sadeghi J et al (2020) Effectiveness of geogrid reinforcement in improvement of mechanical behavior of sand-contaminated ballast. Geotext Geomembr 48(6):768–779
11. Paikaray B et al (2021) Behavior of rectangular footing on geosynthetic reinforced crusher dust. In: Smart technologies for sustainable development. Springer, Singapore, pp 347–357
12. Sheikh IR, Shah MY (2020) State-of-the-art review on the role of geocells in soil reinforcement. Geotech Geol Eng 1–15
13. Lal D et al (2020) Effect of geocell on the behaviour of soil. IOP Conf Ser Mater Sci Eng 998(1)
14. Bayraktar OY (2020) Use of geosynthetics in road construction. Kastamonu Üniversitesi Mühendislik ve Fen Bilimleri Dergisi 6(2):107–113
15. Hegde AM, Palsule PS (2020) Performance of geosynthetics reinforced subgrade subjected to repeated vehicle loads: experimental and numerical studies. Front Built Environ 6:15
16. Sitharam TG, Hegde AM, Kolathayar S (2020) Geocells. Springer Singapore
17. Al-Subari L, Hanafi M, Ekinci A (2020) Effect of geosynthetic reinforcement on the bearing capacity of strip footing on sandy soil. SN Appl Sci 2(9):1–11
18. Kalidas P, Chavan AA (2020) A review on efficiency of geo grid in road pavement construction
19. Guo X, Zhang H, Liu L (2020) Planar geosynthetic-reinforced soil foundations: a review. SN Appl Sci 2(12):1–18

20. Pinto MIM, Cousens TW (2000) Effect of the foundation quality on a geotextile-reinforced, brick-faced soil retaining wall. Geosynth Int 7(3):217–242
21. Fluet JE (1988) Geosynthetics for soil improvement: a general report and keynote address. In: Symposium on geosynthetics for soil improvement at the ASCE convention American Society of Civil Engineers. No. Geotechnical Special
22. Jewell RA (1996) Soil reinforcement with geotextiles. Construction Industry Research and Information Association
23. Koerner RM (1998) Designing with geosynthetics. Prentice Hall, Englewood Cliffs, pp 709–714
24. Bonaparte R, Holtz RD, Giroud JP (1987) Soil reinforcement design using geotextiles and geogrids. In: Geotextile testing and the design engineer. ASTM International
25. Das BM, Maji A, Shin EC (1998) Foundation on geogrid-reinforced sand–effect of transient loading. Geotext Geomembr 16(3):151–160
26. Demir A et al (2014) Experimental and numerical analyses of circular footing on geogrid-reinforced granular fill underlain by soft clay. Acta Geotech 9(4):711–723
27. Mosallanezhad M, Hataf N, Sadat Taghavi SH (2016) Experimental and large-scale field tests of grid-anchor system performance in increasing the ultimate bearing capacity of granular soils. Can Geotech J 53(7):1047–1058
28. Makkar FM, Chandrakaran S, Sankar N (2017) Behaviour of model square footing resting on sand reinforced with three-dimensional geogrid. Int J Geosynthetics Ground Eng 3(1):3
29. Sweta K, Hussaini SKK (2019) Behavior evaluation of geogrid-reinforced ballast-subballast interface under shear condition. Geotext Geomembr 47(1):23–31
30. Hegde AM, Sitharam TG (2015) Three-dimensional numerical analysis of geocell-reinforced soft clay beds by considering the actual geometry of geocell pockets. Can Geotech J 52(9):1396–1407
31. Biswas A, Murali Krishna A, Dash SK (2016) Behavior of geosynthetic reinforced soil foundation systems supported on stiff clay subgrade. Int J Geomech 16(5):04016007
32. Lal D, Sankar N, Chandrakaran S (2017) Behaviour of square footing on sand reinforced with coir geocell. Arab J Geosci 10(15):1–8
33. Ouria A, Mahmoudi A (2018) Laboratory and numerical modeling of strip footing on geotextile-reinforced sand with cement-treated interface. Geotext Geomembr 46(1):29–39
34. Sridhar R, Prathapkumar MT (2017) Behaviour of model footing resting on sand reinforced with number of layers of coir geotextile. Innov Infrastruct Solutions 2(1):1–8
35. Tavangar Y, Shooshpasha I (2016) Experimental and numerical study of bearing capacity and effect of specimen size on uniform sand with medium density, reinforced with nonwoven geotextile. Arab J Sci Eng 41(10):4127–4413

Chapter 9
Coconut Tree Roots as Natural Geosynthetics

Leonardo Souza and Purnanand Savoikar

Introduction

Introduction

In today's world which is focused on sustainable infrastructure, the role of natural geosynthetics is set to play a major role. Geosynthetics interact with soil in various ways to improve the soil properties [1–3]. However, their function fails when they are damaged. Luckily, there is a natural self-repairable geosynthetic that has been used for thousands of years by Traditional Goan Saraswat Bunds. This paper aims to study that effect.

Traditional Goan Saraswat Bunds

On disappearance of River Sarasvati, a group of residents thereof headed by certain Brahmin rishis settled in Goa. They constructed Bunds to reclaim land and settled there [4, 5]. Traditional Goan Saraswat Bunds are unique in that they are reinforced by coconut roots. These roots act as natural geosynthetic as discussed below. They line rivers from which land was reclaimed and fields that were reclaimed (Fig. 9.3). Some of these fields with high salinity are called Khazans [6–9]. They are constructed from soil borrowed from the river stabilized by coconut leaf ash, jaggery molasses, lime added, and to soil in fixed proportions [10–12], and reinforced by coconut tree roots. TGSB has several benefits. They enhance Scenic beauty while reducing pollution. Their fibrous roots compact the soil. Their growth encourages growth of other plants

L. Souza · P. Savoikar (✉)
Department of Civil Engineering, Goa Engineering College, Farmagudi, Goa 403 401, India
e-mail: psavoikar@gmail.com

K. Muthukkumaran et al. (eds.), *Transportation and Environmental Geotechnics*, Lecture Notes in Civil Engineering 298, https://doi.org/10.1007/978-981-19-6774-0_9

and grasses. TGSB can be maintained by anyone, and they also generate income. They are perfect for tropical rural coastal and hilly infrastructure where coconut trees grow easily.

Comparison of Coconut Root with Other Geosynthetics

Today's development focus is on natural products, for example, jute and coir. Sustainable infrastructure with biostabilization, bioremediation, and bio-engineering are driving today's research [13–16]. There exists ample literature on roots, natural fibers, and coir [17–24]. There is a paucity of technical literature dealing with coconut tree roots.

Coconut tree root systems mechanically enhance soil stability by three mechanisms: (a) Root mat on the top in surface layers hold soil particle together thus increases the shear strength of soil (c-ϕ), and the grasses that grow near the roots increase this, (b) water uptake by the roots increases the stability of slope, and (c) tree root systems that penetrate deep have soil reinforcement action as explained below.

Although coconut tree roots are weaker than synthetic geotextiles individually, due to their density of root per meter square they collectively perform better than standard geotextiles having superior penetration, greater reach, and increased anchorage (Fig. 9.1). They are well suited for slope stabilization since roots can penetrate substrata including rocks thus tying together underlying fractured rock-mass together. Additionally, they are self-repairable and grow back when injured or broken thus ensuring continued strength. As their density grows during their lifetime, the factor of safety proportionately increases.

Coconut tree roots grow even after the embankment is built unlike geosynthetics. Due to scar tissue growth, roots get additional strength on strain hardening.

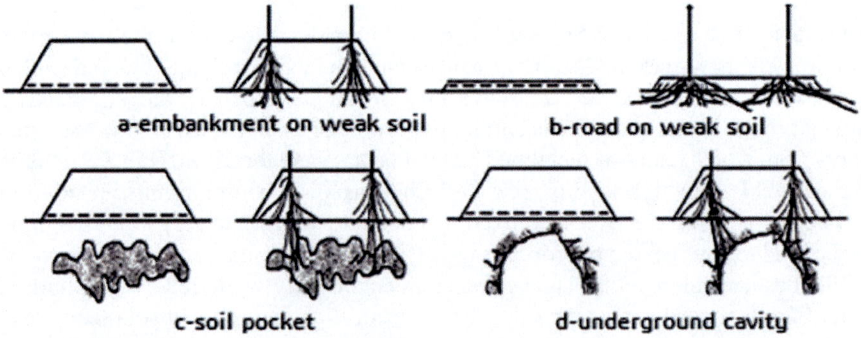

Fig. 9.1 Comparison of weak subsoils reinforced by geosynthetic and coconut roots (showing penetration of subsoil and increased anchorage of coconut roots)

Geosynthetics Action of Coconut Root

Geosynthetics Action of Coconut Root

Geosynthetics are materials that are artificially inserted or incorporated into the soil to improve their properties. Geosynthetics are a major diverse group of materials and include all polymer-based fabrics fibers and cotton, coir, and jute fibers. Geosynthetic materials in general have following main functions: separation of dissimilar materials; reinforcement of soil masses; filtration of soil, controlling solid particles transport within soil; drainage pathways within the soil mass; and impede fluid flow by acting as a containment/flow barrier [25]. Coconut root as a Geosynthetic performs the functions of filtration, and reinforcement involving interactions with the surrounding soil. The roots perform the filtration function by the matric root suction. The soil reinforcement functions (Fig. 9.2, Table 9.1) are as follows: soil nailing action, RE wall action, micropile action, soil anchor action, root pile action, and soil shear property (c-ϕ) improvement action. All these actions occur simultaneously, and hence, the effect is an integrated one.

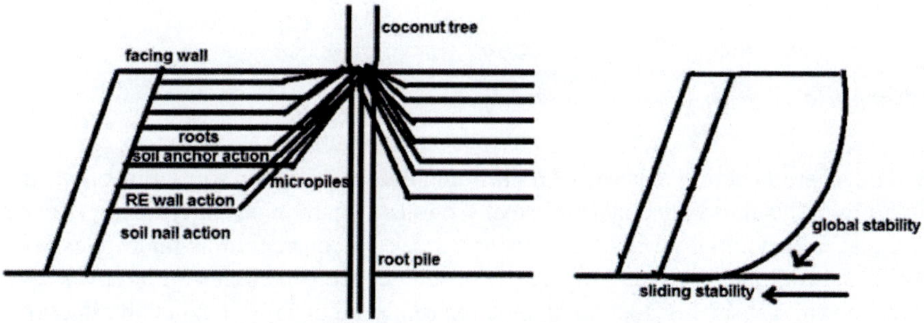

Fig. 9.2 Geosynthetic action of coconut tree roots and types of stability

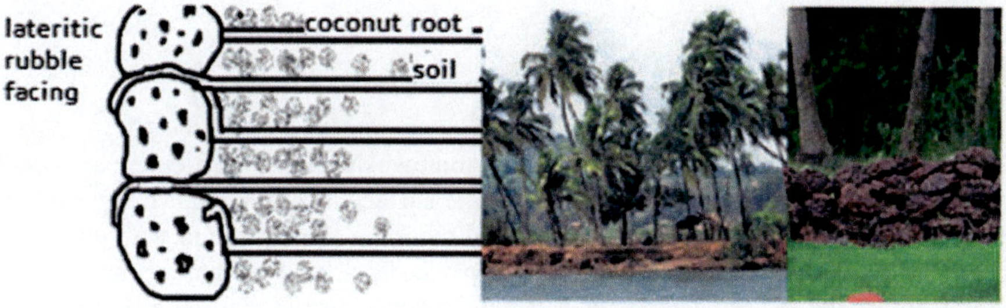

Fig. 9.3 Coconut roots found on TGSB attach to outer rubble layer act as soil nails

Table 9.1 Geosynthetic action of coconut tree roots

No.	Geosynthetic action
1	RE wall
2	Micropile
3	Soil anchor
4	Root pile
5	Soil shear improvement
6	Soil nailing

RE Wall Action

Reinforced wall action consists of reinforcing the soil by using a strip element attached to a panel. Panels normally consist of pre-cast concrete, and the strip consists of steel material compacted into place. A Reinforced Earth Wall is a paneled wall with a reinforcement starting from it and extending inwards crossing the slip circle and failure wedge. Each rubble facing of the wall acts as a mini panel with a couple of roots attached. Each can thus be analyzed just like a regular paneled Reinforced earth wall section.

Micropile Action

Soil micropiling action consists of reinforcing the soil by using a pile-pile elements. Soil micropiles normally consist of steel sections (usually pipes or *H* beams) driven into place. The initial parts of the tree roots can be considered as micropiles. The horizontal root mat is about 2 m deep. It penetrates the slip circle and failure wedge at the top two thirds. Just like reticulated micropiles, they are spread in all directions and can be grouped together for analysis as a set of micropiles 2 m deep at all angles in every direction.

Soil Anchor Action

Soil anchor action consists of reinforcing the soil by using a tension element attached to anchor block in the soil and tightened at the surface. Soil piles normally consist of steel material placed into bored holes, and then, the tip is grouted with concrete to create the anchor. It is then tightened in place. Each root of the coconut tree is tied at both ends. At the facing wall, it gets attached to the wall (Fig. 9.3). They act as two-way anchored cable. They can be analyzed for their soil anchor action considering them as cables anchored to the stem. The root length averages 5–10 m wide in horizontal direction depending on the type of soil. It is longer in sandy and cohesionless soils and shorter in soil with fines.

Root Pile Action

Soil piling action consists of reinforcing the soil by using a pile element. Soil piles normally consist of concrete, timber, or steel material inserted into place. The vertical roots of the coconut tree act as a root pile of approximate 50-cm diameter and 5–10 m deep depending on the seasonal fluctuation depth of water table. Dry arid areas have deeper thinner root pile. Tropical areas have shallower root pile. This pile is deep enough to penetrate the slip circle and failure wedge and anchor the soil just like a normal pile. The additional flexibility of the pile gives a greater safety to a slope reinforced by coconut tree root.

Biostabilization-Soil Shear Property (C-ϕ) Improvement Action

The soil parameters of coconut root reinforce soil are higher than those of soil without roots. The cohesion increases due to the biostabilization effect of the gums (mucilage and Mucigel) secreted by the coconut tree roots. This Mucilage, which is secreted by root cells as the root grows through the soil, consists of polysaccharides containing hexose and pentose sugars and uronic acids. Rhizosphere microbes also secrete Supplementary mucilage. On contact with the soil, mucilage forms a gel, which binds soil particles and microbes together with the root to form a "rhizosheath." This mixture of soil particles with gelatinous material and microorganisms is termed as Mucigel [26] which has a number of beneficial properties. The growth of roots causes soil to press against them causing an increase in the friction angle. The *c*-value improves by almost 50–100% due to roots.

Soil Nailing Action

Soil nailing action consists of reinforcing the soil by using closely spaced reinforcing tension elements. Soil nails normally consist of polymeric or steel material inserted into pre-bored holes and grouted into place (Fig. 9.4). Coconut tree roots attach themselves to the facing stones and at failure act as soil nails (Fig. 9.3).

Soil nail systems have varied and complex failure conditions and thus require careful review and analysis of available data. The Indian standards however do not provide the designer with an adequate design process that can be used in an Indian

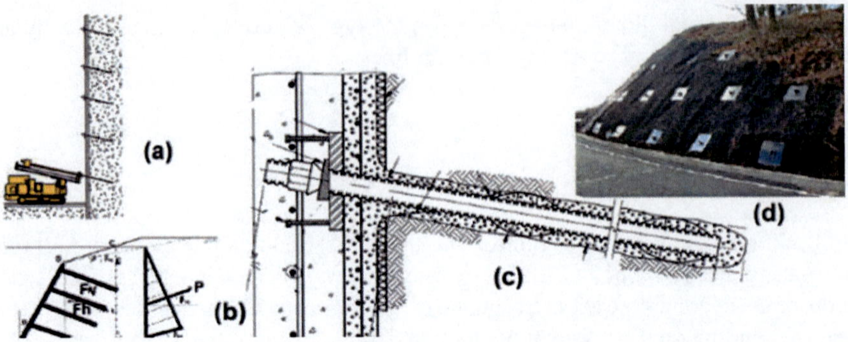

Fig. 9.4 Conventional soil nails **a** driving nails **b** stability diagram **c** soil nail **d** application

context so some reference to foreign codes is necessary. We can refer to a series of foreign codes like the Australian code AS4678-2002 [27], American Code [28], and British codes [29] to arrive at a method to apply in this case. Geotechnical engineers largely prefer soil nailing as an efficient and cost-effective stabilization technique for vertical cuts and slopes. Customary design and process are time-consuming; hence, the possibility of mistakes is there which necessitates a simplified method for designing the soil nail walls [30]. Soil nail provides acceptable tensile strength. This helps in binding the active soil zone in the resistant zone [31]. Soil nailing is an effective, cost-effective, and practicable option to support near vertical even vertical soil cutting which can be used in different slope stability geotechnical applications [32]. Together the soil nails and the facing contribute to the stability of the excavation [28] just like in TGSB coconut tree roots. Soil nail walls are constructed using a "top–down" construction sequence [28], and coconut tree roots also develop in atop down sequence. Due to this, only this action has been analyzed in this study.

Properties of Coconut Roots as Geosynthetics

In a retaining wall with coconut root as soil nails, the material behavior and properties—the base soil, the root reinforcement and the lateritic rubble facing element—and their mutual interactions significantly affect the performance of structure [33]. Soil of the bund was tested and classified in the laboratory of Goa Engineering College as per standard norms and practices [34–36]. Soil was found to be sandy with fines and presence of organic matter. Bund Soil is SM-SW and Root Soil is SM-OI as per IS 1498-1970 classification.

Root adhesion varies with soil shear strength (Fig. 9.5) which was found out from pullout test for roots in a tank in laboratory (Fig. 9.6).

For simplification sake, the root distribution is assumed as uniformly 10 cm c/c in both directions. Roots were tested in laboratory (Fig. 9.6).

Fig. 9.5 Variation of root adhesion with soil shear strength

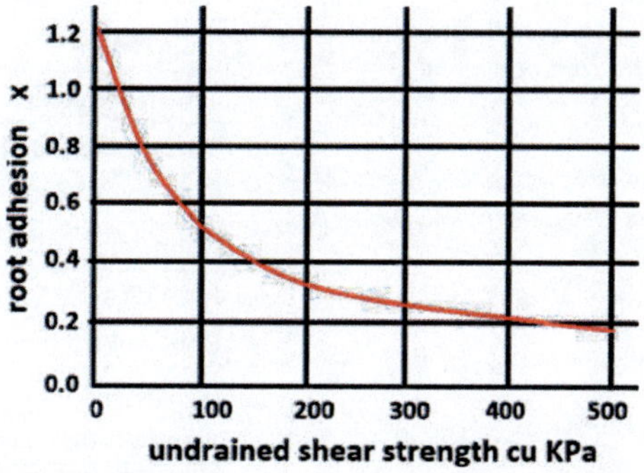

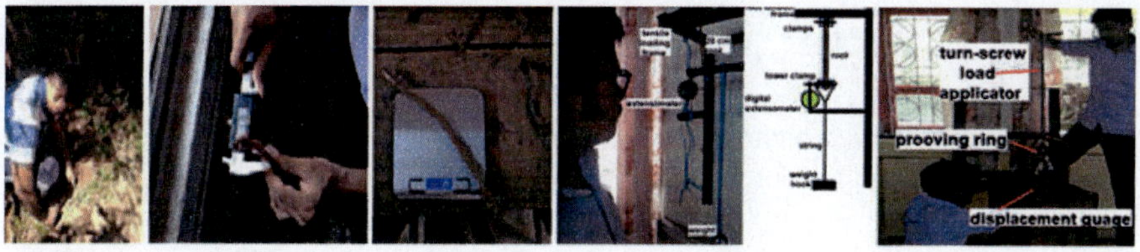

(a) (b) (c) (d) (e)

Fig. 9.6 Testing of roots/measurement for **a** length **b** diameter **c** weight **d** tensile strength **e** adhesion

Table 9.2 Values for soil and root

Soil	Density	Cohesion	Friction
	γ (kN/m^3)	c (kN/m^2)	($\phi°$)
Bund soil	20	11	38
Root soil	15	30	40

For geosynthetic soil nail action, soil properties (Table 9.2) and root properties (Table 9.3) were used and factors of safety were calculated, where q_u is root friction contribution of soil, z is depth of soil, H is height of TGSB (3 m), ϕ is the failure wedge angle, and factor of safety in tension = $\text{FoS}_t = 1.8$.

Factor of Safety for Soil Nailing Action

A detailed design procedure as well as explanation of various terms involved in design methodology for steel soil nails is given in FHWA manual. The appropriate design had been adapted for coconut tree root in this design. The embankment safety due to presence of coconut tree roots can be thought of as soil nail action. Soil is reinforced by the roots as each root acts as a soil nail.

Safe Maximum Tension in longest root

Table 9.3 Root properties for reinforcement

Properties	Notation	Unit	Value
Tensile strength	f_{yr}	kN/m^2	2×10^4
Cross-sectional area	A_r	m^2	2×10^{-5}
Length	l_r	m	5
Diameter	d	m	0.005
Spacing	$s_v = s_h$	m	0.21
Root soil friction	δ		0.67
Adhesion	α		0.25
Skin friction factor	K	Graph	4

$$T_{\max} = \frac{A_r f_{yr}}{\text{FoS}_t} = \left(\frac{1 - \sin \emptyset}{1 + \sin \emptyset}\right)(q_u + \gamma z)s_h s_v \tag{9.1}$$

Find q_u from above equation $q_u = 65.5$ N/mm^2
Maximum tension in root is $T_{\max} = 0.4$ N/mm^2.
Root Pullout Capacity in longest root

$$R_{pr} = (\pi d)l_r q_u \tag{9.2}$$

Root pull-out capacity $R_{pr} = 5.14$
Equivalent root strength for all roots in the failure wedge

$$T_{eq} = \frac{\sum R_{pr}}{s_h} \tag{9.3}$$

Failure wedge weight

$$W = \gamma \left(0.5H^2 \cos \varphi\right) \tag{9.4}$$

Horizontal pressure

$$P = \left(\frac{\gamma H^2}{2}\right)\left(\frac{1 - \sin \emptyset}{1 + \sin \emptyset}\right) \tag{9.5}$$

Factor of safety global with root

$$\text{FoS}_g = \frac{T_{eq} \cot \varphi + \left[W \cot \varphi + T_{eq} \sin \varphi\right] \tan \emptyset}{W \sin \varphi} \tag{9.6}$$

Factor of safety global without root

$$\text{FoS}_g = \frac{[W \cot \varphi] \tan \emptyset}{W \sin \varphi} \tag{9.7}$$

Table 9.4 Factor of safety for soil nail action of coconut roots

Property	Value
H (m)	3
W (kN)	69.11
P (kN)	26.73
T_{eq} (kN)	61.68
FoS_g	1.54
FoS_s	4.23
FoS_p	12.85

Table 9.5 Combined factor of safety for root reinforcement action

Action	Sliding	Global
Soil nailing	4.23	1.54
FoS plain	1.32	1.194
Increase	1.87	0.454
% Increase	320	77.5

Factor of safety sliding with root

$$FoS_s = \frac{q_u L + \tan \text{Ø}(W + P)}{W} \tag{9.8}$$

Factor of safety sliding without root

$$FoS_s = \frac{W \tan \text{Ø}}{P} \tag{9.9}$$

Factor of safety pullout

$$FoS_{pr} = \frac{R_{pr}}{T_{max}} \tag{9.10}$$

Factor of safety pullout is $FoS_p = 12.85$.

The factor of safety against global instability is 1.54 with roots as compared to 1.194 without roots, an increase of 77%. The factor of safety against sliding is 4.23 with roots as compared to 1.32 without roots, an increase of 320% (Tables 9.4 and 9.5).

Conclusions

Traditional Goan Saraswat Bunds use coconut tree roots as natural geosynthetics. The technology was developed by the ancient Saraswats who settled in Goa. Coconut root as a geosynthetic performs the functions of filtration, and reinforcement: soil nailing action, RE wall action, micropile action, soil anchor action, root pile action, and soil shear property (c-ϕ) improvement action. Soil nail action is investigated and found to give a factor of safety against global instability increase of 77%, and a factor of safety against sliding increase of 320% with roots as compared to without roots. This shows that the coconut tree root is a very good natural geosynthetic soil reinforcement. It must be adopted for other small height embankments all over wherever coconut trees can grow.

References

1. Zornberg JG (2011) Advances in the use of geosynthetic in pavement design. In: Proceedings of the second national conference on geosynthetics, geosynthetics India'11, vol 1. India Institute of Technology Madras, Chennai, India, pp 3–21
2. Handbook for Geo-synthetics (2013). www.gmanow.com
3. Haghi AK (2019) Experimental analysis of geotextiles & geofibers composites. WSEA Press
4. Keni C (1998) Saraswats in Goa and beyond. Vasco-da-Gama, Goa: Murgaon Mutt Sankul Samiti, MMS ID 9927485283503681, Van Pelt—Non Circulating DS432.S35 K46
5. Keni C (2008) The Saraswats. Sahyadri Books. 9788190757102 I HB I pp 446 I 2008 I V. M. Salgaocar Foundation
6. D'Silva R, Barreto E (2012) Goa Government's Dilemma. http://www.iosrjournals.org/iosr-jbm/pages/ies-mcsc-volume-2.html
7. Iyer H (2014) The lost Khazans of Goa–India Water Portal. www.indiawaterportal.org.articles.khazans-Goa
8. Kamat N (2004) History of Khazan land management in Goa—ecological, economic and political perspective. In: Seminar on history of agriculture in Goa. Department of History, Goa University
9. Sonak SM (2008) Khazan ecosystems of Goa—building on indigenous solutions to cope with global environmental change. In: Advances in Asian human-environmental research Springer book series (AAHER). ISBN 978-94-007-7201-4
10. Souza L, Naik A, Chanekar TP, Savoikar P (2016) Stability of traditional 'Bundhs'—earthen levees—from Goa. In: Proceedings of Indian geotechnical conference, IGC 2016-15-17 December. IIT Madras, Chennai, India
11. Souza L, Savoikar P (2019) Design of traditional Goan Saraswat bunds. In: Proceedings of Indian geotechnical conference, IGC 2019. Surat, Gujarat
12. Souza L, Savoikar P (2017) Ground improvement using coconut leaf ash. In: Proceedings of the Indian geotechnical conference IGC 2017. Indian Institute of Technology, Guwahati, Assam
13. Carmona JP, Oliveira PJ, Lemos LJ (2016) Biostabilization of a sandy soil using enzymatic calcium carbonate precipitation. In: Proceedings of 3rd international conference on transportation geotechnics, advances in transportation geotechnics, vol 3
14. Pepe G, Baudinelli E, Zanini M, Calcaterra D, Cevascol A, Scarpellini P, Firpo M (2020) Application of bioengineering techniques as geo-hydrological risk mitigation measures in a highly valuable cultural landscape: experiences from the Cinque Terre National Park (Italy). Sustainability 12:8653. https://doi.org/10.3390/su12208653. www.mdpi.com/journal/sustainability
15. Preti F, Giadrossich F (2009) Root reinforcement and slope bioengineering stabilization by Spanish Broom (*Spartium junceum* L). Hydrol Earth Syst Sci 13:1713–1726
16. Polster DF (2003) Soil bioengineering for slope stabilization and site restoration. botanicgardens.uw.edu›uploads›sites›2013/12. In: Mining and the environment III. Laurentian University, Sudbury, Ontario, Canada

17. Ali N, Farshchi I, Mu'azu MA, Rees SW (2012) Soil-root interaction and effects on slope stability analysis. EJGE Bund C 17:319–328
18. Anggraini V (2016) Potential of coir fibres as soil reinforcement. PJSRR Pertanika J Sch Res 2(1):95–106. eISSN: 2462-2028
19. Das D, Kaundinya D, Sarkar R, Deb B (2016) Shear strength improvement of sandy soil using coconut fibre. Int J Civ Eng Technol (IJCIET) 7(3):297–305
20. Fatahi B, Indraratna B, Khabaz H (2007) Soft soil improvement by tree root suction. Aust Geomech J Aust Geomech Soc 42(4):13–18
21. Fathi L (2014) Structural and mechanical properties of wood from coconut palms, oil palms and date palms. PhD Thesis, University of Hamburg, Hamburg, Germany
22. Satyanarayana KG, Pillai CKS, Sukumaran K, Pilllai SGK, Rohatgi PK, Vijayan K (1982) Structure property studies of fibres from various parts of the coconut tree. J Mater Sci 17:2453–2462
23. Gray DH, Ohashi O (1983) Mechanics of fiber reinforcement in sand. J Geotech Eng 109(3):335. Ascelibrary.org (ASCE) 0733-9410
24. Kalita et al (2016) Comparative study of soil reinforced with natural fiber, synthetic fiber and waster material. IJLTET 6. ISSN 2278-621X
25. Gohil DP, Solanki CH, Dessai AK (2009) Application of geosynthetic for ground improvement: an overview. In: Proceedings of Indian geotechnical conference IGC 2009. GeoTide, Guntur, Andhra Pradesh
26. Bending GD (2003) The rhizosphere and its microorganisms in encyclopedia of applied plant sciences. Science Direct
27. Bridges C (2017) Design of soil nailed walls according to AS4678-2002. In: Proceedings of 8th Australian small bridges conference
28. FHWA-NHI-14-007 Soil Nail Walls Reference Manual U.S. Department of Transportation Publication No. Federal Highway Administration FHWA GEC 007 (2015)
29. BS 8006-2: code of practice for strengthened/reinforced soils, Part 2: soil nail design, BRITISH STANDARD (2012)
30. Zand-Parsa K, Zand-Parsa K (2008) The simplified KZP5 method for soil nail design in granular soils. In: Proceedings of the international conference on case histories in geotechnical engineering, 29. https://scholarsmine.mst.edu/icchge/6icchge/session05/29
31. Panigrahi RK, Dhiman G (2019) Design and analysis of soil nailing technique for remediation of landslides. Highw Res J 10(1):24–32
32. Babu SG (2020) Ground reinforcement using soil nailing. NPTEL Course, lecture 33
33. Babu SG, Singh VP (2009) Appraisal of soil nailing design. Indian Geotech J 39(1):81–95
34. IS 2720-Part 8: methods of test for soils—determination of water content—dry density relation. Bureau of Indian Standards, New Delhi (1983)
35. IS 1498: Classification and identification of soils Bureau of Indian Standards BIS, New Delhi, (1970)
36. Soil Classification Basics: Description of USCS, ASTM & AASHTO Soil Classification Basics, PPT, University of Massachusetts Lowell (2015)

Chapter 10
Experimental and Numerical Study on California Bearing Ratio for Two-Layer Soil Systems

Razib Hussain and Amit Kumar Das

Introduction

Roads and highways play a vital role in transportation and economic growth of a nation. Construction of roads and highways require construction of pavement which is a multi-layer structure constructed on the subgrade. The main function of a pavement is to reduce the stresses transmitted due to vehicle loads and distribute the loads through various layers of the pavement structure to the natural soil subgrade. The design of a pavement depends mainly on the strength of the subgrade. Therefore, before construction of pavement, it is necessary to determine the strength of the subgrade. Among many methods developed to determine strength of subgrade, design philosophy based on the California bearing ratio (CBR) gained wide acceptability [1, 2]. Thus, for a road design, it is important to evaluate CBR values of the subgrade soil. Natural deposit in general depicts a layered lithology. Therefore, the knowledge of CBR of top soil layer along with effect of CBR of underneath layers on it is important. Moayed et al. [3] conducted CBR test on unreinforced and reinforced two-layered soil samples consisting of sand as the base layer overlying a cohesive soil. From the test results, it was concluded that the threshold value of sand layer thickness is dependent on the strength of subgrade clayey soil, which is dependent on the compaction moisture and soaking condition. Caicedo and Mendozaa [4] in their study presented the influence of the variability of soil parameters on the California Bearing Ratio tests by means of numerical simulations. From the study, it was observed that the CBR value depends on Young's modulus as well as shear strength parameters such as apparent cohesion and friction angle. Ronad and Rakaraddi [5] attempted to find CBR values in layered soil for both soaked and unsoaked condition. Soils used in this study were black cotton soil and red soil. It was found that

R. Hussain · A. K. Das (✉)
Department of Civil Engineering, National Institute of Technology Silchar, Silchar, Assam 788010, India
e-mail: amitkdas@civil.nits.ac.in

K. Muthukkumaran et al. (eds.), *Transportation and Environmental Geotechnics*, Lecture Notes in Civil Engineering 298, https://doi.org/10.1007/978-981-19-6774-0_10

addition of coarse-grained soil layer over fine-grained soil layer results in increase in CBR value. Osouli et al. [6] in their study attempted to compare results of soaked and unsoaked CBR tests conducted on unbound aggregate samples, and it was found that higher plasticity indices have a negative effect on both soaked and unsoaked CBR values. Determination of CBR of soil from laboratory tests requires time and energy, and there are always chances of human error where complicate situations are considered. Therefore, various researchers [7–14] tried to estimate CBR values of soil through various experimental and statistical approach and established correlations between CBR and various soil parameters.

It is observed that California bearing ratio (CBR) for soils has been studied extensively. Most of these studies are based on experimental determination of CBR. Along with experimental determination, statistical analysis and numerical modeling of CBR are also studied in different research works. Numerical modeling of CBR based on shear strength properties of soil may reduce the effort and time consumption, but numerical model for CBR determination of layered soil is very limited. In addition, contribution of underneath soil layers on CBR of top soil layer is not sufficiently addressed. Therefore, in the present study, an effort has been made to develop a numerical model validated with experimental results for determination of CBR of layered soil.

Experimental Study

Experimental study was performed on soil samples comprising single as well as double layers of soils. Barak river sand and locally available clay from nearby area of NIT Silchar, Assam, India, were used to prepare the samples for CBR test. Various laboratory tests were also conducted on sand and clay soil to determine their physical properties. The same properties were used to develop numerical model for simulation of CBR test in ABAQUS 2018. Physical properties of sand and clay soil determined from various laboratory tests are presented in Table 10.1.

Sample Preparation for CBR Test

CBR test was conducted following IS: 2720 (Part 16)-1987 [15]. For CBR test on single-layered soil, about 5 kg of the soil passing through 20 mm sieve was taken and mixed thoroughly after adding water as per optimum moisture content. The soil was compacted in 3 layers giving 56 number of blows, and unsoaked CBR test is conducted by applying load keeping the rate of penetration at 1.25 mm/min till 12.5 mm penetration was reached. While to conduct soaked CBR test, CBR mold along with the soil is put under water for four days to attain fully saturated soil condition and soaked CBR test is conducted after that.

Table 10.1 Physical properties of sand and clay

Properties	Sand	Clay
Cohesion, *c*	0 kPa	15 kPa
Angle of internal friction, ϕ	30°	6°
Young's modulus, *E*	34 MPa	10 MPa
Maximum dry density, ρ	1.764 g/cc	1.349 g/cc
Coefficient of permeability, *k*	4×10^{-4} m/s	3×10^{-9} m/s
Void ratio, *e*	0.508	1.007
Specific gravity	2.66	2.71
Optimum moisture content	10.51%	26%
Soil classification	SP	CH
Liquid limit		65%
Plastic limit		42%
Plasticity index		23%

For double-layered soil, about 2.5 kg of the sand and 2.5 kg of clay passing through 20 mm sieve was taken and mixed thoroughly with their respective optimum moisture content. Sand was placed at the top layer up to half depth followed by bottom half filled up by clay layer with proper compaction. Sand was compacted and CBR test was conducted by applying loading rate of 1.25 mm/min till 12.5 mm penetration was reached. Same procedure was followed for clay on top layer and sand at the bottom layer. CBR values were calculated from load-settlement curve as per IS: 2720 (Part 16)-1987 [15].

Numerical Modeling

3D numerical model was developed using an axisymmetric model with SWEEP feature in ABAQUS. Thus, a finite element axisymmetric model was developed in the ABAQUS/CAE 2018 by rotating a 2D plane of dimension 0.075 m × 0.1273 m for 360° to simulate soil sample prepared in standard CBR mold. The elements selected for the mesh were of type CAX4P (4-node axisymmetric quadrilateral, bilinear displacement, bilinear pore pressure), and mesh was generated with element size of 0.005 mm as shown in Fig. 10.1.

Soaked or fully saturated samples were generated with steady-state pore fluid response and applying saturation ratio of 1 as predefined field in initial step along with void ratio and geostatic stress. Unsoaked soil samples were simulated defining material properties appropriately. In such samples, sorption was defined as a material property in a tabular form with negative pore pressure and saturation in such a way that a constant partial saturation was maintained in the soil. ABAQUS/Standard assumes fully saturated flow if sorption was not defined. Gravity load was applied to the whole model, and zero pore pressure was applied at the top surface. Boundary condition of

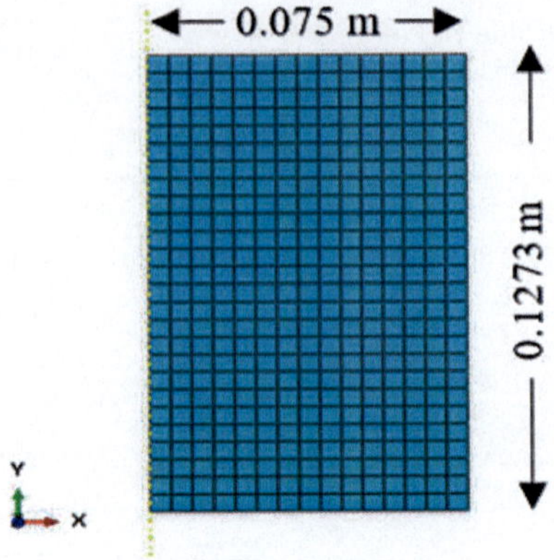

Fig. 10.1 Meshing of soil model

type XSYMM was selected for the left side of the model. Boundary condition of type Displacement/Rotation restricting U1 and U2 was applied at bottom, whereas only U1 was restricted at right side of the model. Surcharge weight of 5 kg was applied at the top, and vertical confinement was attained through boundary conditions. Figure 10.2 shows the boundary conditions applied to the soil model. Both soaked and unsoaked models were finally submitted for the analysis, and output file was generated. All the samples were developed for various combination of soil layers in similar manner.

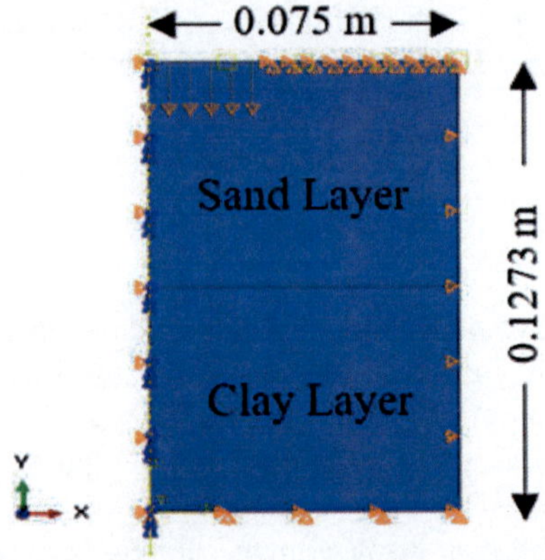

Fig. 10.2 Boundary conditions in soil model

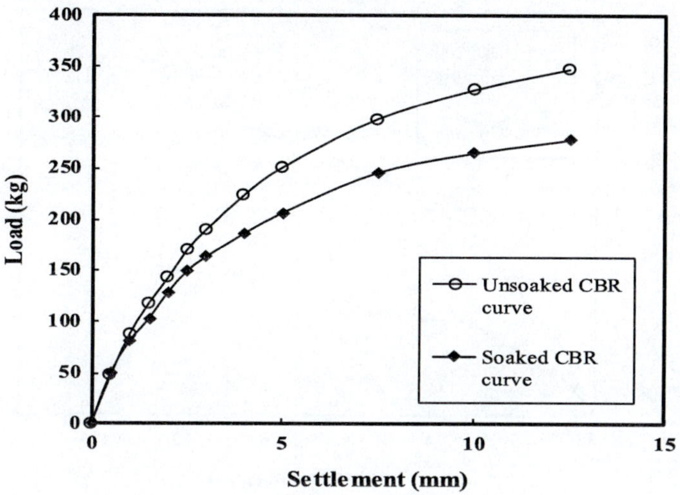

Fig. 10.3 Load-settlement curve for single layer sand

Results and Discussion

Results of Laboratory CBR Test

CBR tests were conducted in laboratory on single- and double-layered soil, and load-settlement curves were plotted for both soaked and unsoaked samples. Figures 10.3 and 10.4 show the load-settlement curves obtained for both unsoaked and soaked conditions when soil sample is single layer sand and clay, respectively. Figure 10.5 shows load-settlement curve for two-layered soil sample with sand as top layer and clay as bottom layer, whereas Fig. 10.6 shows load-settlement curve for layered soil with clay as top layer and sand as bottom layer for both unsoaked and soaked conditions. It is observed that the magnitude of load at 12.5 mm settlement is substantially higher for unsoaked sample than the soaked sample except in case of layered soil with sand as top layer and clay as bottom layer, where it is only marginally higher.

Results of Numerical Analysis

The same samples tested in laboratory were simulated in 3D numerical analysis using ABAQUS 2018. Figures 10.7 and 10.8 show the family of load-settlement curves obtained from numerical analysis for different combination of soil layers in unsoaked and soaked conditions, respectively. It is observed that the unsoaked samples offer better resistance to penetration compared to same soaked samples as in general expected.

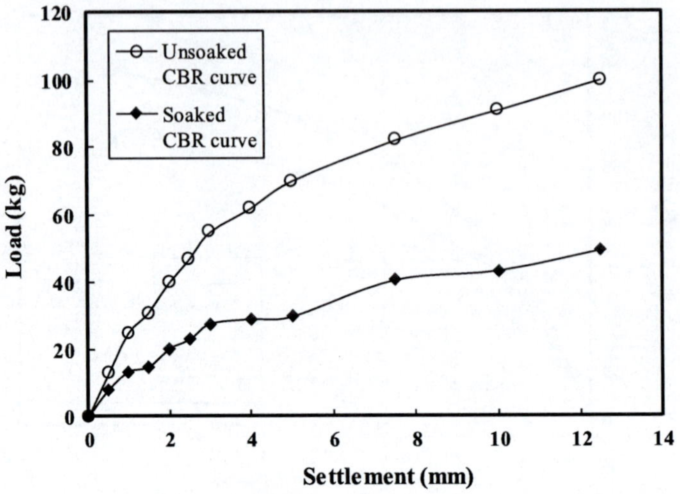

Fig. 10.4 Load-settlement curve for single layer clay

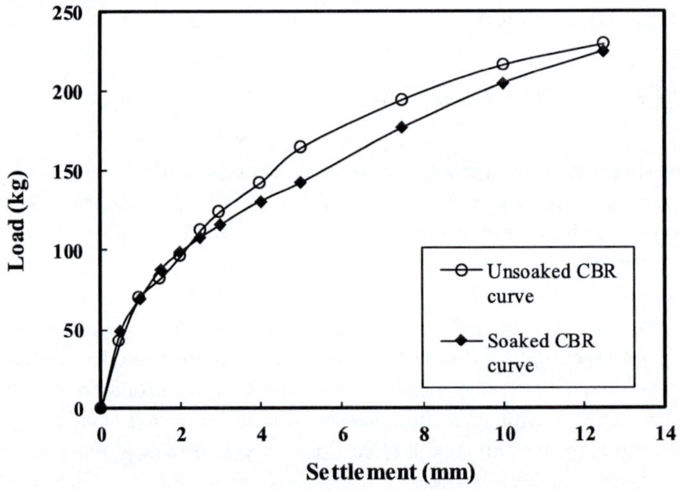

Fig. 10.5 Load-settlement curve for layered soil with sand as top layer and clay as bottom layer

Comparison of CBR Values Between Experimental and Numerical Results

CBR values obtained from experimental study and numerical study are shown in Table 10.2 to check the acceptability of results obtained from numerical analysis. It is observed that the results obtained from experimental study and numerical study are

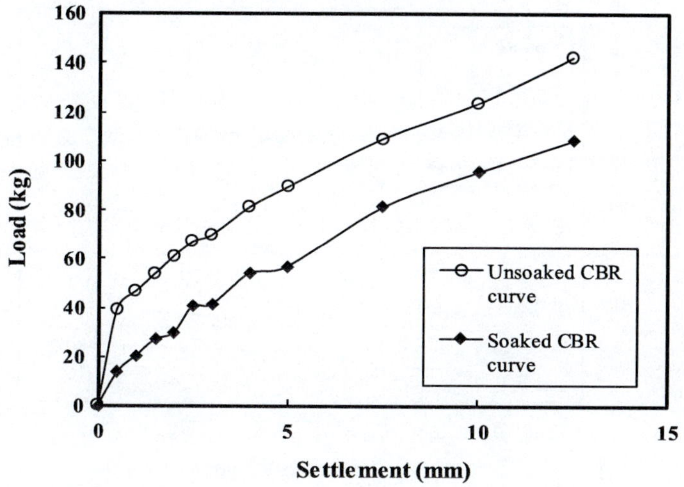

Fig. 10.6 Load-settlement curve for layered soil with clay as top layer and sand as bottom layer

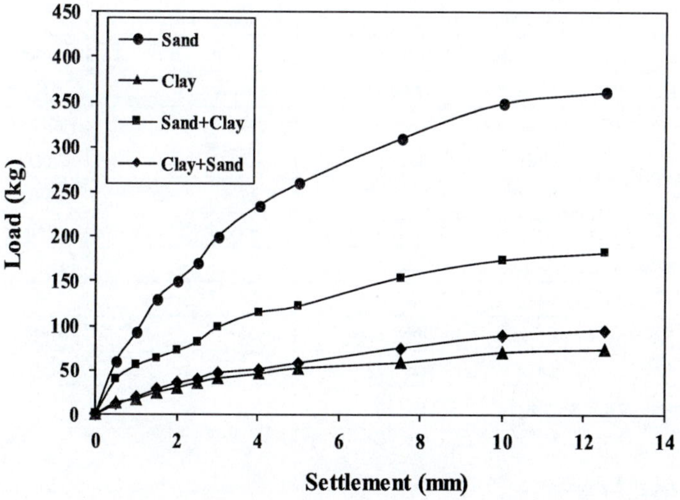

Fig. 10.7 Load-settlement curve for unsoaked condition

in good agreement considering the maximum permissible variation in CBR values is within the permissible limit as per IRC: 37-2012 [16].

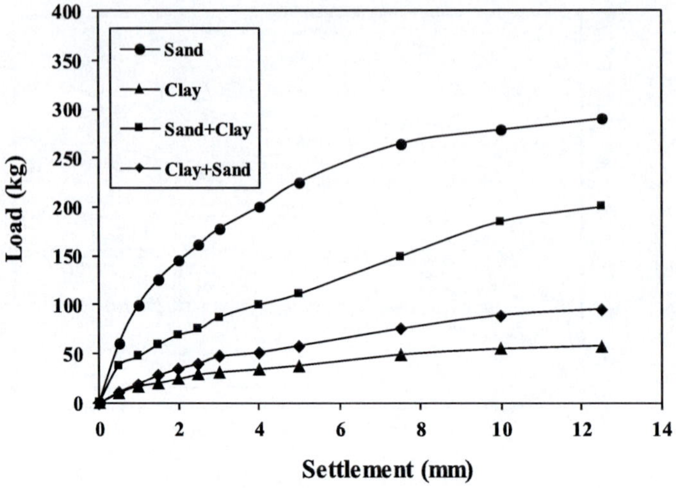

Fig. 10.8 Load-settlement curve for soaked condition

Table 10.2 Comparison between experimental and numerical results

Type	Top layer	Bottom layer	Experimental result	Numerical result
	H/2	*H*/2	CBR (%)	CBR (%)
Unsoaked condition	Sand	Sand	12.37	12.42
	Clay	Clay	3.44	2.60
	Sand	Clay	8.21	5.96
	Clay	Sand	4.89	2.91
Soaked condition	Sand	Sand	10.88	11.75
	Clay	Clay	1.68	2.02
	Sand	Clay	7.91	5.45
	Clay	Sand	2.96	2.30

H = Thickness of CBR sample

Parametric Study

Further study was carried out to investigate stress distribution in two-layer soil systems and assess the effect of top soil layer thickness on CBR value of two-layer soil system using 3D numerical models.

Stress Distribution in Two-Layered Soil System. Stress distribution contours for two-layer soil systems were obtained from numerical analysis at 12.5 mm penetration of the plunger.

Figure 10.9 shows the stress distribution in the two-layer soil system with sand as top layer and clay as bottom layer having equal thickness. Stress contours are obtained for both unsoaked (Fig. 10.9a) and soaked conditions (Fig. 10.9b). In both the conditions, bulk of stress is taken care by the top layer and there is significant change in stress distribution at the interface between two soil layers. It is also observed that there is change in stress distribution pattern in bottom layer between unsoaked and soaked conditions.

Figure 10.10 shows the stress distribution in the two-layer soil system with clay as top layer and sand as bottom layer having equal thickness. Compared to previous two-layer soil system, in this case, both the layers are observed to take care of stress distribution, though the intensity of stress in bottom layer compared to top layer is higher in soaked sample (Fig. 10.10b) than the unsoaked sample (Fig. 10.10a).

Effect of Top Soil Layer Thickness on CBR. In this study, results discussed till previous section are all based on the two-layer soil system, where both the layers having equal thickness. In natural soil deposit, there is always a possibility of existence of soil layers with different thickness. Therefore, a parametric study was performed to assess the effect of change in thickness of top layer soil on CBR of two-layer soil systems, where total thickness of sample is constant. Figure 10.11

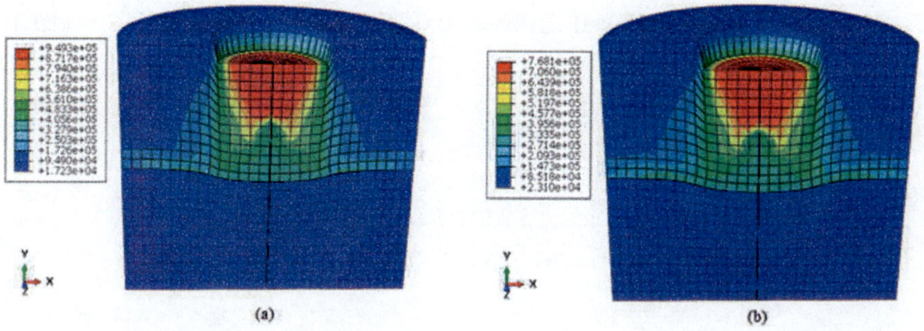

Fig. 10.9 Stress distribution in layered soil with sand as top layer and clay as bottom layer at 12.5 mm displacement: **a** unsoaked **b** soaked

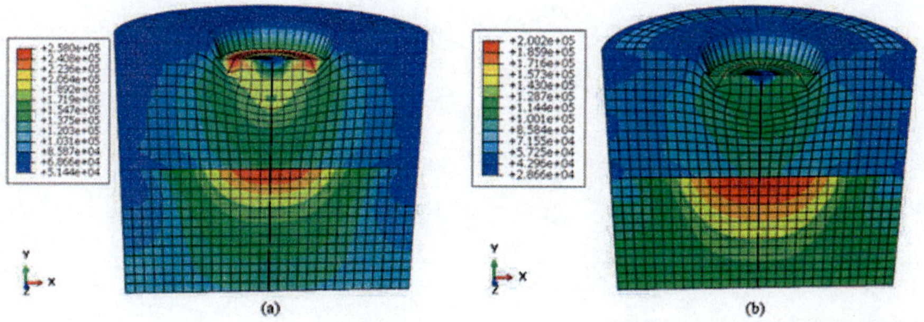

Fig. 10.10 Stress distribution in layered soil with clay as top layer and sand as bottom layer at 12.5 mm displacement: **a** unsoaked **b** soaked

shows the change in CBR value with change in thickness of sand (h_s) as the top layer, where clay is the bottom layer. It is observed that the CBR value increases steadily with increase in thickness of top layer, when top layer is granular material like sand. Using the data presented in Fig. 10.11, linear regression was performed and correlations as Eqs. 10.1 and 10.2 were obtained for unsoaked and soaked CBR, respectively.

$$\text{Unsoaked CBR} = 0.0898h_s + 1.0165 \tag{10.1}$$

$$\text{Soaked CBR} = 0.075h_s + 1.0574 \tag{10.2}$$

Figure 10.12 shows the change in CBR value with change in thickness of clay as the top layer, where bottom layer is sand. It is observed that the CBR value decreases sharply until the thickness of top layer (h_c) is half of the total sample thickness and thereafter rate of change in CBR value decreases with further increase in thickness of top layer, when top layer is cohesive material. Following correlations were obtained for unsoaked and soaked CBR from Fig. 10.12:

$$\text{Unsoaked CBR} = -0.041h_c + 6.0519 \tag{10.3}$$

$$\text{Soaked CBR} = -0.0136h_c + 3.32 \tag{10.4}$$

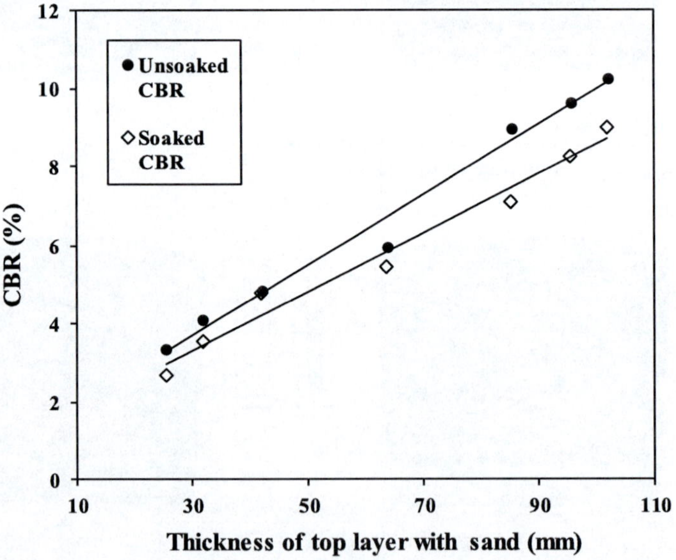

Fig. 10.11 Variation of CBR value with thickness of top layer as sand and bottom layer as clay

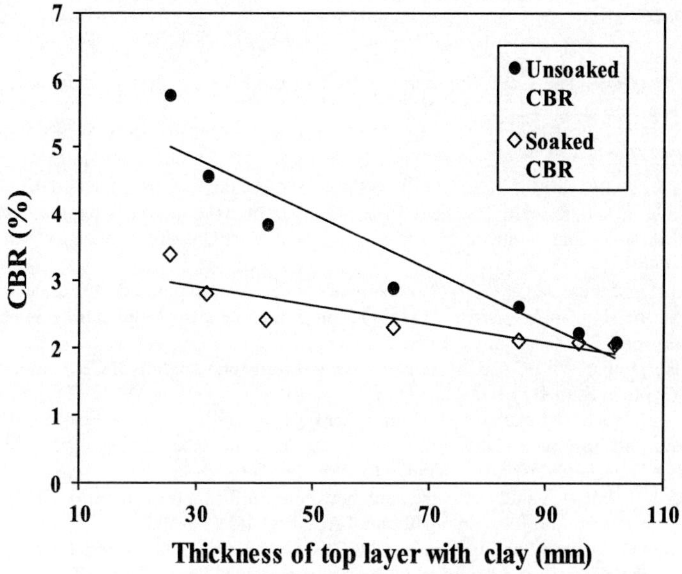

Fig. 10.12 Variation of CBR value with thickness of top layer as clay and bottom layer as sand

Conclusion

In the present study, predictability of numerical models for CBR of single-layer and two-layer systems of soils are established with further development of insights in CBR of layered soil systems. Correlations are proposed to calculate CBR of two-layer soils based on thickness of top layer for both unsoaked and soaked conditions. Following conclusions are drawn from the present study:

1. The numerical model developed in ABAQUS/CAE 2018 to assess the CBR value is efficient enough to predict unsoaked and soaked CBR of layered soil.
2. In the layered soil, with sand as the top layer and clay as the bottom layer, the unsoaked and soaked CBR values increase as the thickness of sand layer increases and decrease as the thickness of clay layer increases.
3. In the two-layer soil systems, with clay at top and sand at bottom, both soaked and unsoaked CBR is higher than CBR of only clay sample.
4. CBR of two-layer soils is dominated by the type of soil at top.
5. Correlations established between CBR and top layer soil thickness in a layered soil can be used to predict soaked and unsoaked CBR values of both single and two-layer soil systems consisting of sand and clay.

However, further study may be carried out to check the accuracy level of the proposed correlations in the field cases with similar soil type and lithology.

References

1. Davis EH (1949) The California bearing ratio method for the design of flexible roads and runways. Géotechnique 1(4):249–263
2. Pandian NS, Sridharan A, Raju PSRN (1999) California bearing ratio test simplified. J Test Eval 27(1):72–75
3. Moayed RZ, Allahyari F, Nazari M (2012) Effect of the sand layer thickness on the CBR values of two layered subgrade. In: Proceedings of 3rd international conference on new developments in soil mechanics and geotechnical engineering. Near East University, Nicosia, North Cyprus, pp 437–444
4. Caicedoa B, Mendozaa C (2015) Geotechnical behaviour of unpaved roads: understanding the CBR test. In: Manzanal D, Sfriso AO (eds) From fundamentals to applications in geotechnics. IOS Press, pp 95–100
5. Ronad HN, Rakaraddi PG (2016) An experimental study on variation of CBR value for layered soils. J Geotech Stud 1(1):1–7
6. Osouli A, Salam S, Othmanawny G, Tutumluer E, Beshears S, Shoup H, Eck M (2017) Results of soaked and unsoaked California bearing rate tests on unbound aggregates with varying amounts of fines and dust ratios. Transp Res Rec 2655(1):13–19
7. Talukdar DK (2014) A study of correlation between California bearing ratio (CBR) value with other properties of soil. Int J Emerg Technol Adv Eng 4(1):59–562
8. Aderinola OS, Oguntoyinbo E, Quadri AI (2017) Correlation of California bearing ratio value of clays with soil index and compaction characteristics. Int J Sci Res Innov Technol 4(4):12–22
9. Al-Amoudi OSB, Asi IM, Wahhab HIA, Khan ZA (2002) Clegg hammer—California-bearing ratio correlations. J Mater Civ Eng 14(6):512–523
10. Purwana YM, Nikraz H (2014) The correlation between the CBR and shear strength in unsaturated soil conditions. Int J Transp Eng 1(3):211–222
11. Sathawara JK, Patel AK (2013) Comparison between soaked and unsoaked CBR. Int J Adv Eng Res Stud 2(1):132–135
12. Al-Hashemi HM, Bukhary AH (2016) Correlation between California bearing ratio (CBR) and angle of repose of granular soil. Electron J Geotech Eng 21(17):5655–5660
13. Lakshmi SM, Subramanian S, Lalithambikhai MP, Vela AM, Ashni M (2016) Evaluation of soaked and unsoaked CBR values of soil based on the compaction characteristics. Malays J Civ Eng 28(2):172–182
14. Yashas SR, Harish SN, Muralidhara HR (2016) Effect of California bearing ratio on the properties of soil. Am J Eng Res 5(4):28–37
15. IS 2720-Part 16: Method of test for soils: Laboratory determination of CBR, Bureau of Indian Standards, New Delhi, India (1987)
16. IRC 37: Guidelines for design of flexible pavement, Indian Roads Congress, New Delhi, India (2012)

Chapter 11
Strength and Durability Characteristics of Cement-Treated Soil for Its Application in Sub-Base and Base Course of Flexible Pavements

Gaurav Gupta

Introduction

Sources of excellent quality aggregates in various regions of the world are becoming scarce with time. Stringent environmental laws have also hampered the future growth of present stone quarries in developing nations, which have required the utilization of inferior levels of aggregates or weak materials in the layers of pavements using appropriate stabilization procedures [1–3]. Lately, the quick speed of pavement construction in India has also entailed significant enthusiasm among pavement engineers to establish the use of marginal materials like aggregates and soils present in the close vicinity of the road construction site. Chemical stabilizers like cement, lime, and fly ash-lime mixture have been effectively utilized to stabilize many soils, resulting in improved strength, durability, stiffness, and workability and upgrading pavement construction quality [4, 5]. Cement-treated base (CTB) courses have been used in developed countries such as Brazil, South Africa, the United States, and Australia for numerous years [6–8]. However, barely a few examples of CTB exist in India [9].

Granular soils are spread over a large area of land across several regions of India. They are employed chiefly in the sub-base course of flexible pavements in their natural condition, provided they fulfill the Indian Roads Congress's criteria. However, these soils frequently fail to fulfill criteria regarding unconfined compressive strength, necessitating modification with appropriate hydraulic binders such as lime and cement for the application of the materials in sub-base course and base course of flexible pavements. Given the importance of cement-treated granular material characterization, thorough research of mechanical characterization of such materials was carried out in this work. It is also crucial to evaluate the durability of cement-treated materials for application in pavement layers. Standard methods

G. Gupta (✉)
Geotechnical Engineering (M. Tech), Delhi College of Engineering, New Delhi, India
e-mail: gaurav007gupta@gmail.com

Technical Director, Dr. Ghuman & Gupta Geotech Consultants, Mohali, Punjab 160055, India

K. Muthukkumaran et al. (eds.), *Transportation and Environmental Geotechnics*,
Lecture Notes in Civil Engineering 298, https://doi.org/10.1007/978-981-19-6774-0_11

to measure the durability of treated materials include wetting and drying test and freezing-thawing test. The limited mass loss stated by regulations or standards is compared to the loss of mass of the treated soil specimen after 12 cycles of wetting and drying and 12 cycles of freezing-thawing.

Literature Review

In the recent decade, numerous studies have been published on the efficiency of granular soil stabilization with cement, fly ash, geopolymer, lime, and a blend of any of the two materials. Fall et al. used repeated load triaxial tests to determine the impact of soil compaction and cement on the resilient modulus of treated gravel lateritic soils obtained from several places in Senegal [10]. They discovered that the amount of cement in the soil is a critical variable that affects its stiffness. The resilient modulus of treated soil ranged from 0.4 to 5.6 MPa at 3% cement. Mengue et al. recently investigated the enhancement of Cameroon's soil [11]. They found that 3 and 6 % cement dosage is adequate for obtaining the treated soil's permissible mechanical characteristics for application as sub-base and base material. Joel and Agbede demonstrated that adding sand to cement improves the stability of soil (poorly graded) for application as a foundation layer [12]. The addition of 45% sand content and 6% dosage of cement blend with the soil resulted in an increase of 270% in UCS (period of curing: 7 days). Portelinha et al. showed that a low dosage of lime and cement could improve the workability of soil in Brazil [13]. The authors discovered that the highest alteration in the plasticity occurred at a 3% dosage of cement. Shankar et al. treated soil with cement and pond ash, and the stabilization process resulted in a 440% increase in the UCS of the soil [14]. Caro et al. recommended stabilization of soil with at least 6% cement dosage to enhance the soil's performance [15]. UCS is commonly used to evaluate the appropriateness of cement-treated soils bound material in pavements due to the ease of the test method procedure. Table 11.1 presents the details of UCS requirements established by several standards of practice. It is worth noting that most investigations on cement-treated materials have used crushed stones or strong aggregates. Therefore, the results of these publications are more suited for comparable materials and these studies cannot be applied to determine the behaviour of other treated materials that use low-strength materials, such as granular soils.

Materials and Methods

The materials, sample preparation, and testing procedures used in this work for the mechanical characterization of cement-treated soil are detailed in the following paragraphs.

Table 11.1 Criteria for the minimum *UCS* of CTB and CTSB layers in flexible pavements

Reference	UCS (MPa)	
	Medium to high volume roads	Low volume roads
NCHRP [16]	> 1.72 (CTSB)	> 1.72 (CTSB)
	> 5.17 (CTB)	> 5.17 (CTB)
IRC 37 [17]	1.5–3.0 (CTSB)	–
	> 4.5 (CTB)	–
IRC SP 89 [18]	–	> 3.0
	–	> 1.7

Table 11.2 Engineering properties of the untreated granular soil

Property	Results
Liquid limit	–
Plastic limit	–
Plasticity index	Non-plastic
MDD (kN/m^3)	21.2
OMC (%)	10.3
Gravel (%)	34.2
Sand (%)	48.5
Silt (%)	17.3
Clay (%)	0

Soil

One granular soil was obtained from in the northern region of India. Representative samples were obtained from a suitable depth of the quarries with sufficient care. Table 11.2 summarizes the engineering properties of untreated granular soil. The soil is categorized as silty sand (SM) in the Indian soil classification system.

Cement

For the stabilization of the granular soil, ordinary Portland cement (OPC) of grade 43 has been used in the present study.

Methodology

The granular soil was treated with 2, 3, 4, 5, 6, 7, and 8% dosage of cement. All the contents were added by the weight of dry soil. The samples were tested after curing samples for 7 days. The compaction characteristics, strength, and durability of the untreated and the treated soil were determined. To establish the compaction characteristics of the untreated and the treated soil, a modified Proctor test was performed according to the test procedure detailed out in IS 2720 Part 8 [19]. To determine the strength behaviour of the untreated and the treated soil, unconfined compressive strength (UCS) test was performed under unsoaked and 96 h water-soaked conditions. As per the recommendations of IRC SP 89 for granular soils, cubical samples of 150 mm side length were prepared with the varying dosage of cement [18]. To prepare a sample at maximum dry density (MDD), the samples were prepared in five layers and 25 blows were given to each layer, comparable to a modified Proctor compaction test. After extraction, the produced samples were covered in polythene sheets to prevent moisture evaporation and stored for curing for 7 days. The samples were tested in a compression testing machine.

To establish the durability of the cement-treated soil, the soils were subjected to 12 cycles of wetting, drying, and freezing and thawing, as per the procedure detailed in IS: 4332 Part-IV [20]. The samples were subjected to curing for 7 days prior to performing the alternate wetting, drying, and freezing and thawing cycles. One wetting and drying cycle consists of soaking the sample in water for 5 h and heating the sample at 72 °C temperature in an oven for 42 h. After completion of every cycle, two firm strokes are given on the complete surface area of the samples with the help of a wired brush. The loss in weight of the samples was calculated after 12 cycles of the test. On the other hand, one freezing and thawing cycle consists of freezing the sample at −23 °C temperature at −23 °C temperature in a freezing chamber for 24 h and thawing the sample at room temperature (23 °C) for 24 h. After completing every cycle, two firm strokes were given on the complete surface area of the samples with the help of a wired brush. The loss in weight of the samples was calculated after 12 cycles of the test.

Strength and Durability Criteria for the Application of Cement-Treated Soil in the Sub-Base Course and a Base Course of Flexible Pavements as Per IRC 37: 2018

To apply cement-treated soil in the sub-base course and base course of flexible pavements in India, the soils must fulfill the strength and durability criteria laid down by IRC 37: 2018 [17] as detailed out in Table 11.1. Regarding the durability criteria, as per clause 8.2.4 of IRC 37:2018, the maximum loss in weight of CTB and CTSB materials should not exceed 14% 12 cycles of the test and drying. Also,

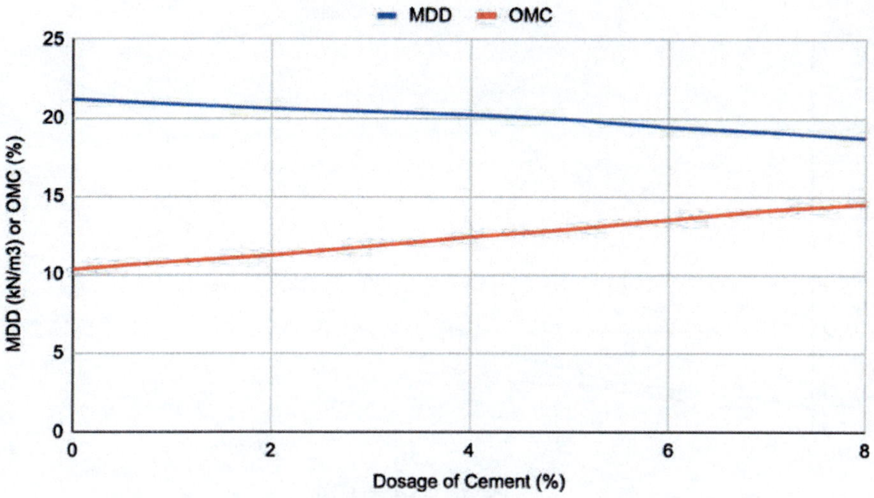

Fig. 11.1 Variation of MDD and OMC of the soil with the varying dosage of cement

as per clause 8.2.4 of IRC 37:2018, the maximum loss in weight of CTB and CTSB materials should not exceed 14% after 12 cycles of the test.

Results and Discussions

Modified Proctor Compaction Test

Figure 11.1 shows MDD and OMC values for the soil with the varying dosage of cement. Results of the research show that with the rise in the dosage of cement, MDD of the soil decreased, and OMC of the soil increased. The decrease in MDD of the soil is due to the flocculation of soil particles. The increase in OMC of the soil is due to the use of water for the hydration of cement.

Unconfined Compressive Strength Test

Figure 11.2 shows UCS values for the soil with the varying cement dosage subjected to soaked and unsoaked conditions. Results of the research show that with the rise in the dosage of cement, UCS of the soil increased. Furthermore, it was seen that the soil treated with 4 and 7% dosage of cement fulfills the criteria of the minimum UCS of CTSB and CTB layers. Therefore, the gain in UCS of the soil is due to the formation of cementitious products due to the hydration of cement.

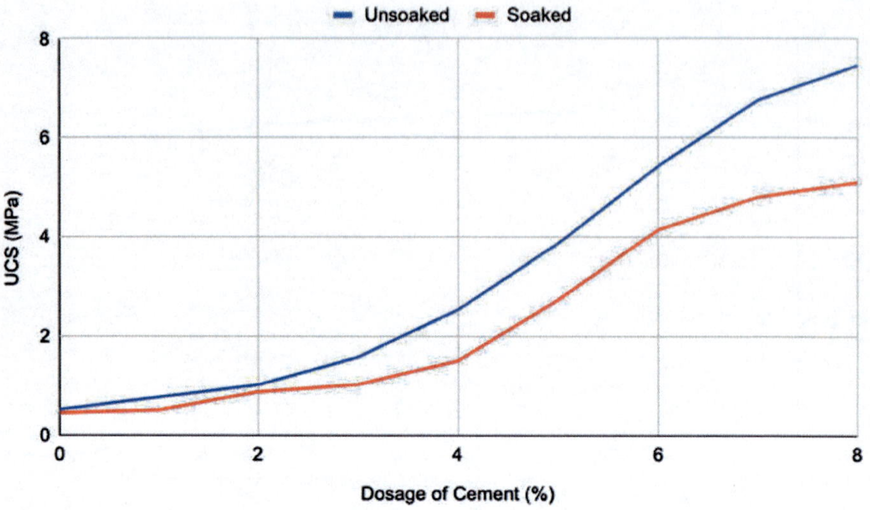

Fig. 11.2 Variation of unsoaked and soaked UCS of the soil with the varying dosage of cement

Wetting and Drying Test

Figure 11.3 shows the loss in weight of the samples for cement-treated soil with the varying dosage of cement subjected to soaked and unsoaked conditions. Results of the research show that with the rise in the dosage of cement, loss in weight of the soil decreased. Furthermore, it was seen that the soil-treated with a 3% dosage of cement fulfills the criteria of the durability of CTSB and CTB layers. The decrease in loss of weight of the soil is due to linking and joining soil particles to each other by cementitious products formed due to cement hydration.

Freezing and Thawing Test

Figure 11.3 shows the loss in weight of the samples for cement-treated soil subjected to 12 cycles of freezing and thawing with the varying dosage of cement subjected to soaked and unsoaked conditions. Results of the research show that with the rise in the dosage of cement, loss in weight of the soil decreased. Furthermore, it was seen that the soil treated with a 4% dosage of cement fulfills the criteria of the durability of CTSB and CTB layers. The decrease in loss of weight of the soil is due to the linking and joining of soil particles by cementitious products formed due to the hydration of cement.

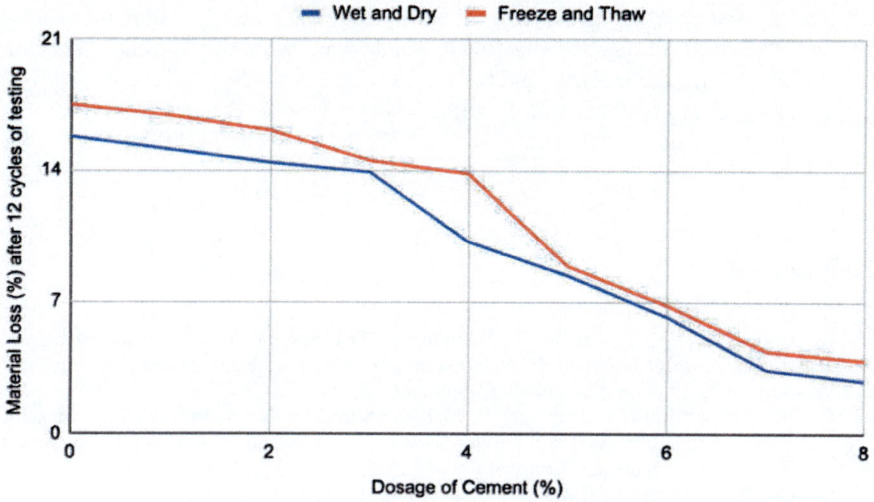

Fig. 11.3 Variation of material loss (%) after 12 wet–dry and 12 freeze–thaw cycles of the soil with the varying dosage of cement

Recommendations

Based on experimental results, and strength and durability criteria for CTSB and CTB layers laid down by IRC 37 2018, 4% dosage of cement for CTSB layer and 7% dosage of cement for CTB layer are recommended [17].

Conclusions

The present research was carried out to evaluate the strength and durability characteristics of granular soil treated with the varying dosage of cement for their application in the sub-base and base courses of flexible pavements. To fulfill the criteria in this regard, several laboratory tests, such as modified Proctor compaction tests, UCS tests, wetting and drying test, and freezing and thawing test, were carried out. The tests were performed on various combinations of the granular soil and cement. The important conclusions arrived at from this study are as follows:

1. With the rise in the dosage of cement, MDD of the granular soil decreased, and OMC of the granular soil increased.
2. With the rise in the dosage of cement, unsoaked and soaked UCS of the granular soil increased. As a result, the minimum dosage of cement fulfilling the criteria for UCS of CTSB and CTB layers was 4% and 7%, respectively.
3. With the rise in the dosage of cement, loss in weight of the granular soil decreased for the samples subjected to 12 cycles of wetting and drying and 12 cycles of

freezing and thawing. Thus, the minimum dosage of cement fulfilling the criteria for CTSB and CTB layers of the loss in weight was 3% for the wetting and drying test and 4% for the freezing and thawing test, respectively.

4. Considering the strength and durability criteria for CTSB and CTB layers, the recommended dosage of cement is 4 and 7% for CTSB and CTB layers.

References

1. Biswal DR, Sahoo UC, Dash SR (2017) Strength and stiffness studies of cement treated granular lateritic soil. In: International congress and exhibition sustainable civil infrastructures: innovative infrastructure geotechnology. Springer, Cham
2. Disfani MM, Arulrajah A, Haghighi H, Mohammadinia A, Horpibulsuk S (2014) Flexural beam fatigue strength evaluation of crushed brick as a supplementary material in cement treated recycled concrete aggregates. Constr Build Mater 68:667–676
3. Liebenberg JJE, Visser AT (2003) Stabilization and structural design of marginal materials for use in low-volume roads. Transp Res Rec 1819(1):166–172
4. Xuan DX et al (2012) Mechanical properties of cement-treated aggregate material—a review. Mater Des 33:496–502
5. Paul DK, Theivakularatnam M, Gnanendran CT (2015) Damage study of a lightly stabilised granular material using flexural testing. Indian Geotech J 45(4):441–448
6. Jitsangiam P, Nusit K, Chummuneerat S, Chindaprasirt P, Pichayapan P (2016) Fatigue assessment of cement-treated base for roads: an examination of beam-fatigue tests. J Mater Civ Eng 28(10):04016095
7. Ma Y, Gu J, Li Y, Li Y (2015) The bending fatigue performance of cement-treated aggregate reinforced with polypropylene filament fiber. Constr Build Mater 83:230–236
8. Paul DK, Gnanendran CT (2016) Characterization of lightly treated granular base materials using monotonic and cyclic load flexural testing. J Mater Civ Eng 28(1):04015074
9. Biswal DR, Sahoo UC, Dash SR (2020) Mechanical characteristics of cement stabilised granular lateritic soils for use as structural layer of pavement. Road Mater Pavement Des 21(5):1201–1223
10. Fall M, Sawangsuriya A, Benson CH, Edil TB, Bosscher PJ (2008) On the investigations of resilient modulus of residual tropical gravel lateritic soils from Senegal (West Africa). Geotech Geol Eng 26(1):13–35
11. Mengue E, Mroueh H, Lancelot L, Eko RM (2017) Mechanical improvement of a fine-grained lateritic soil treated with cement for use in road construction. J Mater Civ Eng 29(11):04017206
12. Joel M, Agbede IO (2011) Mechanical-cement stabilization of laterite for use as flexible pavement material. J Mater Civ Eng 23(2):146–152
13. Portelinha FHM, Lima DC, Fontes MPF, Carvalho CAB (2012) Modification of a lateritic soil with lime and cement: an economical alternative for flexible pavement layers. Soils Rocks São Paulo 35(1):51–63
14. Shankar AU, Suresha SN, Kashinath B (2008) Characterisation of lateritic soil modified with pond ash and cement. Indian Highw 36(6)
15. Caro S, Agudelo JP, Caicedo B, Orozco LF, Patiño F, Rodado N (2019) Advanced characterisation of cement-stabilised lateritic soils to be used as road materials. Int J Pavement Eng 20(12):1425–1434
16. Olidis C, Hein D (2004) NCHRP guide for the mechanistic-empirical design of new and rehabilitated pavement structures materials characterization: is your agency ready. In: 2004 annual conference of the transportation association of Canada
17. IRC: 37 Guidelines for the design of flexible pavements. Indian Road Congress, New Delhi, India (2018)

18. IRC: SP 89 Guidelines for soil and granular material stabilization using cement, lime and flyash. Indian Road Congress, New Delhi, India (2010)
19. IS 2720-Part 8: Methods of test for soils—determination of water content—dry density relation. Bureau of Indian Standards, New Delhi (1983)
20. IS 4332-Part 4: Methods of test for treated soils—wetting and drying, and freezing and thawing tests for compacted soil-cement mixtures. Bureau of Indian Standards, New Delhi (1968)

Chapter 12
Evaluation of Pond Ash Mixed with Coir as a Pavement Material

Ashish Malik, Safeer Ahmed, and S. K. Singh

Introduction

Thermal power facilities in India create massive amounts of coal ash as a waste product. Coal ash production in India is around 225 million metric tons per year, and disposal is a big issue [1]. Thermal power plants that use pulverized coal as a fuel create massive amounts of ash as a waste product. As time goes on, the amount of ash produced is likely to go up and thereby creating challenges because of its disposal and potential of polluting the environment. As a result, engineers are being encouraged to use coal ash in large quantities for developmental activities [2]. It has been noted that most thermal power stations deposit coal ash into neighboring low-lying areas, resulting in an "ash pond." In India, these ash ponds have already taken up 65,000 acres of precious land [3]. Indian coal has 35–45% ash, which contributes to significant amounts of pond ash [4]. The diversion of huge tracts of land to garbage disposal causes relocation concerns, loss of agricultural productivity, grazing land and habitat, as well as other hand use implications. With a population of 1.21 billion people according to the 2011 Census, India is putting a lot of strain on its resources. Thermal power plants (TPPs) use coal as one of the resources they use to generate electricity. Coal is burned to produce steam, fly ash, and bottom ash, which are then used to generate power. Fly ash and bottom ash make about 12–20% of the total weight of coal [5]. After coal is burned, fly ash and bottom ash remain as waste for TPPs. For the disposal of this material, two approaches are used: dry disposal and wet disposal. In India, the wet disposal method is commonly used, in which bottom ash and fly ash are combined, mixed 1:10 with water to produce a slurry, and then carried to ash ponds via dykes [6]. When the slurry reaches the ash ponds, it is

A. Malik (✉) · S. Ahmed · S. K. Singh
Department of Civil Engineering, PEC (Deemed to be University), Chandigarh 160112, India
e-mail: ashishmalik.phdciv@pec.edu.in

S. K. Singh
e-mail: sksingh@pec.edu.in

K. Muthukkumaran et al. (eds.), *Transportation and Environmental Geotechnics*,
Lecture Notes in Civil Engineering 298, https://doi.org/10.1007/978-981-19-6774-0_12

allowed to settle, and the resulting slurry is known as pond ash. Though pond ash is used in a variety of ways, it accounts for about 67% of total ash produced by weight, with 25% going to the cement industry and the rest going to bricks and tiles, ash dyke raising, mine filling, reclamation of low-lying areas, roads and flyovers, and so on, leaving one-third ash unused [1]. As a result, there is an ever-increasing ponding space for storing ash, as well as other environmental concerns. On the other hand, the recent increase in the construction of highways and roads in India necessitates a large amount of natural soil and aggregates [7]. The brutal exploitation of rich soil and natural aggregate is being used to meet this need. When ash deposits dry up without being covered by water or plants, fugitive dust can pollute the air. The majority of the area around huge ash ponds or ash dikes is typically not covered, wetted, or soaked in water. When the ash dries up, it becomes a great source of fugitive dust emissions. As a result, the situation has deteriorated significantly [8]. Pond ash has been used as structural fills and highway embankment construction to address these issues in low-lying locations. However, due to a lack of expertise and confidence, it has yet to gain traction.

Some research has been done to see if compacted pond ash may be used in geotechnical construction such as embankments, retaining walls, and structural fills. However, in order to keep the inherent strength of the compacted pond ash, these structures must be safeguarded from getting wet, which is a tough task in field circumstances [9]. With this in mind, fiber reinforcement in the form of recron-3 s was employed to influence the stress–strain behavior of destabilized material in the pond ash sample [10]. A series of CBR, direct shear, proctor, and UCS were used to examine the influence of fiber reinforcing on the stress–strain behavior and strength parameters of compacted mixtures. Figure 12.1 represents ash pond from which ash was collected from Ropar thermal power plant.

Along with the abundant availability of coir at cheaper costs than other polymetric materials, it is a natural reinforcing material and is better suited for India. Coir has several advantages, including durability, biodegradability, and lack of auxiliary chemicals cheap cost of raw materials. Engineers and scientists are looking for alternative viable resources that may be utilized to address a variety of issues that arise around the world over time. This study aims to utilize waste from different industries, to achieve sustainability.

Materials and Methods

The use of pond ash in geotechnical projects on a large scale will alleviate the challenges that thermal power plants have with its disposal. In this regard, it is necessary to examine the behavior of structures built with pond ash in order to ensure their stability and safe operation. Despite the fact that there is no suitable substitute for full-scale field tests, laboratory tests have the advantage of permitting careful control of many of the variables encountered in field. The trends and behavior patterns discovered in laboratory experiments can be utilized to better understand

Fig. 12.1 Pond ash collected from Ropar TPP's ash pond

how structures operate in the field. Pond ash is the primary material, and coir is like an admixture materials used in the current study:

Pond ash

Pond ash used in this study was collected from the thermal power plant of Ropar, Punjab. The samples were dried at the temperature of 105 °C. The ash sample was screened through 2 mm sieve to separate out the foreign and vegetative matters. Then, the pond ash samples were stored in airtight container so that the moisture does not enter the sample. Figure 12.2 represents pond ash sample used in the study.

Coir

Coconut fiber or coir comes from the husk of the coconut and is classified as a hard structural fiber. The coir fiber is flexible enough to twist without breaking, and it maintains its curl as if it was continually waved. It is a significant commercial commodity used in mattress manufacturing. The diameter of coir fiber has been found using Vernier as shown in Fig. 12.3. The average value of diameter of coir fiber came out as 0.25 mm.

Coir or coconut fiber used in the current study has been taken from a mattress industry in Panchkula. Coir was then treated with water to remove impurities, and it was manually cut with help of scissors with size ranging in length from 1 to 4 cm for various aspect ratios. Also, the coir fiber has been mixed (by weight) randomly

Fig. 12.2 Representative pond ash sample

Fig. 12.3 Diameter measurement of coir using vernier caliper

Table 12.1 Sample denotation for different aspect ratios

Aspect ratio	Sample composition (by weight) (%)	Denotation
50	100 pond ash	PA
	99.5 pond ash + 0.5 coir	$PA0.5CF_{50}$
	99 pond ash + 1 coir	$PA1CF_{50}$
	98.5 pond ash + 1.5 coir	$PA1.5CF_{50}$
	98 pond ash + 2 coir	$PA2CF_{50}$
100	99.5 pond ash + 0.5 coir	$PA0.5CF_{100}$
	99 pond ash + 1 coir	$PA1CF_{100}$
	98.5 pond ash + 1.5 coir	$PA1.5CF_{100}$
	98 pond ash + 2 coir	$PA2CF_{100}$
150	99.5 pond ash + 0.5 coir	$PA0.5CF_{150}$
	99 pond ash + 1 coir	$PA1CF_{150}$
	98.5 pond ash + 1.5 coir	$PA1.5CF_{150}$
	98 pond ash + 2 coir	$PA2CF_{150}$

in the pond ash in percentages of 0.5, 1, 1.5, and 2, respectively. The mix sample has been denoted as per Table 12.1.

Specific gravity of PA sample came out as 2.11 and for coir fiber is 0.91. Specific gravity test has been performed as per IS: 2720 (Part-III): 1980 and was used to calculate the of pond ash. For coarser particles, sieve analysis was used in accordance with IS: 2720 (Part-IV): 1985, while hydrometer analysis was used for finer particles. The moisture content and dry density relationships were determined using the proctor test, as defined by IS: 2720 (Part-VII): 1980. Pond ash was mixed with water and increasing amounts of coir fiber (by dry weight of pond ash) for different aspect ratios, commencing at 0, 0.5, 1, 1.5, and 2%. For specimens compacted to standard proctor density with varying percentages of fiber, the unconfined compressive strength test as per IS: 2720 (Part-X): 1991 of pond ash specimens reinforced with coir fibers of length around 25 mm was performed (i.e., 0.5, 1, 1.5, and 2%). Bearing ratio is one of the most important parameters in evaluating soil subgrades for both rigid and flexible pavement constructions. CBR test has been conducted as per IS: 2720 (Part-XVI): 1991. CBR is also needed in a number of processes such as determination of pavement thickness.

Results and Discussion

The aim of the study is to use pond ash and coir as subgrade and sub-base material for light traffic conditions. To use the two materials in road construction, the strength and geotechnical properties of pond ash (PA) and pond ash–coir mix (PA-CF) in subgrade and sub-base of pavement need to be evaluated. The results for specific gravity, grain

Table 12.2 Geotechnical properties

Properties of pond ash	Value	Properties of pond ash	Value
Specific gravity (G)	2.11	D_{10} (mm)	0.023
Gravels (>4.75 mm)%	0	D_{30} (mm)	0.058
Coarse sand (4.75–2.0 mm)%	0.16	D_{60} (mm)	0.085
Medium sand (2–0.425 mm)	0.82%	D_{50} (mm)	0.075
Fine sand (0.425–0.075 mm)	46.73%	C_u	3.6
Silt (0.075–0.002 mm)	50.52%	C_c	1.7
Clay (< 0.002 mm)	1.77%		

size analysis, shear strength parameters, proctor, and California bearing ratio have been discussed below:

Specific gravity (G) has been found using pycnometer. Both the materials have lower specific gravity when compared to soils in general, but coir fiber is extremely light in weight for a given volume. The value of specific gravity for pond ash (G_{PA}) came out as 2.11 and that of coir fiber (G_{CF}) is 0.91. The lower specific gravity of a material can help civil engineers build lightweight structures. Basic properties of pond ash have been depicted in Table 12.2.

The grain size distribution curve is plotted on a semi-logarithmic scale, with percentage finer on *y*-axis and logarithm of diameter on *x*-axis as shown in Fig. 12.4. The shape of the curve gives important information regarding the presence of different sizes. The sample is poorly graded and silty sand with C_u and C_c values of 3.6 and 1.7, respectively.

Pond ash as a material is very light in weight, and it is also evident from maximum dry density vs water content curves of pond ash (PA) and pond ash–coir fiber (PA-CF) mix composition. There is a slight variation in MDD of pond ash and PA-CF mixtures, but when compared simultaneously, MDD of $PA1CF_{50}$ came out to be greater than rest of the mix composition as shown in Fig. 12.5a. From analysis of Fig. 12.5a, it can be said that adding coir fiber increases MDD up to an increment of 1%, further increases in coir percentage decrease the MDD of mix composition as shown in Fig. 12.5a. The decrease in MDD after increasing coir percentage more than 1% may be attributed to lesser specific gravity of coir fiber. Optimum moisture content (OMC) decreases on addition of coir fiber to pond ash (PA) sample from 41% for PA sample to 34% for $PA2CF_{50}$ sample as shown in Fig. 12.5a. Maximum dry density and optimum moisture content variation has been shown when coir fiber percentage in pond ash was varied at an aspect ratio of 50 in Fig. 12.5a. Maximum dry density of 10.3 kN/m^3 was observed at 37% water content for 1% coir fiber mixed in pond ash. The minimum MDD came for $PA2CF_{50}$ sample, i.e., 9.90 kN/m^3 at an

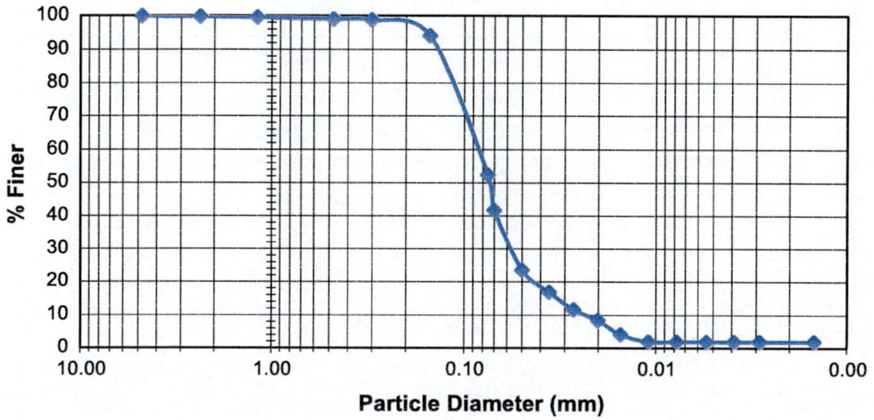

Fig. 12.4 Grain size distribution curve for pond ash sample

optimum moisture content of 34%. Similar results have been obtained for PA-CF mix for aspect ratio 100 and 150 as shown in Fig. 12.5b and c.

The variation of percentage changes in MDD has been plotted against the variation of coir fiber in PA sample as shown in Fig. 12.6a, and percentage changes in OMC have been plotted against the variation of coir fiber in PA sample as shown in Fig. 12.6b. It has been observed that at aspect ratio 100, 1% coir fiber with 99% pond ash gave the highest MDD among all the samples tested. This can be attributed to the better bonding of coir fiber with pond ash for $PA1CF_{100}$ than the rest of the mixes. Optimum moisture content (OMC) decreases on addition of coir fiber to pond ash (PA) sample. The reduction in OMC is of the order of 10–15% as evident from Fig. 12.6b. The decrease may be because of lesser water required by coir fiber to get compacted when compared to pond ash.

It can be established that when coir fiber percentage is increased from 0 to 2%, there is significant increase in value of UCS. The increase may be attributed to the better bonding between coir fiber and pond ash. For pond ash sample, these values of UCS came out as 120 kPa with a downward post peak curve suggesting brittle failure. The values of UCS for PA0.5CF, PA1CF, PA1.5CF, and PA2CF are 170, 240, 250, and 258 kPa, respectively, as shown in Fig. 12.7. It can be said that the nature of failure of UCS samples changes from brittle to ductile as the percentage of coir is increased from zero to 2%. It can also be concluded that on addition of coir fiber to pond ash, the coir fiber imparts plastic behavior to pond ash, i.e., also evident from Fig. 12.7. As the percentage of coir is increased, the curve after attaining peak of UCS starts to attain a flat slope toward PA2CF samples which suggests the nature of failure has changed.

It has been observed that the rate of increase of UCS on addition of coir fiber is more up to coir addition of 1%. After further increase in percentage of coir till 2%, the rate of increase of UCS has remained almost constant as shown in Fig. 12.8. When the percentage of coir fiber is increased from 1 to 2, it has also been found that

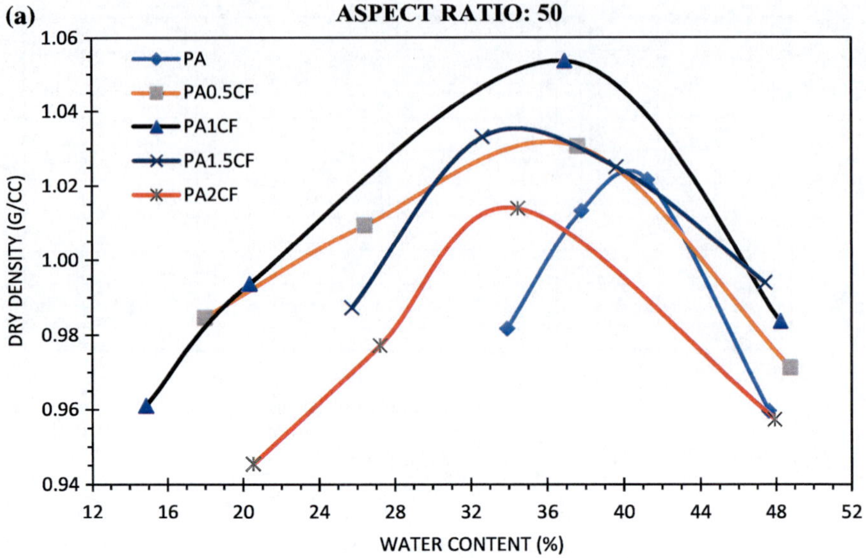

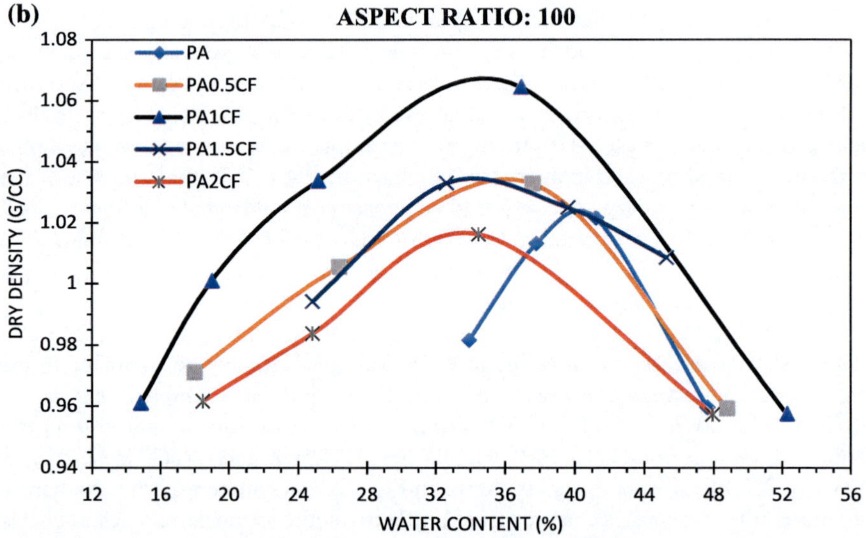

Fig. 12.5 **a** Dry density v/s water content plot for different coir content at AR 50 **b** Dry density v/s water content plot for different coir content at AR 100 **c** Dry density v/s water content plot for different coir content at AR 150

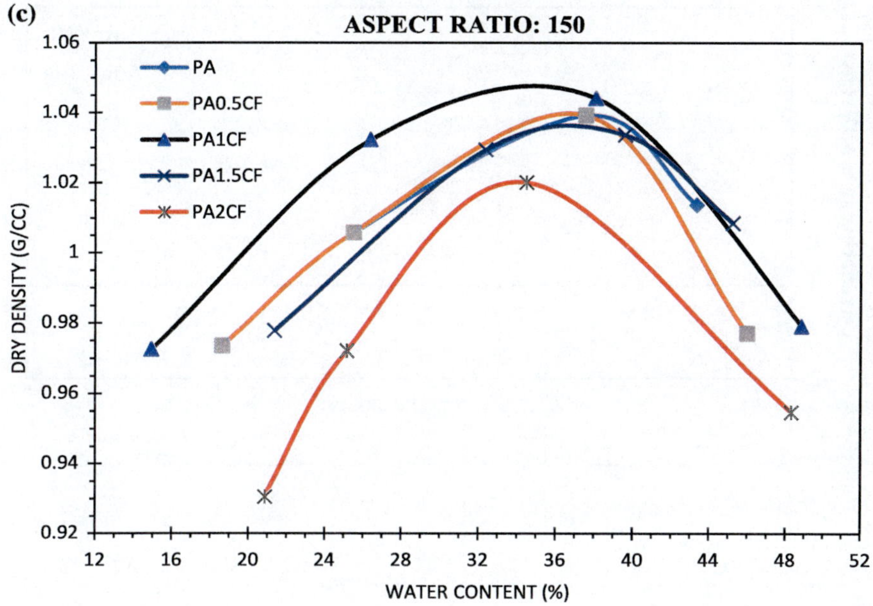

Fig. 12.5 (continued)

the coir fiber imparts more plasticity and the improvement in strength is insignificant and that the reason for lesser change in percentage increase in UCS.

When no coir fiber was added to pond ash, the percentage of failure strain was only 2% and the sample specimen failed as a brittle material. After the percentage of coir fiber was increased to 0.5% (by weight), the strain failure at 3.3% has been shown in Fig. 12.9. Again the percentage was increased from 0.5 to 1%; the values of failure in strain were near about to the 0.5% of coir fiber. After increasing the value of percentage of coir fiber from 1 to 1.5%, the value of strain at failure was 5.2%. After increasing the value of percentage of coir fiber from 1.5 to 2%, the value of strain at failure was obtained 7.2%. At 2% of coir fiber gives the maximum value that shows that if we increase the percentage of coir fiber in pond ash, the percentage strain at failure was also increased. Higher strain values for 1, 1.5, and 2% coir fiber mixtures indicate that the specimen fails after considerable deflections, as shown in Fig. 12.9, which is a favorable sign since brittle failure is undesirable in civil engineering projects.

The CBR test is now conducted for pond ash sample mixed with coir fiber at an aspect ratio of 100. The results are plotted in Fig. 4.9a and b. The CBR value is found to be more for 5 mm penetration as opposed to 2.5 mm. Therefore, two such samples were prepared, and the average value of CBR at 5 mm was taken as the CBR value of the sample. Soaked CBR values for all samples at aspect ratios 50, 100, and 150 are shown in Table 12.3.

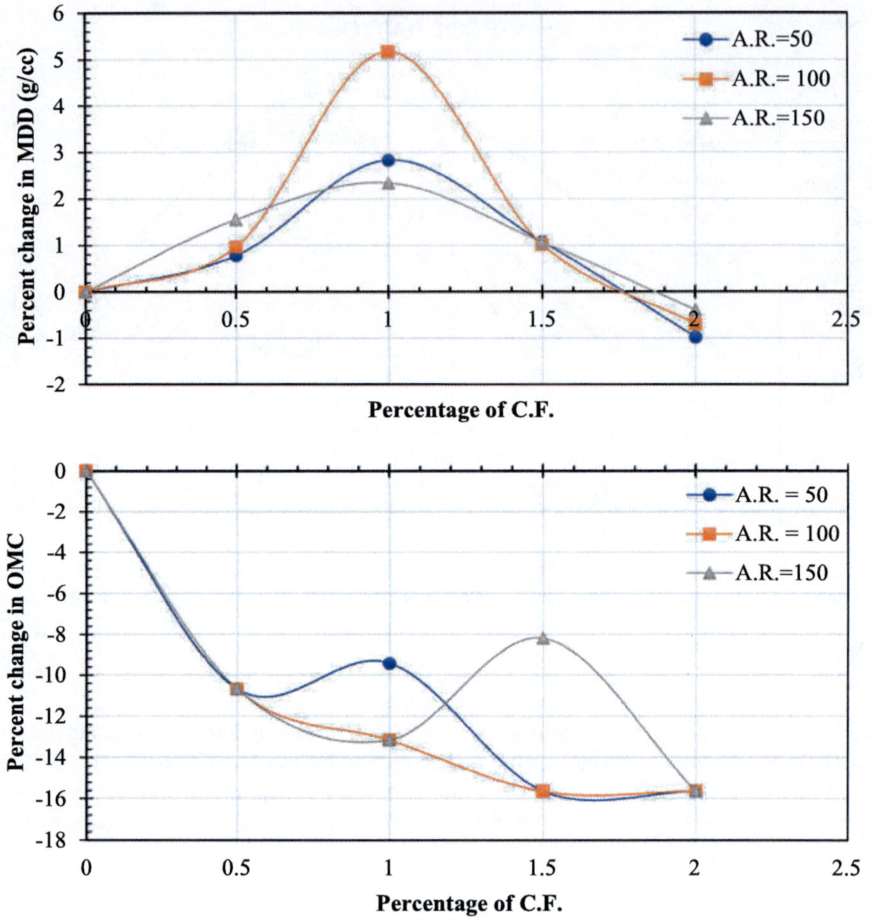

Fig. 12.6 **a** Percentage changes in MDD plotted against the variation of coir fiber in PA sample, **b** Percentage changes in OMC plotted against the variation of coir fiber in PA sample

It can be seen from Fig. 12.10 that the CBR value for a particular aspect ratio increases with increase in coir content up to 1%, and any further increase in percentage of coir starts to decrease the CBR value. Thus, it can be safely said that for all aspect ratios, the optimum value of soaked CBR is attained at a coir content of 1%. The highest CBR value is obtained for 1% coir content at an aspect ratio of 100. The better bond formation of 25 mm length coir fiber with pond ash may account for the higher CBR value at A.R. = 100.

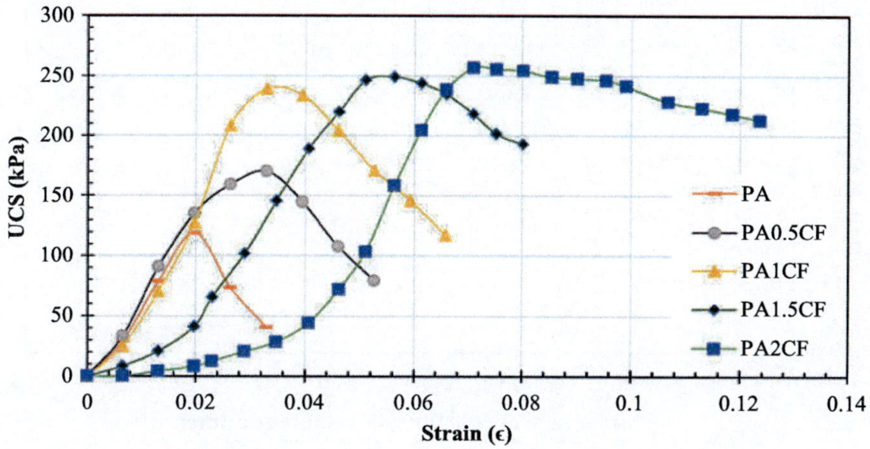

Fig. 12.7 Plot between UCS and strain for PA-CF mix composition

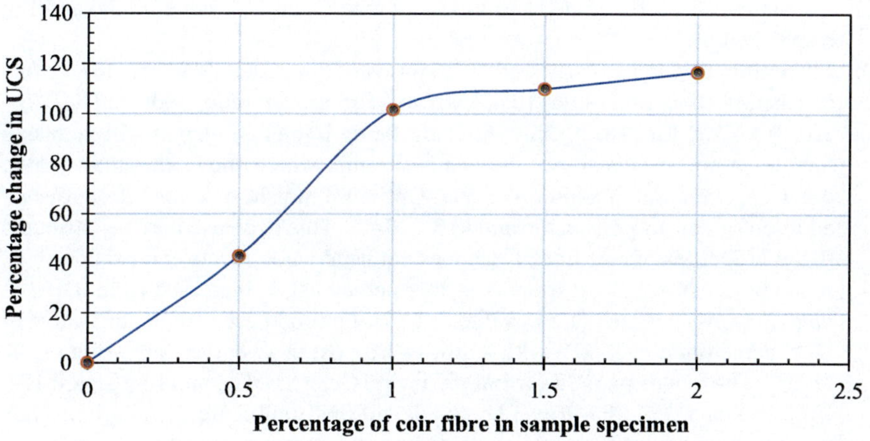

Fig. 12.8 Percentage change in UCS plotted against percentage of coir fiber in sample specimen

Conclusions

The present study was conducted to analyze the effect of percentage variation of coir fiber in pond ash at different aspect ratios. Experiments were performed for each mix composition, and following conclusions have been drawn:

1. The specific gravity of both the materials, i.e., pond ash (2.11) and coir (0.91) was found to be lesser than conventional earth materials. This can help in civil engineering works to become lighter.

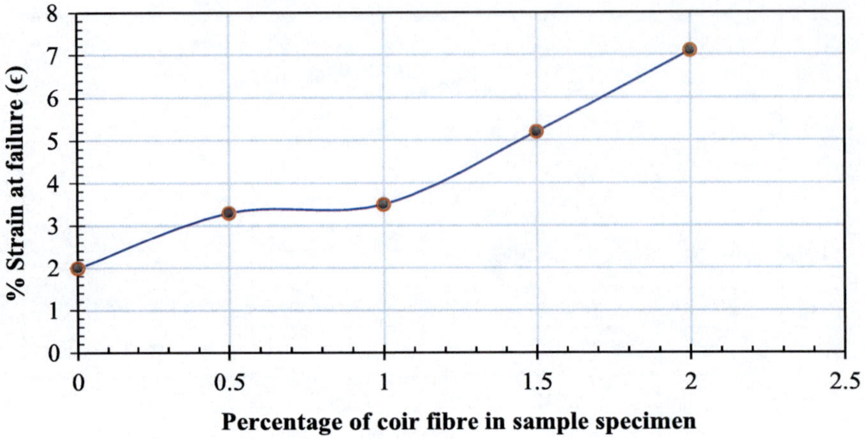

Fig. 12.9 Percentage strain at failure plotted against percentage of coir fiber in sample specimen

2. The sample is poorly graded and silty sand with C_u and C_c values of 3.6 and 1.7, respectively.
3. Maximum dry density showed a slight variation, i.e., (9.90–10.30) kN/m^3 for aspect ratio 50, (9.96–10.47) kN/m^3 for aspect ratio 100, and (10.01–10.28) kN/m^3 for aspect ratio 150. Coir being lighter in weight still increases the maximum dry density of the composite mix which shows that the voids in pond ash decrease on addition of coir. Optimum moisture content decreases on addition of coir to pond ash from 40.3 to 34%. This decrease may be attributed to the properties of coir fiber. Coir does not need water for its compaction.
4. It can be established that when coir fiber percentage is increased from 0 to 2%, there is significant increase in value of UCS. For pond ash sample, this value of UCS came out as 120 kPa with a downward post peak curve suggesting brittle failure. The values of UCS for PA0.5CF, PA1CF, PA1.5CF, and PA2CF are 170, 240, 250, and 258 kPa. It can be concluded that the nature of failure of UCS samples changes from brittle to ductile as the percentage of coir is increased from zero to 2%. As the percentage of coir is increased, the curve after attaining peak of UCS starts to attain a flat slope toward PA2CF samples which suggests the nature of failure has changed.
5. Soaked CBR value of PA and PA-CF mix came at 5 mm for all aspect ratios. CBR value for a particular aspect ratio increases with increase in coir content up to 1%, and any further increase in percentage of coir starts to decrease the CBR value. Thus, it can be safely said that for all aspect ratios, the optimum value of soaked CBR is attained at a coir content of 1%.
6. It can be concluded that pond ash alone can be used as a subgrade material in pavements as it offers a decent CBR value, and addition of coir fiber can increase the plastic behavior of pond ash up to a certain extent.

Table 12.3 **a** Soaked CBR values for all samples at aspect ratio = 50, **b** Soaked CBR values for all samples at aspect ratio = 100, **c** Soaked CBR values for all samples at aspect ratio = 150

(a)

AR = 50

Sample	Sample A		Sample B		Adopted CBR
	CBR (5 mm)	CBR (2.5 mm)	CBR (5 mm)	CBR (2.5 mm)	
PA	9.88	8.06	10.09	8.39	9.88
PA0.5CF100	10.15	8.22	10.44	8.51	10.15
PA1CF100	12.32	10.20	12.43	11.14	12.32
PA1.5CF100	11.48	9.57	11.60	10.56	11.48
PA2CF100	10.58	8.82	10.31	8.56	10.31

(b)

AR = 100

Sample	Sample A		Sample B		Adopted CBR
	CBR (5 mm)	CBR (2.5 mm)	CBR (5 mm)	CBR (2.5 mm)	
PA	9.88	8.06	10.09	8.39	9.88
PA0.5CF100	10.70	8.73	10.73	8.73	10.70
PA1CF100	13.80	11.55	13.19	11.72	13.17
PA1.5CF100	12.22	10.24	11.82	10.61	11.82
PA2CF100	11.69	9.82	10.95	9.14	10.95

(c)

AR = 150

Sample	Sample A		Sample B		Adopted CBR
	CBR (5 mm)	CBR (2.5 mm)	CBR (5 mm)	CBR (2.5 mm)	
PA	9.88	8.06	10.09	8.39	9.88
PA0.5CF100	10.73	8.77	10.65	8.68	10.65
PA1CF100	11.86	9.78	12.06	10.80	11.86
PA1.5CF100	11.01	9.10	11.14	10.13	11.01
PA2CF100	10.77	9.12	10.52	8.73	10.52

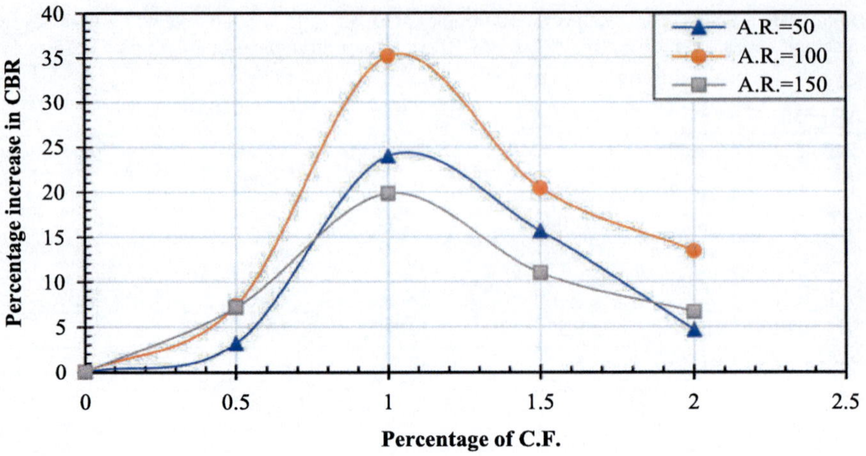

Fig. 12.10 Plot showing variation of strength with coir content for different aspect ratios

References

1. CEA Annual Report (2020) Central electricity authority annual report 2019–20. Government of India
2. Kumar M, Ramesh A (2011) An experimental investigation on engineering behavior of coir fiber and lime fly ash in stabilized mechanistic pavements. Comput Methods Geomech Front New Appl 2:1126–1132
3. Mohanty S, Patra NR (2015) Geotechnical characterization of Panki and Panipat pond ash in India. Int J Geo-Eng 6(1)
4. Kumar PSP (2009) Laboratory investigation of shedi soil stabilized with pond ash and coir. pp 428–430
5. Malik A, Singh SK (2020) Dynamic behaviour of pond ash mixed with crumb rubber. Proc Indian Geotech Conf 2020:95–108
6. Singh J, Singh SK, Alam MA (2019) A study on the geotechnical characterization and water retention characteristic curve of pond ash. Water Sci Technol 80(5):929–938
7. Singh SK, Arif SM (2014) Inclusion of coconut coir fiber in soil mixed with coal ash. Int J Res Eng Technol 3(11):209–213
8. Mishra M, Maheshwari UK, Saxena NK (2016) Improving strength of soil using fiber and fly ash-a review. Int Res J Eng Technol (IRJET) 03(10):1262–1266
9. Maurya S, Sharma DAK, Jain DPK, Kumar DR (2015) Review on stabilization of soil using coir fiber. Int J Eng Res 4(6):296–209
10. Sharan A (2011) Strength characteristics of fibre. M. Tech. dissertation, NIT ROURKELA

Chapter 13
Effect of Soil Properties on the Rutting Performance of Flexible Pavement: A Numerical Study

Barnali Debnath, Plaban Deb, and Partha Pratim Sarkar

Introduction

With the advancement of science and technology, the whole world is expanding and enormous road networks are being constructed [1, 2]. But the massive increase in infrastructural growth is leading to a high traffic volume and traffic density, which is causing serious pavement distresses. Flexible pavements are the most commonly used pavements all over the world, but several pavement distresses are making them worse for riding. Rutting in the flexible pavement is such a kind of major distress, which causes surface undulation and unevenness [3]. Rutting mainly develops as a form of longitudinal surface deflection along the wheel paths in the bituminous concrete layer [4–7]. A typical representation of rutting is shown in Fig. 13.1. Rutting is considered as severe pavement distress as it is a permanent deformation, often leading to the ultimate failure of the pavement [7] due to heavy commercial vehicles, the strength of existing soil as well as a severe environmental condition.

Several researchers have worked on this rutting resistance of flexible pavement, and the researchers have mentioned that the main reasons for rutting are excessive bitumen content, high amount of fine aggregates, excessive moisture content, traffic conditions, temperature, or weak subgrade condition [8]. Many theoretical models are also available to predict rutting depth, which are mainly produced based on the results obtained from laboratory experiments. However, most of the studies have concentrated only on the bituminous concrete layer, and the effect of soil properties/behavior on the rutting resistance is still unknown. Rutting mainly occurs due to either plastic deformation of the surface layer (instability rutting) or may also occur due to the lower layer's permanent settlement (structural rutting). The Indian

B. Debnath (✉) · P. P. Sarkar
Civil Engineering Department, NIT Agartala, Agartala, India
e-mail: brnali540@gmail.com

P. Deb
Civil Engineering Department, Chandigarh University, Mohali, Punjab, India

K. Muthukkumaran et al. (eds.), *Transportation and Environmental Geotechnics*,
Lecture Notes in Civil Engineering 298, https://doi.org/10.1007/978-981-19-6774-0_13

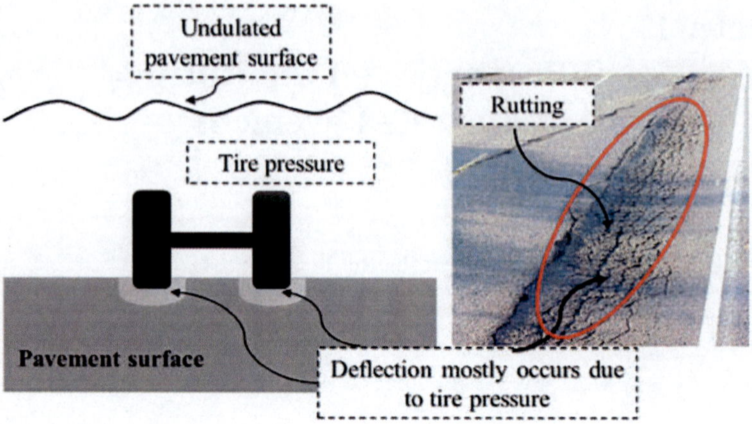

Fig. 13.1 Typical representation of rutting in flexible pavement

Road Congress (IRC) [9] has also mentioned in the standard guideline for flexible pavement design that the rutting might be a part of a weak subgrade or may occur due to subgrade deformation. Therefore, this study mainly focuses on the effect of soil properties or subgrade properties on the rutting resistance of flexible pavement.

In the present study, the rutting resistance of pavement on different soil subgrade conditions is determined with the help of computer-developed finite element-based software "Abaqus" [10–12]. For pavement construction, the important subgrade property is its resilient modulus or subgrade modulus, and hence, the main focus is paid on the subgrade modulus. It is important to note that the other properties of soil such as its dry density, optimum moisture content, CBR, etc. are highly correlated with the resilient modulus, and the resilient modulus alone can express the strength of the subgrade. Along with this subgrade modulus, the thickness of the subgrade is also changed, and a parametric study is conducted. A standard single-axle dual-wheel assembly is prepared for applying the traffic load in the pavement, and the loading is simulated using the repeating loading method. The load repetition is continued up to 20,000 repetitions, and the rutting value, i.e., the surface deflection is checked.

Numerical Modeling

Preparation of Model Geometry

In the present study, a flexible pavement section is chosen for the analysis, where the pavement consists of three different layers. A schematic diagram of the pavement with different layers is shown in Fig. 13.2a. The thickness of the top layer, i.e., bituminous concrete layers is chosen as 60 mm, the thickness of the granular base layer is taken as 150 mm, and the thickness of the compacted subgrade is varied from 300 to 650 mm. Below the subgrade, the embankment is also provided so as to make a similar kind of numerical pavement structure as present in in-situ conditions, although the main focus is paid on the soil subgrade. The embankment height is taken as 1000 mm. The properties of different layers are presented in Table 13.1. For applying the vehicular load, a single-axle dual-wheel assembly is chosen, where the gross weight of the wheels is 80 kN. The model geometry with dimensions, different layers, and circular imprints is shown in Fig. 13.2b. It is well known that the tire imprints of the vehicles are semi-elliptical in nature; however, due to the simplicity of geometry preparation, it is converted to a circular contact area. The radius of the circular tire imprint is chosen as 107 mm, and the horizontal distance between two wheels is taken as 343 mm. A single-axle dual-wheel assembly generally consists of two pairs of wheels giving a total 4 number of wheels. So, the applied load on each wheel will be 20 kN. In the present study, half a portion of the pavement is selected for the analysis due to the symmetry of the pavement section and dual-wheel assembly. The detailed model geometry along with the model dimensions is shown in Fig. 13.3.

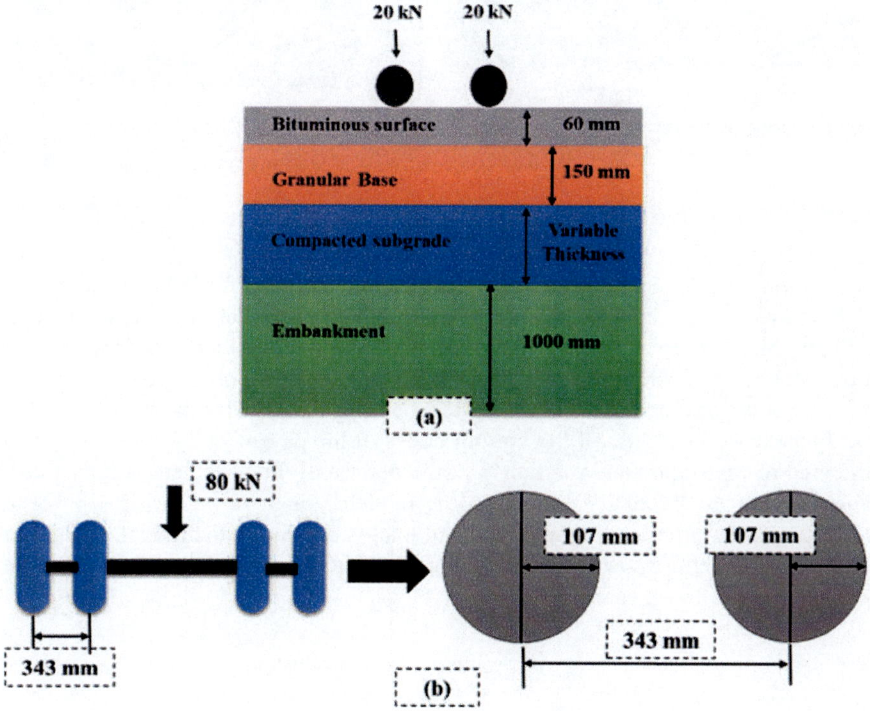

Fig. 13.2 Schematic diagram of the pavement layers and wheel assembly

Table 13.1 Properties of different layers

Pavement layers	Resilient modulus	Poisson's ratio
Surface layer	400 MPa	0.3
Granular base layer	200 MPa	0.3
Compacted subgrade	Variable	0.3
Embankment	50 MPa	0.3

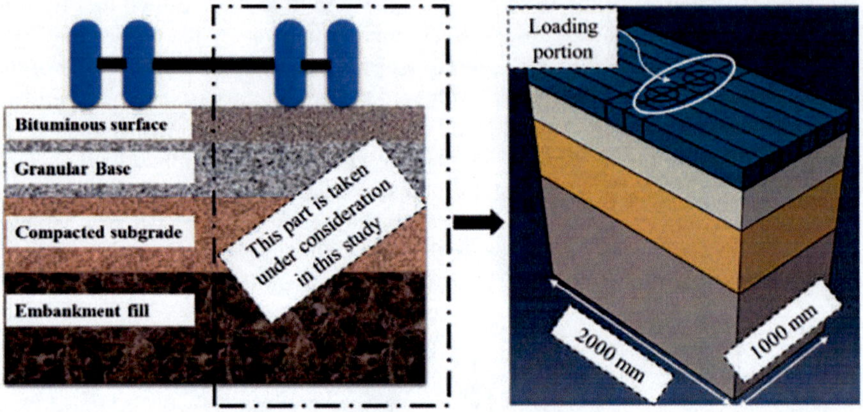

Fig. 13.3 Detail model geometry along with the model dimensions

Selection of Element Type, Mesh, and Boundary

For dividing the whole model geometry into a finite number of small-size elements, an 8-nodded hexahedral brick element (C3D8R) is chosen. For meshing the model, different bias is tried and a very fine mesh is provided at the vicinity of the loading area, whereas comparatively coarser mesh is applied on the other areas. The mesh refinement is also checked, and the selected mesh for the model is shown in Fig. 13.4. For boundary conditions, all the bottom nodes of the pavement are restricted from any kind of translation and rotation, i.e., the bottom of the pavement is fully fixed. The detail of the boundary applied in the model is also shown in Fig. 13.4. For simulating the traffic loading, the repeated load is applied in the model. A total of 20,000 load repetitions is provided, where the frequency is kept as 10 Hz, i.e., the load is repeated 10 times in 1 s.

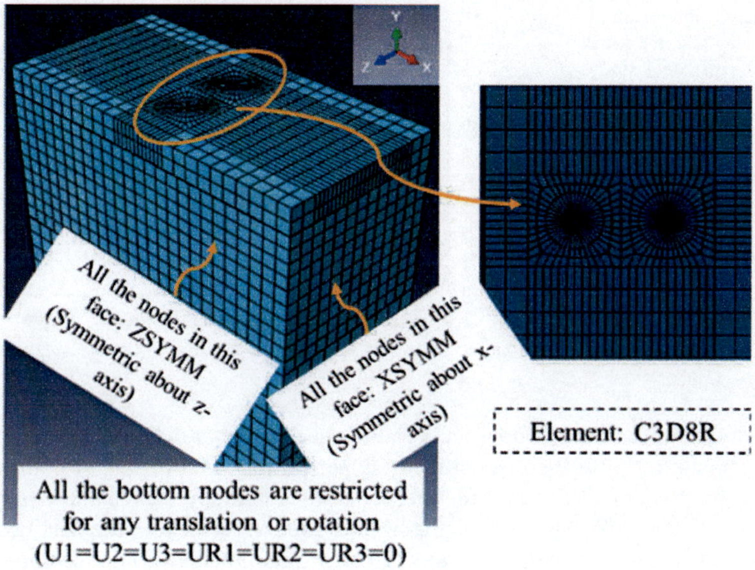

Fig. 13.4 Mesh and boundary applied in the numerical model

Table 13.2 Parametric study

Variables	Adopted values
Subgrade modulus (MPa)	40, 50, 60, 70, 80
Subgrade thickness (mm)	300, 350, 400, 450, 500, 550, 600, 650

Parametric Study

For the parametric study, it is already mentioned that the subgrade modulus and the thickness of the subgrade are changed. The detailed parametric study conducted in this research is mentioned in Table 13.2.

Validation of the Model

The prepared numerical model needs to be validated or compared with other numerical simulations or experimental data so as to confirm the suitability or the accuracy of the developed model. In the present study, validation is done with a numerical study conducted by Ali et al. [6]. As the dimensions and properties of the present study and the reference study are not the same, an exact mimic of the reference study is prepared here using the model development procedure as mentioned earlier.

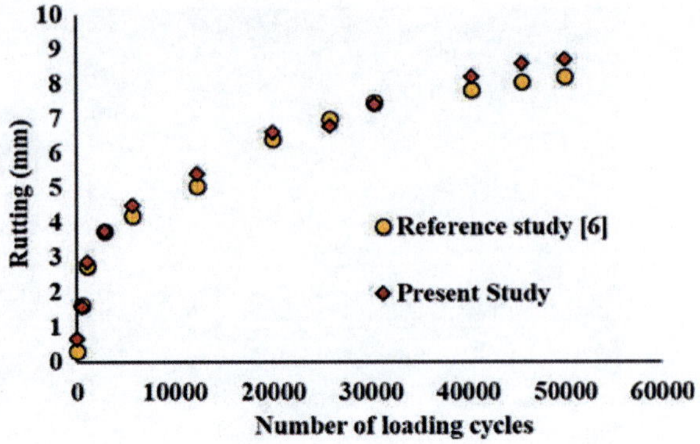

Fig. 13.5 Comparison of results obtained in present study and reference study

The main purpose of this validation is just to check whether the model development process as followed in the present study is performing well or not. A comparison of the rutting value observed in the present study and the reference study is shown in Fig. 13.5, and it is very clear that there is a good agreement in the data obtained in both cases. The standard deviation of the results obtained in the present study and reference study varies from 0.03 to 0.2, whereas the variation in the results lies between −5 and + 6%. This confirms that the process used for developing the numerical model in this study is satisfactory and can be used for further analysis.

Result and Discussion

Rutting Performance of Surface Layer of Flexible Pavement

The rutting behavior of the flexible pavement after applying repeated load is shown in Fig. 13.6, and it is observed that the rutting of the surface layer continuously increases with the increase in the loading cycle. However, it is a very common phenomenon, and similar results are also observed by other researchers [5, 13–16]. A visual representation of rutting behavior can also be observed from the numerical software, and it is presented in Fig. 13.7. The visual inspection of stress developed in the pavement is also shown in Fig. 13.8.

Effect of Subgrade Modulus on Surface Rutting

The effect of subgrade modulus, i.e., the resilient modulus of the subgrade is shown in Fig. 13.9, and it is clear that there is a significant effect of subgrade modulus on the rutting performances of bituminous pavement. It is important to mention that the rutting value is taken at final loading cycle, i.e., after application of 20,000 loading

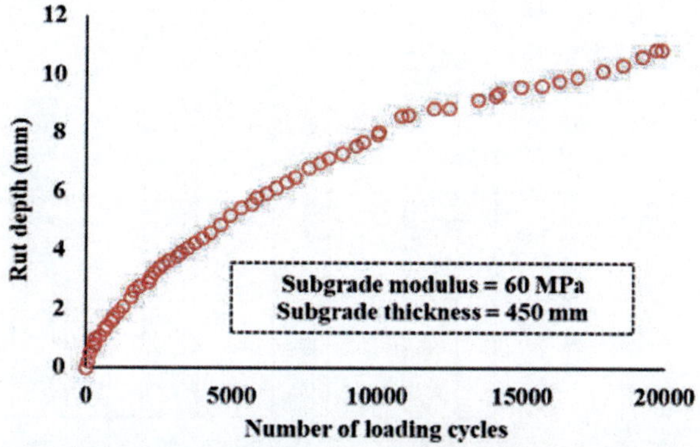

Fig. 13.6 Variation of rut depth with the number of loading cycle

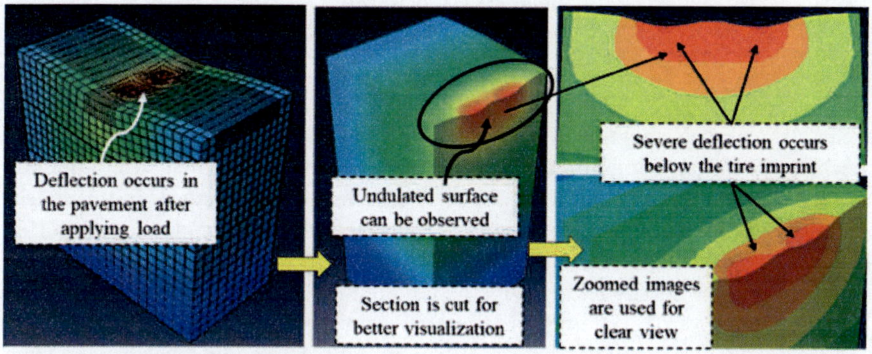

Fig. 13.7 Visual inspection of rutting

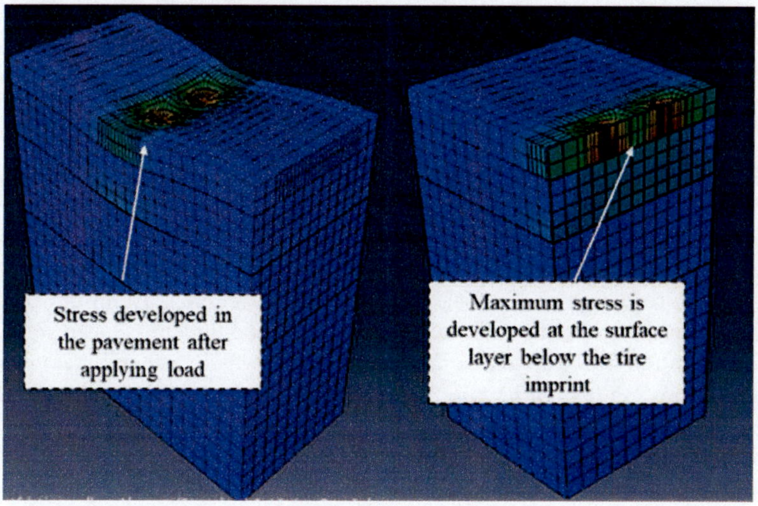

Fig. 13.8 Stress developed in the pavement

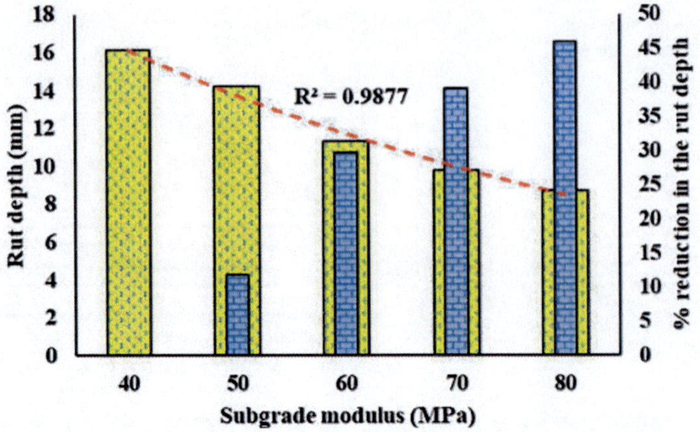

Fig. 13.9 Variation of rut depth with subgrade modulus

cycles. It is already mentioned earlier that the rutting can be simply identified by checking the deformation or the deflection occurred in the pavement after applying a certain amount of vehicular load. From this figure, a sharp decrease in the deflection/rutting value can be observed with the increase in the subgrade modulus. This decrease in the rutting of the surface layer mainly occurs due to the enhancement in the strength of the subgrade. If the resilient modulus of the subgrade is improved, a lesser amount of vertical strain gets generated at the top of the subgrade layer, which leads to a reduction in surface rutting. When the subgrade modulus is changed from 40 to 80 MPa, there observed almost 45% reduction in the rut depth in the surface layer.

Effect of Subgrade Thickness on Surface Rutting

Figure 13.10 is showing the effect of subgrade thickness on the rutting of the bituminous concrete layer. For the parametric study, the subgrade thickness is started from 300 mm and it is continued up to 600 mm. It is observed that the increased subgrade thickness reduces the rutting to a certain extent; however, it is also found that the rate of rutting reduction continuously decreases with the increase in the subgrade thickness. This phenomenon indicates that the subgrade thickness has a significant effect on the rutting value up to a certain limit, beyond which further increment in the subgrade thickness may not be effective enough. It is clear from the figure that the rut depth has been reduced by about 32%, when the subgrade thickness is increased from 300 to 500 mm. However, with the further increment, i.e., increasing thickness from 500 to 600 mm, the rut depth is almost the same or there is a very negligible reduction in the rut depth.

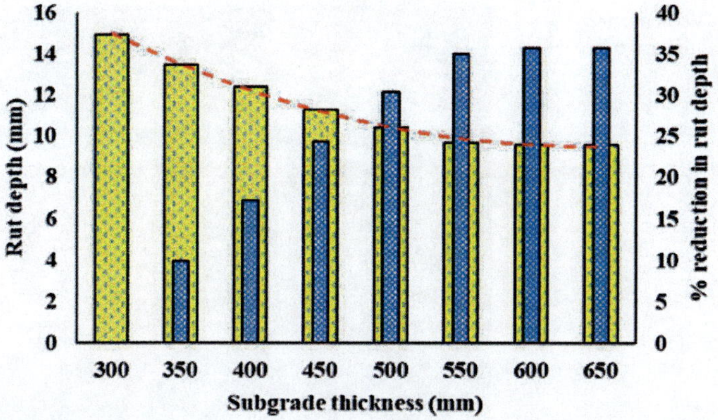

Fig. 13.10 Variation of rut depth with subgrade thickness

Vertical Strain Developed at the Top of the Subgrade and Measurement of Subgrade Rutting

IRC 37-2018 [6] has developed a flexible pavement design principle based on the subgrade rutting criteria and the vertical strain developed at the top of the subgrade layer. The subgrade rutting life can be expressed as

$$N_R = 4.17 \times 10^{-8} \times \left(\frac{1}{\varepsilon_v}\right)^{4.53} \tag{13.1}$$

where N_R is the subgrade rutting life and ε_v is the vertical strain developed at the top of the subgrade. In present study also, the vertical strain in the subgrade is obtained as shown in Fig. 13.11. With the improvement of subgrade properties, i.e., the subgrade modulus, the developed strain tends to reduce, and hence, the rutting life of the pavement goes on increasing. This can be better understood from Fig. 13.12, where the rutting life is presented according to subgrade thickness for several subgrade modulus. In this figure, MR represents the subgrade modulus, and the graph is plotted for a constant rut depth of 10 mm. The figure represents that the rutting life of the pavement can be greatly enhanced by improving the properties of subgrade.

Conclusion

This study mainly focuses on the behavior of subgrade material on the rutting resistance of the flexible pavement. The subgrade modulus or the resilient modulus of the subgrade soil is changed, and the effect of this subgrade modulus on the rutting of flexible pavement is checked. Moreover, the subgrade thickness is also changed, and

Fig. 13.11 Vertical strain developed at the top of the subgrade

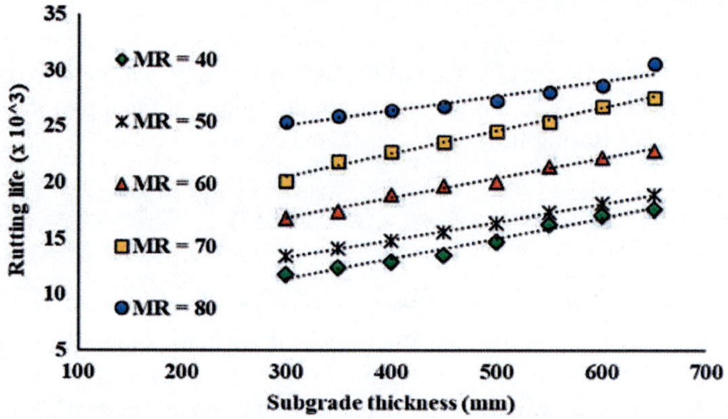

Fig. 13.12 Variation of rutting life with subgrade modulus and thickness

rutting resistance is observed. From the overall results, it is found that the subgrade modulus has a severe effect on the surface rutting of the flexible pavements. It is also visualized from the numerical models that the pavement experiences a surface rutting at the location of the tire imprint, and this rutting is dependent upon the subgrade modulus as well as subgrade thickness. The numerical results indicated that the surface rutting can be reduced up to 45% by increasing the subgrade modulus by two times. Similarly, the subgrade thickness also can be increased for the reduction of surface rutting. However, it is also found from the numerical results that the high subgrade thickness may not be helpful for controlling the rutting depth.

References

1. Debnath B, Sarkar PP (2020) Characterization of pervious concrete using over burnt brick as coarse aggregate. Constr Build Mater 242:118154
2. Erlingsson S (2012) Rutting development in a flexible pavement structure. Road Mater Pavement Des 13(2):218–234
3. Mirzapour Mounes S, Karim MR, Khodaii A, Almasi MH (2014) Improving rutting resistance of pavement structures using geosynthetics: an overview. Sci World J 8:2014
4. Du Y, Dai M, Deng H, Deng D, Wei T, Li W (2020) Evaluation of thermal and anti-rutting behaviors of thermal resistance asphalt pavement with glass microsphere. Constr Build Mater 10(263):120609
5. Du Y, Chen J, Han Z, Liu W (2018) A review on solutions for improving rutting resistance of asphalt pavement and test methods. Constr Build Mater 20(168):893–905
6. Ali B, Sadek M, Shahrour I (2009) Finite-element model for urban pavement rutting: analysis of pavement rehabilitation methods. J Transp Eng 135(4):235–239
7. Imaninasab R, Bakhshi B, Shirini B (2016) Rutting performance of rubberized porous asphalt using finite element method (FEM). Constr Build Mater 1(106):382–391
8. Archilla AR, Madanat S (2000) Development of a pavement rutting model from experimental data. J Transp Eng 126(4):291–299
9. IRC 37: Guidelines for the design of flexible pavements, Indian Road Congress (2018)
10. Deb P, Pal SK (2019) Numerical analysis of piled raft foundation under combined vertical and lateral loading. Ocean Eng 15(190):106431
11. Deb P, Pal SK (2021) Structural and geotechnical aspects of piled raft foundation through numerical analysis. Mar Georesour Geotechnol 12:1–24
12. Deb P, Pal SK (2019) Analysis of load sharing response and prediction of interaction behaviour in piled raft foundation. Arab J Sci Eng 44(10):8527–8543
13. Zhang J, Zhu C, Li X, Pei J, Chen J (2017) Characterizing the three-stage rutting behavior of asphalt pavement with semi-rigid base by using UMAT in ABAQUS. Constr Build Mater 1(140):496–507
14. Zhang Y, Luo X, Deng Y, Hou S, Shi X, Lytton RL (2020) Evaluation of rutting potential of flexible pavement structures using energy-based pseudo variables. Constr Build Mater 30(247):118391
15. Sirin O, Kim HJ, Tia M, Choubane B (2008) Comparison of rutting resistance of unmodified and SBS-modified Superpave mixtures by accelerated pavement testing. Constr Build Mater 22(3):286–294
16. Kim WJ, Lee HJ, Phan HT (2017) Calibration and validation of a rutting model based on shearstress to strength ratio for asphalt pavements. Constr Build Mater 15(149):327–337

Chapter 14
Feasibility of Utilizing MSW Incinerated Bottom Ash for Subgrade and Sub-Base Applications in Flexible Pavements

Deepesh Bansal, Manoj Datta, and G. V. Ramana

Introduction

Industrialization, urbanization, improved living standard accompanied by economic growth, and increasing population have led to the generation of large quantities of municipal solid waste (MSW). A World Bank report (2018) states that the annual global MSW generation is about 2.01 billion tons in 2016 and is estimated to be around 2.59 billion tons per annum by 2030 [1]. MoHUA (2020) report suggested that India with an MSW generation of about 60 million tons per annum will double by 2030. In other words, India will generate around 130 million tons of MSW annually with an average waste generation rate of 0.74 kg/person/day [2]. In India, most of this waste is dumped in an unscientific manner at dumpsites, and these sites have locked around 25,000 acres of land in urban areas, which also poses a threat to the environment [2]. Waste recycling and reduction at source is a problem in our country, and hence, waste-to-energy (WtE) plants provide a feasible option to reduce the quantity of waste. In WtE facilities, MSW is incinerated at a high temperature of around 850–1000 °C, which minimizes the waste quantity by about 75–80% (by weight) and produces electricity [3]. WtE plants generate about 20–25% residue, mainly bottom ash (80–85%) and remaining fly ash (15–20%). In India, currently, these residues are dumped in landfills. At present, India has seven operational WtE plants with a capacity of 92.5 MW. Besides these, more than 50 such projects are underway. The completion of these projects will ensure the utilization of around 30,000 tons of MSW per day with a power generation of about 400–410 MW [4, 5]. However, these WtE facilities will result in the generation of MSWI BA, which, if left unutilized, needs to be dumped in landfills.

One of the primary objectives of Sustainable Development Goals (2018) is to shift the focus on utilizing secondary/recycled material for construction activities to

D. Bansal (✉) · M. Datta · G. V. Ramana
Indian Institute of Technology, New Delhi, Delhi 110016, India
e-mail: ce.deepeshbansal@gmail.com

K. Muthukkumaran et al. (eds.), *Transportation and Environmental Geotechnics*, Lecture Notes in Civil Engineering 298, https://doi.org/10.1007/978-981-19-6774-0_14

reduce the onus on naturally available materials [6]. Therefore, it would be helpful if the MSW incinerated bottom ash (MSWI BA) generated can be utilized in bulk quantities so that there is no or little material left for dumping in landfills. Globally, there are studies available regarding the characterization of bottom ash for its utilization purpose [7–13]. However, in India, there are not many studies available on the utilization potential of MSWI BA, and it is still an evolving field. Also, it has been acknowledged that bottom ash properties vary depending on waste composition and the combustion process [14, 15]. Thus, there is a need for a comprehensive study of the utilization potential of MSW incinerated bottom ash in the country. The present study investigates the physical and mechanical characteristics of MSWI BA for utilization in subgrades and sub-base courses in flexible pavements.

Material Collection and Methodology

MSWI BA samples were collected from two waste-to-energy plants in Delhi. These plants are referred to as Plant A and Plant B. These facilities combinedly receive around 4000 tons of MSW per day, constituting approximately 30% of total waste reaching landfills in Delhi [2, 5]. The waste so received is converted to Refused Derived Fuel (RDF) with pre-processing activities to improve its calorific value. Pre-processing activities include size segregation with trommels, air-ballistic separators, or manual sorting. Further, windrow composting is performed for biodegradation of organics, and it also reduces the incoming moisture content of the waste. Finally, the processed material is fed into WtE units to generate electricity. The daily electricity production capacities of Plant A and Plant B are about 16 MW and 24 MW, respectively. The incineration of waste reduces the waste quantity by approximately 70–75% (by mass) and 85–90% (by volume). In addition, the incineration process results in the generation of two types of ashes, bottom ash and fly ash. These ashes are collected from different locations in the plant and are dumped in landfills separately. These plants combinedly account for the daily production of around 700–750 tons of bottom ash. If a significant portion of this ash can be utilized in road works, it will help reduce the material reaching landfills and reduce the onus on natural material.

Material Collection

MSWI BA samples were collected from the bottom of the reactor after water quenching. Gupta et al [16] studied the heterogeneity of material based on seasonal variation, and it was found that the material has almost similar properties for different seasons [16]. Therefore, it was ascertained to collect ten days' average material as a one-time representative sample. The collected bottom ash had a moisture content of 15–20%, and it was naturally air-dried by spreading under a shed for 5–7 days to make it workable, as shown in Fig. 14.1a. After air drying, the material was appropriately

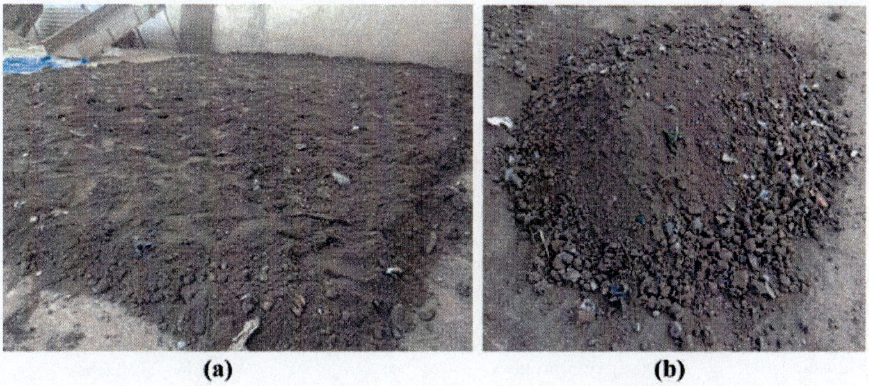

Fig. 14.1 **a** MSWI BA spread for air-drying; **b** Representative sample of MSWI BA

mixed, and a representative sample was obtained through conning and quartering, as shown in Fig. 14.1b. The present study focuses on particle size minus 35 mm, constituting more than 90% of the total MSWI BA generated.

Methodology

The present study focuses on assessing the feasibility potential of MSWI BA for utilization in the subgrade and sub-base course in flexible pavements in an unbound form. A comprehensive laboratory testing was performed for determining the physical and mechanical characterization of MSWI bottom ash, and Table 14.1 summarizes the testing program. In addition, a comparative assessment was performed for characteristics obtained with prescriptive limits in standards: MoRD specifications (2014), MoRTH specifications (2013), and IRC: 37 (2018) [17–19].

Result and Discussion

Grain Size Distribution and Plasticity

Wet sieve analysis was performed to determine the grain size distribution curve of MSWI BA. Figure 14.2 shows the grain size distribution curve of bottom ash from the two plants. It is found that MSWI BA is a well-graded material that consists primarily of sand (53–55%) and gravel (35–40%) size fractions with low fine content (7–10%) and, hence, is classified as silty sand with gravels. The study focuses on particle size

Table 14.1 Required physical and mechanical properties

Property	Type of test	Subgrades	Sub-base course	Test standards
GSD	Sieve analysis	✓	✓	IS: 2720 (Part 4)-1975 ASTM D6913-2017
Specific gravity and Porosity	Specific gravity and water absorption tests	✓	✓	IS: 2720 (Part 3)-1981 (Finer fraction*) IS: 2386 (Part 3)-1963 (Coarser fraction**) ASTM D854-2014
Plasticity	Atterberg limit tests	✓	✓	IS: 2720 (Part 5)-1985 ASTM D4318-2017
Compaction characteristics	Modified proctor test	✓	✓	IS: 2720 (Part 8)-1983 ASTM D1557-2012
Denseness	Relative density test	✓	–	IS: 2720 (Part 14)-1983 ASTM D4253-2016
CBR value	CBR test	✓	✓	IS: 2720 (Part 16)-1987 ASTM D1883-2016
Abrasion (hardness)	LA abrasion test	–	✓	IS: 2386 (Part 4)-1963 ASTM C131-2020
Toughness	Impact test	–	✓	IS: 2386 (Part 4)-1963

* Finer fraction: < 4.75 mm, ** Coarser fraction: 4.75–35 mm

minus 35 mm, constituting more than 90% of the total fraction. In addition, MSWI BA is found to be a non-plastic material.

Specific Gravity and Water Absorption

Specific gravity and water absorption values are reported in Table 14.2. MSWI BA has lower specific gravity values than the natural material (typically 2.65–2.68) and has higher water absorption values compared to natural aggregates (typically 1–3%) [21]. These values indicate the material has intraparticle voids and is lightweight, as suggested in the literature [7–13].

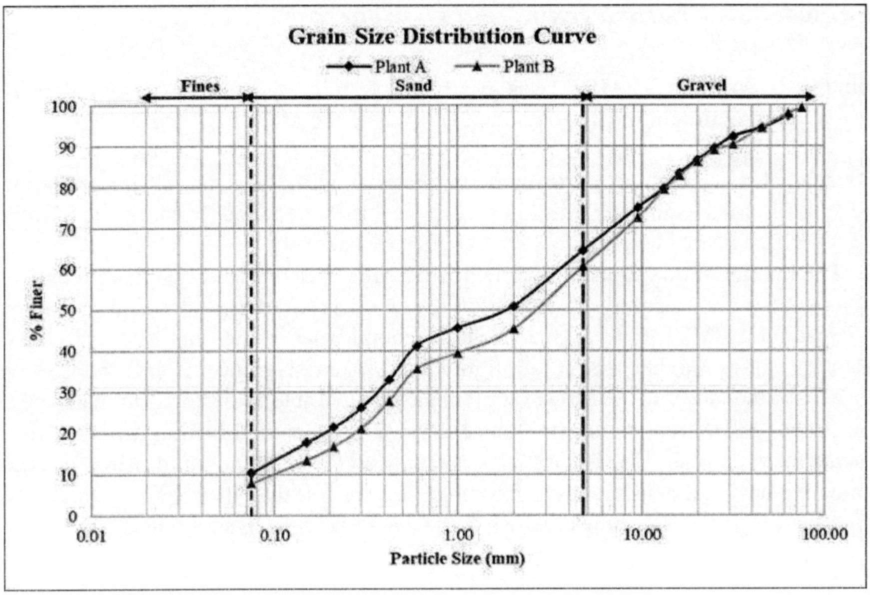

Fig. 14.2 Grain size distribution curve of MSWI BA

Table 14.2 Physical and mechanical properties of MSWI BA

Sl. no	Test	Property	Plant A	Plant B
1	Specific gravity test	SG for coarser fraction	2.00 ± 0.05	1.96 ± 0.05
		SG for finer fraction	2.46 ± 0.04	2.43 ± 0.04
2	Water absorption test	WA for coarser fraction (%)	6.38 ± 0.5	6.64 ± 0.4
		WA for finer fraction (%)	4.82 ± 0.5	4.96 ± 0.3
3	Atterberg limit tests	Plasticity	Non-plastic	Non-plastic
4	Modified proctor test	MDD (kN/m^3)	16.69 ± 0.09	16.63 ± 0.04
		OMC (%)	13.00 ± 0.2	13.20 ± 0.3
5	Relative density test	Max. dry density (kN/m^3)	15.14 ± 0.1	15.27 ± 0.2
		Min. dry density (kN/m^3)	12.42 ± 0.2	12.82 ± 0.2
6	CBR test	CBR unsoaked (%) @ OMC and MDD	52.7 ± 3.7	59.6 ± 5.3
		CBR soaked (%)	39.7 ± 2.6	48.4 ± 4.8
7	LA abrasion test	LA abrasion value (%)	52.2–56.5	53.6–59.2
8	Impact test	Impact value (%)	53.3 ± 0.5	57.3 ± 0.5

Compaction Characteristics and Denseness

Modified proctor test (MPT) was used to determine the maximum dry density (MDD) and optimum moisture content (OMC) of the MSWI BA. MDD and OMC of MSWI BA were found to be 16.63–16.69 kN/m^3 and 13–13.2%, respectively. Relative density can be performed on material with fines (< 75 μ) less than 15% [20]. Maximum and minimum dry densities obtained from the test were 15.14–15.27 kN/m^3 and 12.42–12.82 kN/m^3, respectively.

The maximum dry density obtained from MPT is more than that in the relative density test, and this may be attributed to the crushing of particles due to impact compaction in MPT, contrary to vibratory compaction in the relative density test. Density values indicate that it is a lightweight material. Hence, it will induce low in-situ stresses and will have lower settlement values. Table 14.3 shows standard specifications for different conditions, and MSWI BA meets the MoRD (2014) requirements for rural roads. MSWI BA also meets MoRTH (2013) requirements for road embankments, but density values for subgrades are less than the required standards. However, higher CBR values make it favorable to be used in subgrades.

CBR Value

CBR test was performed by preparing specimens at MDD and OMC as determined from MPT for unsoaked conditions. CBR values in the unsoaked state were 52.7–59.6%. The samples were soaked for four days to assess CBR values (soaked CBR) under critical situations, and these values are 39.7–48.4%. CBR values indicate that MSWI BA has appropriate bearing strength and met design standards for subgrade and sub-base courses, as shown in Table 14.3.

LA Abrasion Value and Impact Value

Los Angeles abrasion value and impact value determined were 52.2–59.6% and 53.3–57.3%, respectively. These values indicate that MSWI BA undergoes more than permissible particle breakdown and, hence, does not meet the sub-base course standard specifications for roads, as shown in Table 14.3. The crushing of particles might be because of the intraparticle voids in the material [7–13].

Table 14.3 Standard specifications for material used in flexible pavements

Sl. no	Property	MoRTH (2013)			MoRD (2014)			Present study
		Embankment	Subgrade	Sub-base	Embankment	Subgrade	Sub-base	
1	Maximum particle size (mm)	75	50	75	75	50	75	35
2	Liquid limit (%)	50	50	25	50	50	25	NA
3	Plasticity index (%)	25	25	6	25	25	6	Non-plastic
4	Density (kN/m^3)	> 15.2 kN/m^3 for height < 3 m	> 17.5 kN/m^3	–	> 14.4 kN/m^3 for height < 3 m	> 16.5 kN/m^3	–	16.63–16.70 kN/m^3
		> 16 kN/m^3 for height > 3 m		–	> 15.2 kN/m^3 for height > 3 m		–	
5	CBR (%)	5[a]	5[a]	30	5[a]	5[a]	20	Soaked CBR-39.7–48.4%
6	LA abrasion value (%)	–	–	40	–	–	–	52.2–59.2%
7	Impact value (%)	–	–	40	–	–	50	53.3–57.3%

a IRC 37 (2018) recommends a minimum CBR of 5% for subgrades and embankments for 450 commercial vehicles per day

Conclusion

MSWI BA samples were collected from two WtE plants in Delhi to evaluate their feasibility in the subgrade and sub-base course in flexible pavements. Based on comprehensive laboratory testing on MSWI BA samples, the following conclusions were made:

- MSWI BA is well-graded silty sand with gravel-like material and is non-plastic in nature.
- It is a lightweight material, and hence, it will induce low in-situ stresses and undergo less settlement.
- Maximum dry density from MPT is more than that in the relative density test due to particle crushing under impact compaction. This may be attributed to the intraparticle voids in the material.
- The properties of MSWI BA are in line with standard specifications for subgrade material in flexible pavements. CBR value indicates that the material will offer a higher bearing capacity to pavement layers.
- Significant particle breakdown is also indicative from abrasion and impact tests, reconfirming the effect of the intraparticle voids in the material. This also limits its direct application in sub-base courses in flexible pavements. Hence, its properties need to be improved, possibly by blending with natural aggregate or cement.

The scope of the present study is limited only to the physical and mechanical characterization of MSWI BA. However, before field application, its geoenvironmental properties also need to be ascertained.

References

1. Kaza S, Lisa Y, Perinaz BT, Frank VW (2018) What a waste 2.0: a global snapshot of solid waste management to 2050. In: Urban Development Series. Washington, DC: World Bank
2. Ministry of Housing and Urban Affairs (MoHUA): Annual Report (2020).
3. Bureau EI (2005) Reference document on the best available techniques for waste incineration. European Union, Brussels
4. Ministry of New and Renewable Energy (MNRE): Power generation from municipal solid waste, 20th Report (2016).
5. Tripathy U (2018) A 21st century vision on waste to energy in India: a win-win strategy for energy security and Swachh Bharat Mission (Clean India Mission). In: A report by the United Nations Centre for Regional Development in collaboration with the Ministry of Housing and Urban Affairs (MoHUA), Government of India and Ministry of the Environment, Government of Japan, India
6. Sustainable development goals, SDGs Homepage, https://sdgs.un.org/goals. Last accessed 21 April 2021
7. Becquart F, Bernard F, Abriak NE, Zentar R (2009) Monotonic aspects of the mechanical behavior of bottom ash from municipal solid waste incineration and its potential use for road construction. Waste Manage 29(4):1320–1329

8. Chang HL, Jau WC, Li KC, Lin CF (2004) The evaluation of the feasibility of utilizing incineration bottom ash as subbase material. Environmental informatics archives, vol 2. ISEIS Publication, pp 1033–1047
9. Dou X, Rena F, Nguyena MQ, Ahameda A, Yina K, Chana WP, Changa VW (2017) Review of MSWI bottom ash utilization from perspectives of collective characterization, treatment and existing application. Renew Sustain Energy Rev 79:24–38
10. Hu Y, Li G, Zhong Y (2010) Utilization of municipal solid waste incineration bottom ash as road construction materials. In: International conference on mechanic automation and control engineering, IEEE, pp 1370–1373
11. Izquierdo M, Lopez-Soler A, Ramonich EV, Barra M, Querol X (2002) Characterization of bottom ash from municipal solid waste incineration in Catalonia. J Chem Technol Biotechnol 77(5):576–583
12. Vizcarra GOC, Casagrande MD, De Motta LMG (2014) Applicability of municipal solid waste incineration ash on base layers of pavements. J Mater Civ Eng 26(6):06014005
13. Wiles CC (1996) Municipal solid waste combustion ash: state-of-the-knowledge. J Hazard Mater 47(1–3):325–344
14. Tang P, Florea MVA, Spiesz P, Brouwers HJH (2015) Characteristics and application potential of municipal solid waste incineration (MSWI) bottom ashes from two waste-to-energy plants. Constr Build Mater 83:77–94
15. Van Gerven T, Van Keer E, Arickx S, Jaspers M, Wauters G, Vandecasteele C (2005) Carbonation of MSWI-bottom ash to decrease heavy metal leaching, in view of recycling. Waste Manage 25(3):291–300
16. Gupta G, Datta M, Ramana GV, Alappat BJ (2018) Feasibility of reuse of bottom ash from MSW waste-to-energy plants in India. In: The 8th international congress on environmental Geotechnics, Springer, Singapore, pp 344–350
17. Ministry of Road Transport and Highways (MoRTH): Specifications for road and bridge works. Published by Indian Road Congress, 5th Revision (2013)
18. Ministry of Rural Development (MoRD): Specifications for rural roads. Published by Indian Road Congress, 1st Revision (2014)
19. IRC: 37-Guidelines for Design of Flexible Pavements. Published by Indian Road Congress, 4th Revision (2018)
20. ASTM D4253–16 Standard test methods for maximum index density and unit weight of soils using a vibratory table. ASTM International, West Conshohocken, PA (2016)
21. Lynn CJ, Ghataora GS, Obe RKD (2017) Municipal incinerated bottom ash (MIBA) characteristics and potential for use in road pavements. Int J Pavement Res Technol 10(2):185–201

Chapter 15
Influence of Existing Ganga Sand on Subgrade Characteristics of Rural Road in South Bengal Districts

Ansar Ali, Tufan Dutta, and B. C. Chattopadhyay

Introduction

Large-scale constructions for development of infrastructures and rural roads are being made in the third world countries for necessary growth and development. In India under Pradhan Mantri Gram Sadak Yojana (PMGSY), rural road constructions are going in for massive scale with the aim to connect remotest villages with population of 500 people with existing roads enabling connection of the villages with nearby towns and growth centers. For such roads, local soils are used to house them on engineered subgrade, generally made of the existing in-situ soil at the location itself. Thus, strength of this subgrade controls the design of the road structures in both the cases of rigid or flexible pavement. In most of the countries, California bearing ratio (CBR) in soaked condition of the subgrade is used to represent the strength of the subgrade and used for design of the higher layers like sub-base, base course, and wearing course of the flexible pavement. Higher the value of soaked CBR of this subgrade, thicknesses of the above layers, required will be lesser and cost-effectiveness of the road construction will increase.

The soils from surface to large depth in flood plains of Bengal are all alluvial. Such soils are extremely erratic and also have variation in properties both in horizontal and vertical direction [1]. The roads constructed in southern districts of Bardhaman, Hooghly, Howrah, and East Medinipur on the west side of river Hooghly and Murshidabad, Nadia, North and South 24 Parganas on the east side of the river are all being constructed on such alluvial soils. These soils are basically fine-grained soil with some Gangetic silver sand in varying proportion [2]. As strength of fine-grained soils are relatively unaffected by the grain size distribution in such soils, presence of coarse grain particles seems to be most influencing factor changing the strength properties of the prepared subgrade made of such soils.

A. Ali (✉) · T. Dutta · B. C. Chattopadhyay
Department of Civil Engineering, Meghnad Saha Institute of Technology, Kolkata 700150, India
e-mail: meeansar@gmail.com

K. Muthukkumaran et al. (eds.), *Transportation and Environmental Geotechnics*, Lecture Notes in Civil Engineering 298, https://doi.org/10.1007/978-981-19-6774-0_15

Since only limited number of CBR (soaked) tests are performed in actual practice as per IRC recommendation (IRC: 36-1970) and as the existing soils below proposed roads in the areas show wide non-uniformity, relation between soaked CBR value of such soils with amount of silver sand present may give easy possible mapping of soaked CBR along the length of the roads by checking sand percentage in the soil at intermittent locations, which can be done at very low cost and time during constructions.

Brief Review of Literature

Some researchers in the recent past have presented the effect of adding sand in different percentages on the engineering properties like liquid limit (LL), plastic limit (PL), compaction characteristics, and soaked California bearing ratio of cohesive soils. Addition of inert materials like sand may be good solution to modify swelling and shrinkage properties of problematic soils [3]. Louafi and Bahar [4] presented in detailed experimented study on effect of addition of fine, medium, and coarse sands in bentonite. The results of the tests showed a decrease in LL, PL, and plasticity index (PI) with increasing sand content and decrease swelling pressure with increasing addition of all three sands.

Roy [5] also studied the effect of three types of sands in different proportions to cohesive soils. Proportion of sand mixed in the study varied from 0 to 15% by weight. The results showed decrement in the value of LL and PI though effect on PL was insignificant. Such observation was also supported by findings of Ravishankar (2012). Maximum dry density (MDD) values of the alluvial soil show significant increase with addition of all the three types of sands used, while optimum moisture content (OMC) values indicated gradual decrease. For strength property, unsoaked CBR at maximum dry density showed high increase in value with increase in sand content. Cabalar and Mustafa [6] investigated the effects of sand content on the behavior of sand–clay mixtures. The result of the tests showed a decrease in unconfined compressive strength (UCS) and on increase in CBR value with increase in sand content. Further, an increase in dry unit weight and decrease in respective OMC with increase in sand content were also observed.

Kollaros and Athanasopoulou [7] used various proportion of sand from 0 to 100% with soft clayey soil. It is reported that LL, PL, and PI all decrease with increase in sand percentage. The OMC of the mixed sand–clay decreases with percentage of sand, while MDD increased with increase in percentage of sand. The result of UCS tests at OMC for the sand–clay mixture shows that up to sand percentage of 60% UCS value increases from 1211 kPa to a high value of 2798 kPa but the strength decreases gradually to a lower value with further increase in sand percentage. Alam and Majumdar [8] reported on the basis of laboratory and field determination of CBR value of soft clayey soils stabilized with granular materials that there is marked improvement due to addition of granular materials.

Yadav and Suman [9] investigated the influence of the Ganga sand on characteristics of cohesive soil when it is used as subgrade materials. Variation of LL, PL, and PI with addition of sand in the range of 0–100% was studied. It was observed that beyond value of sand mixing 20%, the soil behaves like a non-plastic material. However, within 0–20%, LL decreases with addition of sand, PL also decreases, but decrease in LL is more, while that in PL is not that significant, and similar variation is related in PI also. In the compaction characteristics, MDD is increasing initially, reaches a maximum, and then decreases.

Similarly, OMC is seen to be decreasing initially with increasing sand content, and thereafter, certain value of sand content is increasing. CBR value in soaked and unsoaked conditions gradually increases with increasing sand percentages, and after reaching the maximum value, it starts decreasing with further increase of sand content.

Collection of Data on Different Soil Properties

In South Bengal, districts of Murshidabad, Nadia, North 24 Pargana, and South 24 Pargana are on east side of river Hooghly, while the districts of Bardhaman, Hooghly, Howrah, and East Medinipur are on west side of River Hooghly. All these districts are part of alluvial flood plain of River Ganga. The soils in these districts from surface to a great depth are all alluvial and vary erratic both in vertical and horizontal direction. Near-surface soils are basically fine-grained cohesive soils (CI) with some percentage of fine Ganga silver sand. The percentage of sand in these districts generally varies from 0 to 20%.

For construction of rural road generally, near-surface soils are used in the construction of the subgrade. Large numbers of rural roads have been constructed in the recent past for which necessary detailed project reports were submitted by concern authorities in the districts for onward processing. These reports contain results of test on such soil along the routes of the proposed roads at an interval of 500 m. As a result, huge amount of data were collected for these districts for detailed study on effect of varying proportion of fine Ganga silver sand on different engineering properties of such soils. The details of the data on such a results have been presented elsewhere [10].

Methodology for Analysis of Data

From the review of literature, it has been observed that presence of sand in cohesive soils modifies the liquid limit, plastic limit, compaction characteristics, and soaked CBR value to varying extent depending on the percentage of sand mixed and the cohesive soil. For analyzing the collected data of near-surface alluvial soils in eight districts of South Bengal, data of each districts was taken separately and plotted for liquid limit, plastic limit, and plasticity index against percentage of sand present in these soils. Thereafter, OMC and $\gamma_{d\max}$ value of the same soil were plotted against percentage of sand present. Then, reported values of soaked CBR of the same soils were plotted against percentage of sand present in the soil. The process was repeated for remaining seven districts one after another.

Result and Discussions

Variations of Atterberg's Limits with Percentage of Sand

The results of the Atterberg's limit of soils in Murshidabad districts are plotted in Figs. 15.1, 15.2 and 15.3, against percentage of sand present. It is seen from Fig. 15.1 with sand percentage increasing, the liquid limit is gradually decreasing. The variation seems to be linear one, and such linear variation for LL against sand percentage will be convenient for adoption of infield practice.

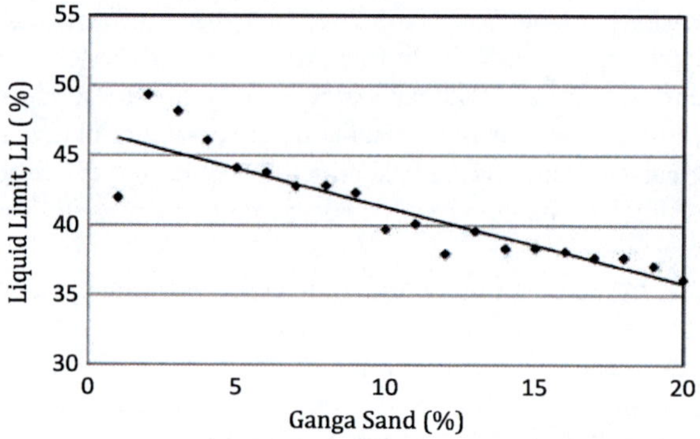

Fig. 15.1 Variation in LL with existing sand percentages at various location in Murshidabad (surficial soils)

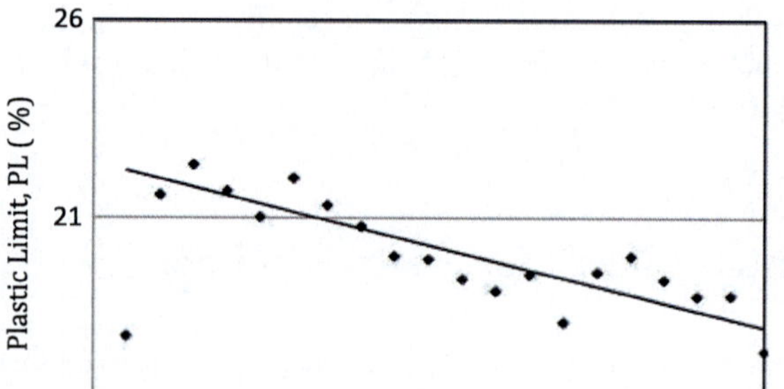

Fig. 15.2 Variation in PL with existing sand percentages at various location in Murshidabad (surficial soils)

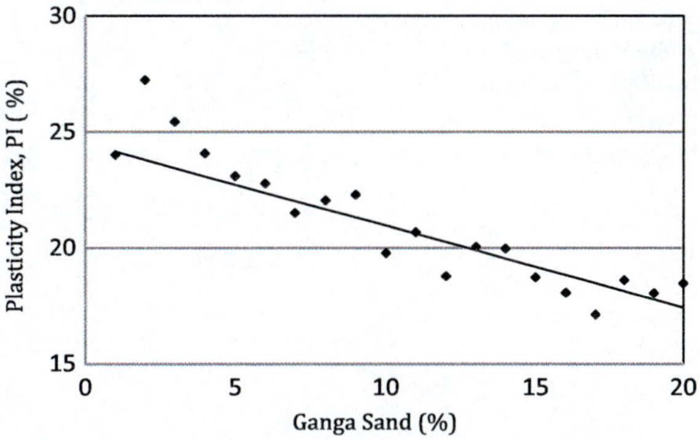

Fig. 15.3 Variation in PI with existing sand percentages at various location in Murshidabad (surficial soils)

From Fig. 15.2, it is seen that plastic limit decreases with increasing percent of sand content, and similar observation for plasticity index was seen in Fig. 15.3.

For others districts, similar observations were made. Detailed presentation on variation of LL, PL, and PI for others seven districts is presented elsewhere [10].

Variation of Compaction Characteristics with Percentage of Sand

The results of compaction characteristics of surficial soils along different routes of proposed roads have reported under PMGSY projects in Murshidabad districts that are presented in this section. Optimum moisture content of the soils under standard compaction tests is presented against percentage of sand in Fig. 15.4. It is observed that the value of OMC is maximum when sand content is nil, but the value decreases gradually with increasing sand content within of limit of 0–20% for sand content. Similar observations were also made for other seven districts.

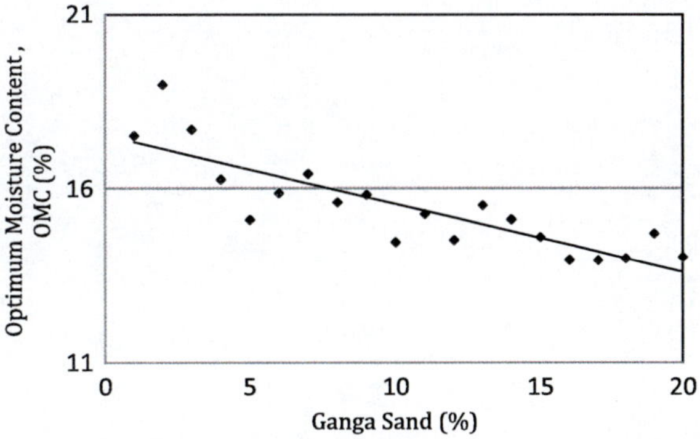

Fig. 15.4 Variation in OMC with existing sand percentages at various location in Murshidabad (surficial soils)

Fig. 15.5 Variation in MDD with existing sand percentages at various location in Murshidabad (surficial soils)

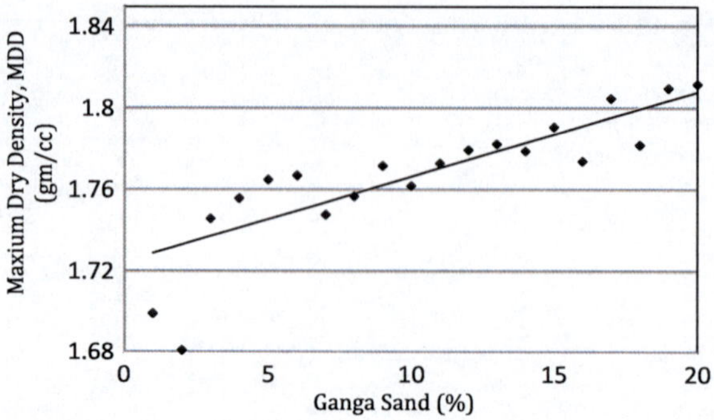

Resulting maximum dry density of these same soils found from the standard compaction test is plotted against percentage of sand present in Fig. 15.5. It is observed that γ_{dmax} shows a distinct increasing trend with increasing sand percentage. A linear relationship between the two parameter is assumed.

Variation of CBR (Soaked) with Percentage of Sand

For district of Murshidabad, values of soaked CBR conducted at optimum moisture content for surficial soils for various routes of the districts are plotted in Fig. 15.6 against percentage of sand present in the soil.

Here also, a distinct increasing trend for soaked CBR was seen with increasing present sand percentage in the soil. Similar trend was observed for soils of seven other districts which are presented in elsewhere [10].

Fig. 15.6 Variation in soaked CBR with existing sand percentages at various location in Murshidabad (surficial soil)

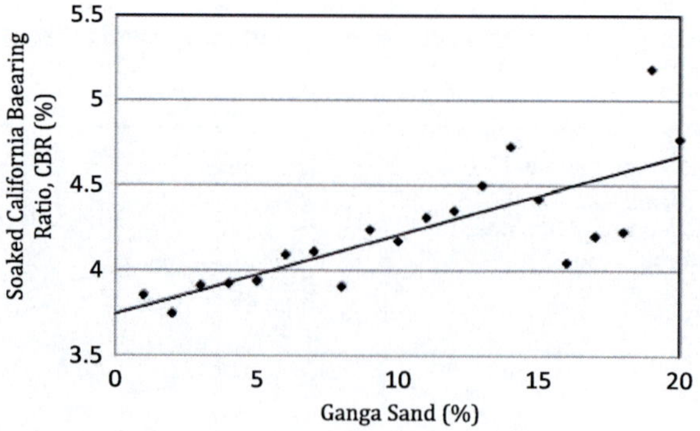

Correlation

Correlation of LL, PL, and PI with Percentage of Sand Present

Liquid limit, plastic limit, and plasticity index of fine-grained soils can be correlated with percentage of sand present in the soil by linear equation,

$$Y = m.X + C \tag{15.1}$$

where Y is LL/PL/PI in percentage, X is percentage of sand present, m is rate of variation, and C is LL/PL/PI of the fine-grained soil with no sand content.

For district of Murshidabad of South West Bengal, the values of m and C were determined for LL, PL, and PI for Figs. 15.1, 15.2, and 15.3. For other seven districts, values of these factors were determined similarly. All such values are presented for all the concerned districts in Table 15.1.

From Table 15.1, it is seen that rate of decrease of LL, PL, and PI with increase in sand percentage is not same for all the districts but ranges within −0.41 to −0.56, −0.18 to −0.25, and −0.18 to −0.38, respectively. Average values can be taken as −0.50, −0.21, and −0.30, respectively, for LL, PL, and PI. Similarly, LL, PL, and PI of the cohesive soil of the eight districts without any sand content lie in as a close range of 41% to 51%, 18% to 27%, and 18% to 26%. And the average values are 46.5%, 23%, and 22%, respectively for LL, PL, and PI.

Correlations of OMC, MDD, and soaked CBR with Percentage of Sand Present

Optimum moisture content, maximum dry density, and soaked California bearing ratio of fine-grained soils can be correlated with percentage of sand present in the soil by linear equation,

$$Y = m.X + C \tag{15.2}$$

Table 15.1 Value of m and C for LL, PL, and PI for eight districts of South Bengal

Name of the district	LL		PL		PI	
	M	*C*	*m*	*C*	*m*	*C*
Murshidabad	−0.55	48	−0.21	22	−0.35	24
Nadia	−0.41	44	−0.22	18	−0.18	21
North 24 Pargana	−0.41	47	−0.19	26	−0.24	20
South 24 Pargana	−0.56	41	−0.20	22	−0.38	18
Bardhaman	−0.55	51	−0.18	27	−0.38	24
Hooghly	−0.54	47	−0.25	23	−0.28	23
Howrah	−0.51	44	−0.23	23	−0.28	20
East Medinipur	−0.48	49	−0.21	24	−0.29	26
Average value	−0.50	46.5	−0.21	23	−0.30	22

Table 15.2 Value of *m* and *C* for OMC, MDD, and soaked CBR for eight districts of South Bengal

Name of the district	OMC		MDD		CBR (soaked)	
	M	*C*	*M*	*C*	*M*	*C*
Murshidabad	−0.19	17	0.0042	1.72	0.046	3.7
Nadia	−0.29	20	0.0048	1.61	0.039	3.1
North 24 Pargana	−0.20	21	0.0048	1.60	0.038	2.8
South 24 Pargana	−0.24	18	0.0039	1.66	0.048	3.2
Bardhaman	−0.26	18	0.0048	1.74	0.043	3.1
Hooghly	−0.20	21	0.0042	1.65	0.039	3.0
Howrah	−0.20	19	0.0046	1.67	0.054	3.4
East Medinipur	−0.18	17	0.0038	1.72	0.048	3.7
Average value	−0.22	19	0.0044	1.67	0.044	3.25

where *Y* is OMC/MDD/CBR in percentage, *X* is percentage of sand present, *m* is rate of variation, and *C* is OMC/MDD/CBR of fine-grained soil with no sand content.

For districts of Murshidabad of South West Bengal, the values of *m* and *C* were determined for OMC, MDD, and CBR for Figs. 15.4, 15.5, and 15.6. For others seven districts, values of these factors were determined similarly. All such values are present for all the concerned districts in Table 15.2.

From Table 15.2, it is seen that rate of decrease of OMC, MDD, and CBR with increase in sand percentage is not same for all the districts but ranges within −0.18 to −0.26, 0.0038 to 0.0048, and 0.038 to 0.054, respectively. Average value can be taken as −0.22, 0.0044, and 0.044. Similarly, OMC, MDD, and CBR of the cohesive soil of the eight districts without any sand content lie in as a close range of 17% to 21%, 1.60% to 1.74%, and 2.8% to 3.7%. And the average values are 19%, 1.67%, and 3.25%, respectively, for OMC, MDD, and CBR.

Comparison of Results with Those of Previous Researchers

Several researchers had presented result of experimental study of variation LL, PL, compaction characteristics, and CBR value of fine-grained soils with variation of sand content in the soil.

Yadav and Suman [9] concluded on the basis of exhaustive experimental study on LL, PL, $\gamma_{d\max}$, OMC, and CBR of cohesive soil on mixing different percentage of fine sand from 0 to 100%. Addition of sand in increasing amount resulted decrease in LL, PL, and OMC. However, corresponding $\gamma_{d\max}$ and soaked CBR value gradually increased with increase in sand content. But rate of decrease of LL, PL, and PI with increase in sand content was −0.48, −0.12, and −0.34, respectively, while corresponding average value of South West Bengal districts is −0.50, −0.21, and −0.30, respectively.

Similarly, the rate of decrease of OMC from result of Yadav and Suman [9] gives a value of −0.16, while corresponding average value for South Bengal districts is −0.22. Similarly, observation can be made on rate of increases of MDD and CBR with increasing sand content. From result of Yadav and Suman [9], the rate of increase of MDD was seen to be 0.0045, while average value of South Bengal districts is 0.0044. Similarly for soaked CBR, rate of increase from Yadav and Suman [9] study is 0.27, while for South Bengal districts are 0.044 (present study).

Thus, experimental study made by Yadav and Suman [9] and [5] on variation for different properties of fine-grained soils with adding sand in increasing percentages shows similar finding as made in the present study qualitatively, while showing slight quantitative variation. These variations are expected because components of cohesive soil are different in those cases.

Conclusions

On the basis of present study, following conclusions can be made:

1. Liquid limit, plastic limit, and plasticity index of cohesive soils are seen to be decreasing with increasing percentage of present Ganga sand.
2. The variation of LL, PL, and PI with increasing sand percentage can be approximated with linear variation and can be expressed as, $LL = m_1.X + C_1$, $PL = m_2.X + C_2$, and $PI = m_3.X + C_3$

   Where X is percentage of sand, C_1 is LL of soil without sand, C_2 is PL of soil without sand, and C_3 is PI of soil without sand, m_1, m_2, and m_3 indicate the value of coefficient of variation for LL, PL, and PI.
3. The average values of m_1, m_2, and m_3 for eight South Bengal districts can be taken as $m_1 = -0.50$, $m_2 = -0.21$, and $m_3 = -0.30$.
4. Average values of liquid limit, plastic limit, and plasticity index without any sand content for these eight districts may be taken as 46.5%, 23%, and 23.5%, respectively.
5. Optimum moisture content decreases with increase in sand percentage, while maximum dry density shows a slight increase in the value with increase in sand percentage. The average coefficient of decrease of OMC with increase in sand percentage is −0.22, while average value of OMC of this soil without any sand content is 19%.

   Similarly, coefficient of variation of MDD with increase in sand percentage is 0.0044, and average value of MDD without sand content is 1.67 gm/cc.
6. With variation in sand percentage, soaked CBR of soils of all the districts is seen to increase linearly and average value of coefficient of variation can be taken as 0.044. The average value of soaked CBR of this soil without sand content can be taken as 3.25%. However, these values are slightly different for each of the districts as coefficient of variation varies between 0.038 and 0.054. Similarly, soaked CBR value of soils without sand varies from 2.8 to 3.7% for above districts.

References

1. Das BM (1999) Principle of foundation engineering, P.W.S publishing, USA, p 81
2. Maity J, Chattopadhyay BC (2015) Prediction of CBR value on gangetic alluvial soil. Int J Innov Eng Technol (IJIET) 5(3):96–99
3. El-Sohby MA, Rabba EA (1981) Some factors affecting swelling of clayey soils. Geotech Eng J 12:19–39
4. Louafi B, Bahar R (2012) Sand: an additive stabilization of swelling clay soils. Int J Geosci 3:719–725, Published online September 2012. http://www.SciRP.org/journal/ijg
5. Roy TK (2013) Influence of sand on strength characteristics of cohesive soil for using as subgrade of Road. In: 2nd Conference of transportation research group of India (2nd CTRG), Proc Soc Behav Sci, vol 104. Science Direct, pp 218–224
6. Cabalar AF, Mustafa WS (2015) Behaviour of sand-clay mixtures for road pavement subgrade. Int J Pavement Eng 1–13
7. Kollaros G, Athanasopoulou A (2016) Sand as a soil stabilizer. In: Proceedings of the 14th international conference, Thessaloniki, Bulletin of the geotechnical society of Greece, vol L. pp 770–777
8. Alam A, Majumdar K (2017) Laboratory and field performance of stabilized soft subgrade with granular material and polymer. Int J Sci Res Develop (IJSRD) 5(05)
9. Yadav NK, Suman SK (2019) Influence of the Ganga sand on subgrade characteristics of cohesive soil. Tech Pap Indian Highw 47(4):32–38
10. Ansar A (2021) Influence of existing Ganga sand on subgrade characteristics of rural road in south Bengal districts. In: Thesis submitted for degree of master of geotechnical engineering at Maulana Abul Kalam Azad University of Technology
11. IRC 36: Recommendation practice for the construction of earth embankment for road works, New Delhi (1970)
12. Shankar AU (2012) Experimental investigation on lithomargic clay stabilized with sand and coir. Res Gate, Tech Pap, Indian Highw 21–31. https://www.researchgate.net/publication/298169903

Chapter 16
Relationship Between Index Properties and CBR of Soil and Prediction of CBR

Jitendra Khatti and Kamaldeep Singh Grover

Introduction

The index properties of soil are properties that help to identify and classify the soil for engineering purposes. The coarse-grained and fine-grained soils are non-cohesive soils and cohesive soils [1]. The particle shape, particle size distribution, consistency, relative density, and clay and clay mineral contents are the index properties of coarse-grained soil. Similarly, consistency, clay and clay minerals content, water content, and particle shape and orientation are the index properties of fine-grained soil. These index properties affect the strength parameters of the soil. The strength of the subgrade of the road is determined by the California bearing ratio (CBR). The CBR ratio is nothing but the ratio of test load to standard load. The CBR of subgrade or soil is determined as per IS 2720: 1979 (P-16). The determination of CBR of subgrade or soil is a cumbersome and time-consuming task. Escaping from the tedious task, numerous researchers and investigators proposed different methods and methodologies to determine the CBR of subgrade or soil.

Kamrul Alam et al. [2] proposed GEP, ANN, and Krigging models to predict the CBR value of fine-grained soil. The published study was carried out for the Bengal basin. The authors developed AI models using G, Cu, Cc, LL, PL, PI, OMC, and MDD parameters of soil. The experimental study was carried out to validate the proposed models. The GEP, ANN, and Krigging models were trained and tested by published data. The laboratory data of soaked and unsoaked CBR was predicted using GEP, ANN, and Krigging models. The authors concluded that the GEP, ANN, and Krigging methods can predict the soaked and unsoaked CBR of soil by the index properties of soil. Montes-Arvizu et al. [3] developed regression models to

J. Khatti (✉) · K. S. Grover
Department of Civil Engineering, RTU, Kota, Rajasthan 324010, India
e-mail: jitendrakhatti197@gmail.com

K. S. Grover
e-mail: ksgrover@rtu.ac.in

K. Muthukkumaran et al. (eds.), *Transportation and Environmental Geotechnics*,
Lecture Notes in Civil Engineering 298, https://doi.org/10.1007/978-981-19-6774-0_16

predict the CBR of soil. The authors developed four models using GW-GM and SM, SP, GW, GW, or GP. The authors developed these four models with gravimetric and volumetric parameters of the soil. The correlation coefficient of geotechnical properties of GW-GM soil was 0.8312, 0.9145, and 0.8821 for CBR vs dry density, water content, and void ratio, respectively. The correlation coefficient for SP soil was 0.7293, 0.7030, 0.7283, and 0.7283 for CBR vs dry density, water content, void ratio, and porosity, respectively. Similarly, the correlation coefficient of the geotechnical properties of GW was 0.9682, 0.7999, 0.9666, and 0.9676 for CBR vs dry density, density, void ratio, and porosity, respectively. The correlation coefficient of GP soil was 0.8556, 0.7118, 0.8527, and 0.8556 for CBR vs dry density, density, void ratio, and porosity, respectively. The correlation coefficient of CBR vs dry density, density, void ratio, and porosity was 0.9317, 0.8522, 0.9346, and 0.9384, respectively, for GW and GP soils. The authors concluded that the proposed Eq. one is more precise for compacted samples at the optimum level, and Eq. 16.4 is more important for experimental results. Shakir Al-Busultan et al. [4] evolved an ANN model to predict the CBR value using soil properties. The ANN model was developed using LL, PI, So_3, soluble salts, gypsum and organics, etc., soil properties. Thirty-five soil datasets validated the model, and the performance of the ANN model was 0.7773 determined. The authors concluded that the ANN model has the potential to predict the CBR value. The authors also concluded that soluble salt is an important parameter in predicting the CBR of soil. Ji et al. [5] predicted the California bearing ratio of soil–rock using the discrete element method and CT scan technique. The authors concluded that the CBR deviation between numerical simulation and laboratory test was 7.4% which shows high reliability. The effect of particle size, large-size particle content, and soil–rock ratio on the CBR of SRM-*G* were investigated. The authors stated that the CBR increases with increasing particle size and content and decreases with increasing the soil content. Finally, the authors concluded that the proposed SRM-*G* model has the potential to predict the CBR of soil. Ravi Chandra et al. [6] suggested simple linear regression models predict the CBR of soil. The proposed models were developed using LL, OMC, PI, and MDD of soil to predict the soaked CBR. The performance of LL, OMC, PI, and MDD simple linear regression models was 0.955, 0.866, 0.924, and 0.967, respectively. The authors concluded that the MDD simple linear regression model has better performance than other models and may be used to predict the soaked CBR of soil. Judes Sujatha et al. [7] proposed multiple regression models using ANN to predict the CBR of soil. The performance of the proposed model was 0.9510. The authors concluded that the proposed model may be used to predict the CBR of soil. Gudeta et al. [8] mapped a study on the prediction of CBR value from soil index properties. The authors used the equations published by Yildirim et al., Nugroho et al. to predict the CBR of soil. The authors suggest that the coefficient of determination is more than 0.9, between 0.7 and 0.89, 0.4 and 0.69, 0.2 and 0.39 and less than 0.2 shows excellent, good, fair, poor, and very poor relationship. The authors also developed three models, (i) using LL and PI, (ii) using Cu, and (iii) using Cu and MDD. The performance of the model (i), (ii), and (iii) was 0.9, 0.8, and 0.8, respectively. The authors concluded that the proposed model is capable of predicting the preliminary soaked CBR of soil. Rani et al. [9]

proposed simple linear regression and multiple linear regression models to predict the CBR of soil. The performance of simple regression and multiple linear regression models was 0.765 and 0.692, respectively. The authors concluded that the proposed models are able to predict the CBR of soil. Sreelekshmypillai et al. [10] developed the empirical correlation to predict the California bearing ratio of fine-grained soil. The published research work was carried out using parameters of reduced, standard, and modified proctor tests. The authors concluded that the LL and PL affect the CBR, and the toughness on CBR increases with increasing applied compaction energy. Ahmad Taha Abdulsadda et al. [11] proposed a multi-layer perceptron neural network model to predict the CBR value using soil index properties. The performance of the proposed MLPNN model was 0.5339. Rehman et al. [12] proposed equations using multiple regression models of LL and PI and Cu and MDD to predict the CBR of fine-grained soil and coarse-grained soil. The authors concluded that the proposed equation may be used for the preliminary prediction of soaked CBR. Rakaraddi et al. [13] developed a simple and multiple regression model to predict the CBR of soil. The performance of the simple regression model of PI, OMC, MDD, and LL was 0.848, 0.867, 0.887, and 0.927, respectively. The performance of the multiple linear regression model was 0.931–0.961. The authors concluded that the multiple linear regression model outperformed simple regression models.

Chattopadhyay et al. [14], Datta et al. [15], and Taskiran [16] also reported that the statistical and artificial intelligence approaches are a powerful tool to predict the CBR (soaked and unsoaked) using index parameters of soil. The PI and dry density play an essential role in the prediction of CBR of soil.

Aim of present research work

The presently available literature survey shows that the statistical and AI approaches have the potential to predict the CBR of soil. Based on the study of the literature survey, the following gaps have been mapped.

- Most of the research works were carried out using simple linear regression and multiple linear regression to predict the CBR of soil. Very few research works were published on the prediction of CBR of soil using ML and DL approaches.
- Most of the research works were carried out with different soil datasets since the quality and quantity of datasets affect the model's performance.
- The published studies were carried out with different datasets and methodologies. Still, the best method or methodology is doubtable.
- The reported studies were carried out with different statistical and artificial intelligence models, which shows that the best architecture model is still doubtable.

Based on the following statements, it has been decided to develop the statistical and artificial intelligence models to predict the soaked CBR of soil. The proposed present study has been carried out by developing simple regression, multiple regression, support vector regression, Gaussian process regression, decision tree, random forest, and artificial neural network approaches. A total of 266 datasets of soil has been used to develop the models of adopted approaches. The best method, methodology, and architecture model has been identified by comparing the performance of models.

Methodology

The simple regression, multiple regression, support vector regression, Gaussian process regression, decision tree, random forest, and artificial neural network AI approaches have been used to predict soaked CBR of soil. Linear approaches have developed simple and multiple regression models. The support vector regression, Gaussian process regression, decision tree, and random forest approaches are associated with machine learning. The artificial neural network is associated with deep learning. The models of simple linear regression, multi-linear regression have been developed in Microsoft Excel 2019 using Data Analysis Tool. The models of support vector regression, Gaussian process regression, decision tree, and random forest have been developed in MATLAB R2020a using Regression Learning Tool. The artificial neural network model has also been developed using the NNTool of MATLAB R2020a. Different hyperparameters have been used to develop these models, as given in Table 16.1.

Data Analysis

A total of 266 datasets of soil has been used to carry out the present research work. The datasets have been collected from different research articles, i.e., Ramasubbarao et al. 2013 [17], Shirur et al. 2014 [18], Patel et al. 2010 [19], Talukdar 2014 [20], Yadav et al. 2014 [21], Leliso 2013 [22], Bassey et al. 2017 [23], Araujo et al. 2016 [24], and Fernando et al. 2020 [25] to develop the artificial intelligence models. The statistics of datasets of soil are given in Table 16.2.

The liquid limit, plasticity index, plastic limit, optimum moisture content, and maximum dry density of soil have been used as input parameters to develop the AI models. The California bearing ratio is affected by LL, PI, PL, OMC, and MDD soil parameters. Hence, relationships have been determined between input parameters and CBRs. The relationship has been determined by the correlation coefficient (r) and shown in Table 16.3 as the Pearson matrix of soil datasets.

The relationship of LL, PL, PI, OMC, and MDD with CBRs (soaked CBR) is 0.56, 0.40, 0.51, 0.58, and 0.65, respectively. G. N. Smith; 1986 [26] suggested that if the correlation coefficient value is more than 0.8, between 0.2 and 0.8 and less than 0.2, it shows strong, good, and weak correlation between two variables. The relationship or correlationn coefficient shows that the input and CBRs have a good correlation, as shown in Fig. 16.1.

Table 16.1 Hyperparameters of AI models [11]

Hyperparameters of machine learning models				
Hyperparameters	SVR	GPR	RF	DT
Basic and kernel function	Auto	Auto	–	–
Kernel scale	Auto (1)	Auto (9.11)	–	–
Signal std. deviation	–	13.133	–	–
Sigma/epsilon	Auto (0.397)	Auto	–	–
Standardize	Enable	Enable	–	–
Optimized numerical parameters	–	Enable	–	–
Box constraint	Auto (3.96)	–	–	–
Ensemble method	–	–	Auto	–
Min. leaf size	–	–	Auto (8)	Auto (1)
Number of learners	–	–	Auto (30)	–
Learning rate	–	–	0.1	–
No. of predictors to sample	–	–	All (1)	–
Max. surrogates per nodes	–	–	–	10

Optimizer options for machine learning models	
Optimizer	Bayesian optimization
Acquisition function	Expected improvement per second plus
Iterations	30
Max. training time in seconds	300
Number of grid divisions	10

Hyperparameters of deep learning (ANN) models	
Hidden layers	Two
Neurons	Ten neurons on each layer
Backpropagation algorithm(s)	Levenberg–Marquardt
Normalizing function(s)	Min–max for input and log for output parameters
Activation function(s)	Sigmoid at hidden layers, linear at the output layer
Train: validation ratio	70: 30
Epochs	1000 (default)
Network type	Feed-forward backpropagation
Network class	Multi-layer perceptron class (MLP)
Mu	0.001
Max. fail	6
Min. gradient	10e-7

Table 16.2 Statistics of datasets of soil

Parameters	LL	PL	PI	OMC	MDD	CBRs
Minimum	22.00	0.00	8.29	10.50	1.25	0.80
Maximum	94.00	40.00	58.00	35.40	2.30	5.86
Mean	49.59	24.60	24.99	20.02	1.68	2.94
Mode	48.00	32.00	22.00	17.00	1.78	2.10
Median	49.00	26.00	24.00	19.00	1.63	3.00
Standard deviation	16.51	9.41	10.62	5.85	0.20	1.25
Confi. level (95%)	4.96	2.83	3.19	1.76	0.06	0.38

Table 16.3 Pearson matrix of soil datasets

(*r*)	LL	PL	PI	OMC	MDD	CBRs
LL	1.00					
PL	0.75	1.00				
PI	0.90	0.38	1.00			
OMC	0.80	0.67	0.67	1.00		
MDD	0.74	0.68	0.59	0.89	1.00	
CBRs	0.56	0.40	0.51	0.58	0.65	1.00

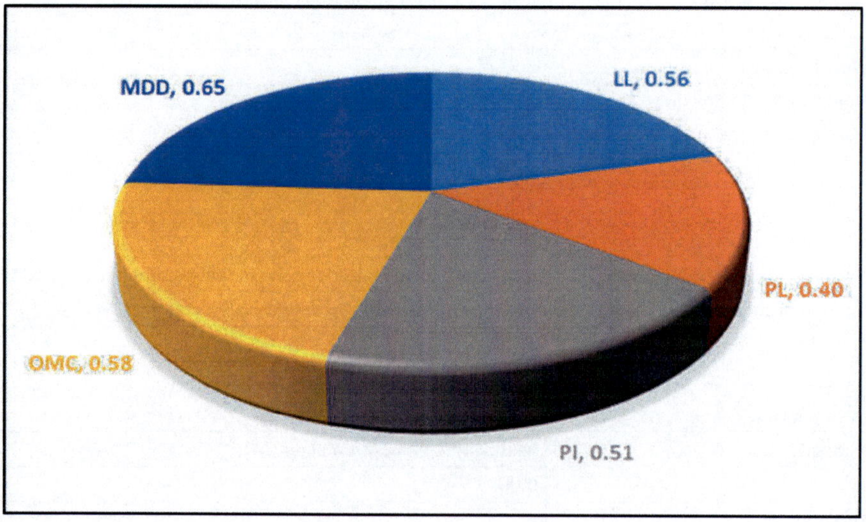

Fig. 16.1 Correlation between input variables and CBRs

Data Preprocessing

The process of cleaning, removing outliers, and missing data are known as data pre-processing. Before using the datasets of soil, the outliers and missing datasets have been removed from the datasets. The min–max and log normalization functions have been used to normalize the dataset in the present research.

Training and Testing Datasets

The present study has been carried out using 266 datasets of soil. These datasets have been divided into 212 training and 53 testing datasets. The proposed AI models such as regression model, SVM, GPR, DT, and RF have been trained and tested by 212 and 53 datasets. The 212 training datasets have also been subdivided into 148 training and 64 validation datasets for the artificial neural network model. The artificial neural network model has been tested on 53 datasets.

Results and Discussions

The California bearing ratio of soil has been predicted using regression analysis (simple linear regression, multi-linear regression), support vector regression, decision tree, random forest, Gaussian process regression, and artificial neural network models. The analyzed results and performance have been discussed below.

Simple Regression Analysis (SRA)

The California bearing ratio of soil has been predicted by simple linear regression analysis. The five SRA models, SRA1_CBRs, SRA2_CBRs, SRA3_CBRs, SRA4_CBRs, SRA5_CBRs, have been developed using LL, PL, PI, OMC, and MDD input parameters. During the development of SRA models, the following simple linear regression equations have been proposed.

$$\mathrm{SRA1\_CBRs} = 40.226 - 0.6138 * \mathrm{LL} \tag{16.1}$$

$$\mathrm{SRA2\_CBRs} = 35.336 - 0.9251 * \mathrm{PL} \tag{16.2}$$

$$\mathrm{SRA3\_CBRs} = 27.9 - 0.7684 * \mathrm{PI} \tag{16.3}$$

Table 16.4 Performance of SRA models

Designation of SRA models	Training performance			Testing performance		
	RMSE	*R*	MAE	RMSE	*R*	MAE
SRA1_CBRs	14.9482	0.6068	9.7835	14.2879	0.5398	9.7397
SRA2_CBRs	16.6775	0.4461	10.3748	15.9455	0.3363	10.1890
SRA3_CBRs	15.6425	0.5563	10.1098	14.1707	0.5430	9.1443
SRA4_CBRs	14.4455	0.6675	9.7872	13.9018	0.5991	10.3310
SRA5_CBRs	13.5593	0.6982	9.1155	12.4661	0.6929	8.9567

$$\mathrm{SRA4\_CBRs} = 38.791 - 1.504 * \mathrm{OMC} \tag{16.4}$$

$$\mathrm{SRA5\_CBRs} = 50.33 * \mathrm{MDD} - 74.763 \tag{16.5}$$

The training and testing performances of SRA models have been determined and given in Table 16.4.

The SRA5_CBRs model has been developed using the maximum dry density parameter of soil, and this parameter is also a compactive parameter. The performance of the SRA5_CBRs model is 0.6929, which is comparatively better than other SRA models.

Multiple Regression Analysis (MRA)

A multiple regression analysis has been performed to predict the California bearing ratio of soil. The multi-linear regression model has been predicted the California bearing ratio. The multi-linear regression (MLR) model has been developed by input parameters such as LL, PL, PI, OMC, and MDD of soil. An equation has been proposed by multi-linear regression analysis, and the equation is –

$$\mathrm{MLR\_CBRs} = 0.0589 * \mathrm{LL} - 0.441 * \mathrm{PI} + 0.5174 * \mathrm{OMC} + 54.424 * \mathrm{MDD} - 84.603 \tag{16.6}$$

Equation 16.6 has been used to predict the soaked CBR of soil. The training and testing performances of the MLR_CBRs model have been determined and given in Table 16.5.

From Table 16.5, it has been observed that the MLR_CBRs model has predicted soaked CBR of soil with 0.7466 and is comparatively better than SRA5_CBRs (0.6929).

Table 16.5 Performance of MLR model

Designation of MLR models	Training performance			Testing performance		
	RMSE	*R*	MAE	RMSE	*R*	MAE
MLR_CBRs	13.4876	0.6857	9.6414	11.8773	0.7466	8.2460

Table 16.6 Performance of SVR model

Designation of SVR models	Training performance			Testing performance		
	RMSE	*R*	MAE	RMSE	*R*	MAE
SVR_CBRs	13.7980	0.6708	9.6849	11.7722	0.7502	7.7162

Table 16.7 Performance of GPR model

Designation of GPR models	Training performance			Testing performance		
	RMSE	*R*	MAE	RMSE	*R*	MAE
GPR_CBRs	7.2607	0.9220	4.0697	5.3741	0.9468	3.6551

Support Vector Regression (SVR)

The support vector regression is based on the kernel function. The kernel functions are nothing but it is mathematical formulas or algorithms. The support vector regression model has been developed using LL, PL, PI, OMC, and MDD to predict the soaked CBR of soil. The performance of the SVR_CBRs model has been determined and given in Table 16.6.

The performance of SVR_CBRs models is 0.7502, which is comparatively better than the MLR_CBRs model.

Gaussian Process Regression (GPR)

The Gaussian process regression method is also based on the kernel function. The Gaussian process regression model has been developed using LL, PL, PI, OMC, and MDD. The performance of the GPR_CBRs model has been determined and given in Table 16.7.

Random Forest (RF)

The random forest method is also named a random tree forest. A random forest is a group of decision trees. The random forest method has been used to predict the soaked

Table 16.8 Performance of RF model

Designation of RF models	Training performance			Testing performance		
	RMSE	*R*	MAE	RMSE	*R*	MAE
RF_CBRs	8.5486	0.8888	4.2534	6.6151	0.9292	4.1810

Table 16.9 Performance of DT model

Designation of DT models	Training performance			Testing performance		
	RMSE	*R*	MAE	RMSE	*R*	MAE
DT_CBRs	9.1140	0.8718	4.6915	7.7630	0.9052	4.4645

CBR of soil in the present research work. The random forest model (RF_CBRs) has been developed by LL, PL, PI, OMC, and MDD parameters. The performance of RF_CBRs has been determined and given in Table 16.8.

Decision Tree (DT)

The decision tree is also a type of supervised machine learning which splits the data continuously as per specific parameters. The decision tree (DT_CBRs) model has been developed using LL, PL, PI, OMC, and MDD of soil to predict the soaked CBR of soil in the present research work. The training and testing performance of the DT_CBRs model has been determined and given in Table 16.9.

Artificial Neural Networks (ANN)

The artificial neural network is a type of supervised deep learning. The artificial neural network (LMNN_CBRs) model has been developed using LL, PI, PL, OMC, and MDD to predict the soaked CBR of soil. The ANN model has been developed by using two hidden layers and the Levenberg–Marquardt algorithm. The performance of the LMNN_CBRs model has been determined and given in Table 16.10.

The performance of SRA, MLR, SVR, GPR, RF, DT, and ANN has been compared to determine the best architecture AI model to predict soil's soaked California bearing

Table 16.10 Performance of ANN model

Designation of ANN models	Training performance			Testing performance		
	RMSE	*R*	MAE	RMSE	*R*	MAE
LMNN_CBRs	0.3998	0.8943	0.7194	4.4619	0.9736	3.2093

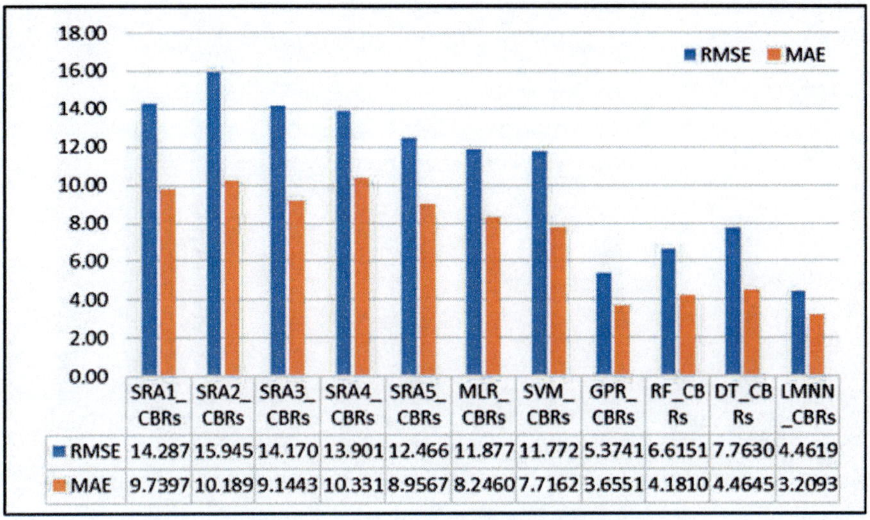

Fig. 16.2 Comparison of RMSE and MAE performances of AI models

ratio. The RMSE and MAE performances of SRA, MLR, SVR, GPR, RF, DT, and ANN models have been compared and shown in Fig. 16.2.

From Fig. 16.2, it has been observed that the LMNN_CBRs model has been predicted soaked CBR of soil with 4.4619 RMSE and 3.2093 MAE. The performance (*R*) of SRA, MLR, SVR, GPR, RF, DT, and ANN has also been compared and shown in Fig. 16.3.

From Fig. 16.3, it has been observed that the LMNN_CBRs model has been predicted soaked CBR of soil with 0.9736, which is comparatively better than other AI models. The LMNN_CBRs model has been identified as the best architecture model based on the RMSE, MAE, and *R* performance. The LMNN_CBRs model has been used to predict the soaked CBR of 53 soils. The predicted results of soaked CBR have also been compared with laboratory test results of soaked CBR and shown in Fig. 16.4.

From Fig. 16.4, it has been observed that the LMNN_CBRs model has predicted the soaked CBR approximately equal to the laboratory test results of soaked CBR. The effect of the model has been presented by plotting the actual vs. predicted curve against the null model and shown in Fig. 16.5.

From Fig. 16.5, the coefficient of correlation of actual vs. predicted soaked CBR is 0.9731. The correlation coefficient shows a strong correlation between actual and predicted soaked CBR of soil. Hence, the LMNN_CBRs model may be used to predict the soaked CBR of soil.

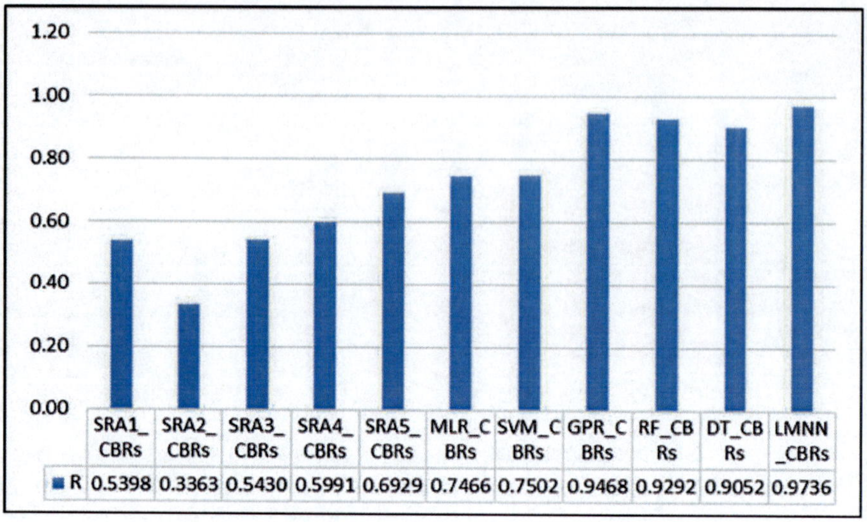

Fig. 16.3 Comparison of performance (*R*) of AI models

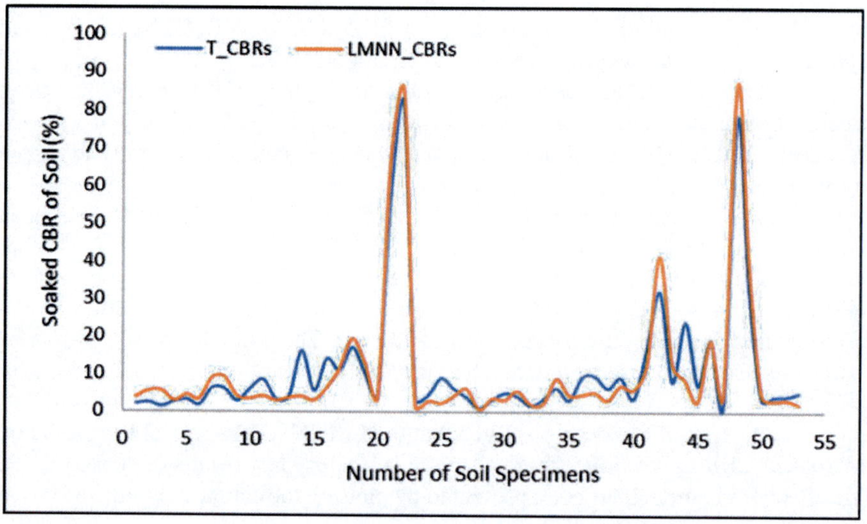

Fig. 16.4 Comparison of laboratory test CBR (T_CBRs) and predicted CBR using LMNN_CBRs

Comparison with Literature Survey

The performance of the LMNN_CBRs model has been compared with presently available models in the literature survey and given in Table 16.11.

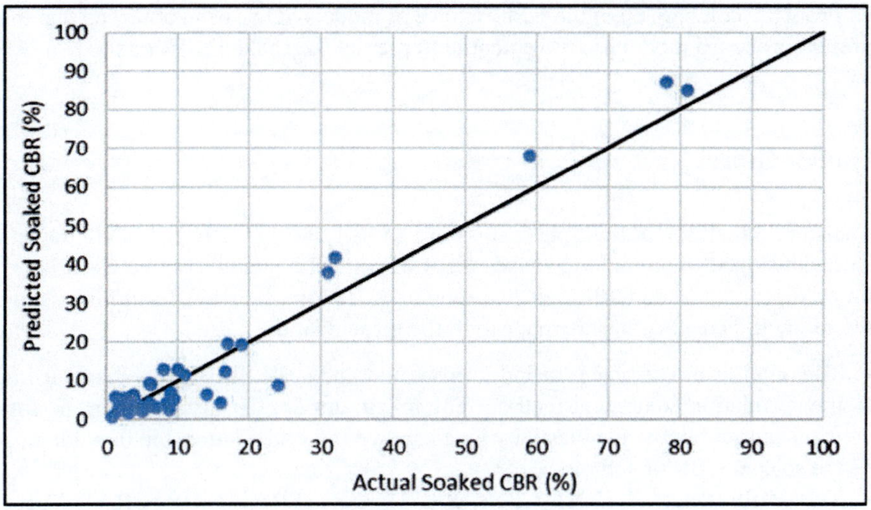

Fig. 16.5 Actual (T_CBRs) vs. predicted CBR plot

Table 16.11 Comparison of performance of models

Authors	Performance	Approaches
Montes-Arvizu et al. [3]	0.7030–0.9682	Simple regression analysis
Al-Busultan et al. [4]	0.7773	Artificial neural network
Ravi Chandra et al. [6]	0.866–0.967	Simple regression analysis
Sujatha et al. [7]	0.9515	Artificial neural network
Gudeta et al. [8]	0.8–0.9	Simple regression analysis
Rani et al. [9]	0.768, 0.692	Simple and multiple regression
Abdulsadda et al. [11]	0.5339	Artificial neural network
Rakaraddi et al. [13]	0.848–0.927	Simple regression analysis
Rakaraddi et al. [13]	0.931–0.961	Multiple regression analysis
Khatti et al. [Present study]	0.9736	Artificial neural network

From the comparison of the performance of models, it has been observed that the present proposed model has the potential to predict the soaked CBR of soil.

Conclusions

The simple regression, multiple regression, SVR, GPR, DT, RF, and ANN models were developed to predict the soaked CBR of soil. The relationship between index properties and soaked CBR was also established. From the results and discussion, the following conclusions are mapped in the present study.

- The relationship between the index properties and soaked CBR was drawn using the correlation coefficient method. The maximum dry density, optimum performance, liquid limit, and plasticity index showed a good relationship in predicting the soaked CBR of soil.
- The performance of SRA1_CBRs, SRA1_CBRs, SRA1_CBRs, SRA1_CBRs, SRA1_CBRs, SRA1_CBRs, MLR_CBRs, SVR_CBRs, GPR_CBRs, RF_CBRs, DT_CBRs, and LMNN_CBRs was 0.5398, 0.3363, 0.6430, 0.5991, 0.6929, 0.7466, 0.7502, 0.9468, 0.9292, 0.9052, and 0.9736, respectively. The LMNN_CBRs model was identified as the best architecture AI model based on the performance of models.
- The best architecture AI model (LMNN_CBRs) performance was compared with presently available prediction models of CBR, and it was reported that the proposed LMNN_CBRs model can predict the soaked CBR of soil.

Finally, the proposed LMNN_CBRs model can predict the soaked CBR of soil, and it may be used to predict the CBR of soil for published research work.

References

1. Arora KR (2004) Soil mechanics and foundation engineering. Standard Publisher Distributors, New Delhi, India
2. Kamrul Alam SK, Mandal A, Shiuly A (2020) Prediction of CBR value of fine-grained soil of Bengal basin by Genetic expression programming, artificial neural network and Krigging method. J Geol Soc India 99:190–196
3. Montes-Arvizu ME, Chavez-Alegria O, Rojas-Gonzalez E, Gaxiola-Camacho JR, Millan-Almaraz JR (2020) CBR predictive models for granular bases 8 using physical and structural properties. Appl Sci 10(4):1414
4. Al-Busultan S, Aswed GK, Almuhanna RRA, Rasheed SE (2020) Application of artificial neural networks in predicting subbase CBR values using soil indices data. In: 3rd International conference on engineering sciences, IOP conference series, materials science and engineering, p 671
5. Ji X, Li J, Cui Z, Li S, Xiong Y, Hu J, Jiang Y (2020) A prediction method for the California bearing ratio of soil-rocks mixture based on the discrete element method and CT scanning. Adv Civil Eng Hindawi 1–12, Article ID 9794756

6. Ravichandra AH, Shivakumar K, VinayKumar H, Md. Khalid S, Basavaraj B(2019) Prediction of CBR value by using index properties of soil. Int Res J Eng Technol 6(7): 3740–3747
7. Judes Sujatha S, Arul Franco P, Angelin Niji N, Sumil Kumaran A (2019) Prediction of CBR from index properties of soil through ANN modelling. J Emerg Technol Innovative Res 6(2):287–296
8. Gudeta AD, Patel AV (2018) Prediction of CBR value from index properties of different soil: review. Int J Creative Res Thoughts 6(2):1190–1197
9. Rani S, Nagaraj (2017) Prediction of CBR value with soil index properties: case study on Yadadri, region. Int J Latest Eng Manage Res 2(7):09–12
10. Sreelekshmypillai G, Vinod P (2019) Prediction of CBR value of fine-grained soils at any rational compactive effort. Int J Geotech Eng 13(6):560–565
11. Abdulsadda AT, Jaleel DA (2017) Predicting CBR value from index properties of soil using expert system. Global J Res Eng E Civil Struct Eng 17(1):23–28
12. Rehman ZU, Khalid U, Farooq K, Mujtaba H (2017) Prediction of CBR value from index properties of different soils. Tech J Univ Eng Technol Taxila Pak 22(2):17–26
13. Rakaraddi PG, Gomarsi V (2015) Establishing the relationship between CBR with different soil parameters. Int J Res Eng Technol 4(2):182–188
14. Chattopadhyay BC, Maity J (2012) Prediction of CBR of different groups of alluvial soils for the design of flexible pavements. Proceedings of the international symposium of engineering under uncertainty: safety assessment and management 2012:833–847
15. Datta T, Chattopadhyay BC (2011) Correlation between CBR and index properties of soil. Proc Indian Geotech Conf Dec 15–17:131–133
16. Taskiran T (2010) Prediction of California bearing ratio of fine-grained soils by AI methods. Adv Eng Softw 41(6):886–892
17. Ramasubbarao GV, Siva SG (2013) Predicting soaked CBR value of fine-grained soils using index and compaction characteristics. Jordan J Civil Eng 7(3):354–360
18. Shirur NB, Hirmath SG (2014) Establishing relationship between CBR value and physical properties of soil. IOSR J Mech Civil Eng 11(5):26–30
19. Patel RS, Desai MD (2010) CBR predicted by index properties of Alluvial soils of south Gujrat. In: Indian geotechnical conference 2010, IGC Mumbai chapter & IIT Bombay, Dec 16–18, In: 2010 Proceedings of the Indian geotechnical conference, Mumbai, pp 79–82
20. Talukdar DK (2014) A study of correlation between California bearing ratio (CBR) value with other properties of soil. Int J Emerg Technol Adv Eng 4(1):559–562
21. Yadav D, Jain PK, Kumar R (2014) Prediction of soaked CBR of fine-grained soils from classification and compaction parameters. Int J Adv Eng Res Stud 3(4):119–121
22. Leliso Y (2013) Correlation of CBR value with soil index properties for Addis Ababa subgrade soil, Addis Ababa University (Thesis)
23. Bassey OB, Attah IC, Ambrose EE, Etim RK (2017) Correlation between CBR values and index properties of soils: a case study of Ibiono, Oron and Onna in Akwa Ibom State, Resour Environ 7(4):94–102
24. Araujo W, Ruiz G (2016) Correlation equations of CBR with index properties of soil in the city of Piura. In: 14th LACCEI international multi-conference for engineering, education and technology, engineering innovations for global sustainability, pp 1–7
25. Fernando JW, Burbano JLA, Sierra EEC (2020) Subgrade soils characterization data, for correlation of geotechnical variables on urban roads in northern Colombia. Data Brief 32(106095):1–7
26. Smith GN (1986) Probability and statistics in civil engineering: an introduction. Collins, London

Chapter 17
Numerical Evaluation of Performance of the Geosynthetic-Reinforced Walls with Poorly Graded Reinforced Fill

Mainak Majumder

Introduction

Geosynthetic-reinforced soil (GRS) walls are currently one of the most used technologies as the alternative solution of earth retaining walls. In recent times, several experimental and numerical works have been conducted to investigate the factors affecting the behavior of the GRS Walls. Useful reviews on the GRS walls using numerical modeling can be found out in the literature by Bathurst and Hatami [1–3], Hatami and Bathurst [4]. Basic mechanism of the GRS wall is the mobilized friction between the geogrid and good quality of the reinforced backfill materials. However, the criteria of the reinforced backfill material as per different standards are very difficult to achieve due to the unavailability of the borrow area soil and transportation costs. The criteria of the different standard guidelines [5] have been tabulated in Tables 17.1 and 17.2, and as per the standards, allowable fines in the backfill soil have very stringent criteria. The effective cohesion must be almost negligible as per all the standards but scarcity of such required backfill criteria is difficult to fulfill at the construction site. This enhanced to use of the poor low permeable native soil as the reinforced backfill material in the reinforced zone of GRS walls. Adverse climatic condition as well as poor founding soils may cause catastrophic failure if frictions between geogrid and backfill soils are not fully mobilized. Not only in heavy rainfalls or high-water table condition even in dry-state condition also the poorly graded fill can lead to the failure of GRS walls. On the other hand, the effective usage of low-quality low permeable (with high fines) soil can reduce the overall construction cost by 30–42% [6]. For this purpose, an attempt has been made to understand the GRS wall behavior with poorly graded backfill soil as reinforced soil material.

M. Majumder (✉)
Engineering Research Manager, Research & Development Department, Transportation Infra IC, RREC, Larsen & Toubro, Mumbai 400093, India
e-mail: lncs@springer.com

K. Muthukkumaran et al. (eds.), *Transportation and Environmental Geotechnics*,
Lecture Notes in Civil Engineering 298, https://doi.org/10.1007/978-981-19-6774-0_17

Table 17.1 Specification of reinforced soil as per different standards

Properties	Units	Reinforced soil		
		FHWA: NHI-10–024	IRC: SP 102–2014	BS: 8006: 2008
Fines content	%	0–15	0–15	0–25
Coefficient of uniformity	-	4–20	> 2	> 4
Liquid limit	%	NA	NA	≤ 45
Plasticity index	%	≤ 6	≤ 6	≤ 6
Permeability		Free–draining soil	Free–draining soil	Free– draining soil
Effective cohesion	kPa	≈ 0	≈ 0	≈ 0
Effective friction angle	°	30–34	30–34	30–40

Numerical Model Development

To evaluate the utility of the poorly graded backfill material in GRS wall, a three-dimensional (3D) finite element model has been developed in the commercial finite element package name ABAQUS (V6.14) [7]. The 3D numerical modeling is best fitted for modeling GRS walls with discrete or discontinuous reinforcements. In this study, firstly GRS wall has been designed to satisfy the internal and external stabilities as per [5, 8, 9].

The height of the GRS wall model has been designed to have proportional and plain geometry condition. The height (H) of the GRS wall model has been assumed as 7.5 m above the foundation soil for base model. Different other heights of the wall have been considered for the purpose of analysis and comparison with the base model. Total eight layers of geogrid (GG) reinforcements have been given at a vertical spacing of 0.8 m, starting from 0.2 m height from the base of the wall. The reinforcement length has been fixed as 5.75 m (0.7H) in all the layers. The vertical spacing between the GG has been kept 0.8 m, whereas horizontal spacing has been varied, depending on the types of reinforcement has been modeled. The foundation soil has been modeled with a height of 7.5 m (1*H*), width 7.5 m (1*H*), and length of 22.5 (3*H*) (Fig. 17.1).

To simulate the embankment loading, a surcharge of 40 kPa has been applied on the top of the GRS wall. This loading has been kept constant throughout the analyses. The boundary conditions has been modeled as the vertical sides of the GRS wall and foundation soil are free in all the directions. The bottom of the foundation soil has been considered as the fixed to eliminate the influence of foundation deformations on the behavior of the GRS wall. The left side of the GRS wall and foundation soil have been modeled as the roller boundary to simulate the vertical deformation due to the progressive surcharge of the construction of reinforced backfill zone. The influence of the pore water pressure has not been considered in the analysis. In all the cases,

Table 17.2 Specification of reinforced soil fill gradation as per different standards

	FHWA: NHI-10–024		IRC: SP 102- 2014		BS: 8006- 2010		Turkish guidelines 2013		NCMA 2006	
	Sieve size (mm)	Percent finer (%)	Sieve size (mm)	Percent finer (%)	Sieve size (mm)	Percent finer (%)	Sieve size (mm)	Percent finer (%)	Sieve size (mm)	Percent finer (%)
Gradtion	102	100	75	100	125	100	125	100	102	100–75
	0.0425	0–60	4.75	85–100	90	80–100	75	85–100	4.76	100–20
	0.075	0–15	0.425	60–90	75	65–100	12.5	25–100	0.425	0–60
					37.5	45–100	2	15–100		
			0.075	< 15	10	15–60	0.60	10–65	0.075	0–35
					5	10–45				
					0.6	0–25	0.075	< 15		
					0.063	0–12				

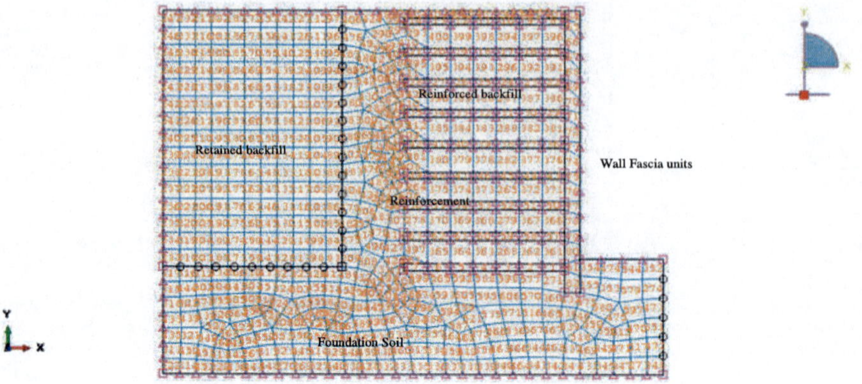

Fig. 17.1 Cross-sectional view of the FE model

the retained backfill material, foundation soil, and fascia units have been assumed to have same material to avoid complexity of the results.

For the numerical analyses, the backfill (both reinforced and retained) and foundation soils have been modeled as the 3D homogeneous, deformable continuum primarily quad-shaped brick elements. For modeling the zones immediately behind the facing, finer mesh sizes have been used.

Elastoplastic drained analysis has been carried out for the various cases to understand the behavior of the poorly graded backfill material in the GRS walls. Therefore, time of construction does not have substantial effects on the behavior of the numerical models. Consolidation of soil has not been considered in the present study.

Material Properties

Properties of different material and their constitutive behavior have been described in this section.

Reinforcement Properties

Geogrid (GG) with ultimate tensile strength of 50 kN/m, 100 kN/m, 250 kN/m, and 300 kN/m, respectively [4, 10], has been used as the reinforcement in the analysis. In the present study, same type of GG has been used for a particular model. Afterward, strength of the GG has been varied to understand the influence of reinforcement strength on the fascia movement. In Table 17.3, different properties of geogrids have been mentioned. The interaction between geogrid and backfill material has been modeled as frictional contact in ABAQUS [7]. Geogrids are placed in linear manner

Table 17.3 Reinforcement material properties

Parameter	Name	Unit	Geogrids-1	Geogrids-2	Geogrids-3	Geogrids-4
Geogrid strength	T	kN/m	50	100	250	300
Material model	Type	–	Linearly elastic	Linearly elastic	Linearly elastic	Linearly elastic
Length of geogrid	L	m	5.75	5.75	5.75	5.75
Vertical spacing	S_v	m	0.8	0.8	0.8	0.8
Horizontal spacing	S_h	m	0.75	0.75	0.75	0.75
Width	W	m	0.08	0.08	0.08	0.08

with c/c horizontal spacing of 0.75 m. In the FE model, geogrids have been modeled as elastic slender elements with an axial stiffness, and they can sustain only tensile forces.

Backfill and Foundation Soil Properties

Soil materials, namely reinforced backfill, retained backfill, and foundation soils, have been modeled as Mohr–Coulomb soil model with linearly elastic perfectly plastic behavior. The drained analyses have been performed for all types of model. Different parameters have been tabulated in Table 17.4. In the base model analysis, the parameters have been assumed as per the standard criteria. In Table 17.5, different cases considered for the poorly graded soil behavior have been mentioned. Foundation soil has been assumed to be stiffer as compared to backfill soil so that no ground improvement has not been provided for founding soil.

The GRS wall has been modeled with segmental panel-type wall facing. The wall facing has been designed as a rectangular precast concrete member with modular dimension of 1.5 m width, 1 m height, and 150 mm thickness. The wall face has been modeled as the plate element in ABAQUS [7]. The plate has been modeled as linear elastic material which can resist forces in tension and compression. The unit weight of 24 kN/m^3, elastic modulus of 35 GPa, and Poisson's ratio of 0.15 have been used for the wall units.

Results and Discussions

The results of the present paper have been discussed in three different subsections. In the first part, validation of the present model has been presented. In the second part

Table 17.4 Material properties of backfill and foundation soil (base model)

Parameter	Name/symbol	Unit	Reinforced backfill	Retained backfill	Foundation soil
Material model	–	–	Mohr–Coulomb	Mohr–Coulomb	Mohr–Coulomb
Unit weight	γ	kN/m^3	18	18	20
Peal angle of internal friction	ϕ_{Peak}	°	30	25	30
Residual angle of internal friction	ϕ_{residual}	°	0.8	0.8	0.8
Dilation angle	ψ	°	0.75	0.75	0.75
Cohesion	c	kPa	2	2	5
Poisson's ratio	υ	-	0.3	0.3	0.3

Table 17.5 Range of parameters used for poorly graded backfill materials

Parameters	Units	Range of values
Peak angle of internal friction	°	{30, 28, 26}
Cohesion	kPa	{1, 5, 10}
Strength of geogrid	kN/m	{50, 100, 250, 300}

of the work, the influence of marginal soil on the fascia displacement of the GRS wall has been discussed. Later, connection load and lateral thrust on the facing units have been presented.

Validation of the FE Model

For the validation purpose, laboratory experiment conducted by Hatami and Bathurst [4] has been considered. The present numerical model has been simulated with the full-scale GRS wall as developed by Hatami and Bathurst [4]) in working stress conditions. The fascia units have been constructed with solid modular concrete column. Full-scale tests have been conducted on 3.6 m height GRS wall with modular block facing at a batter of 8 degrees from vertical. The tests have been conducted at the Royal Military College of Canada (RMCC). The uniform surcharge load of 115 kPa has been applied on the GRS wall at the end of construction (EOC). The backfill soil used in the full-scale test has been RMC sand, a subangular, poorly graded clean sand. The unit weight of the soil has been assumed as 16.8 kN/m^3. The biaxial polypropylene geogrid reinforcement has been used in RMCC tests. For modeling purpose in present model, linear elastic truss (wire) tensile elements with no bending strength has been considered. The "wall 1 (the control wall) which has six layers of polypropylene geogrid has been considered here for the validation purpose. EOC

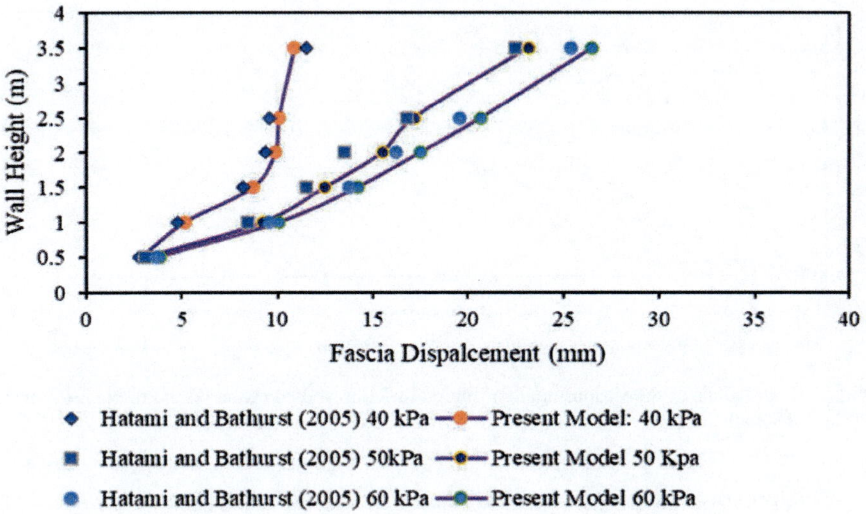

Fig. 17.2 Validation of fascia displacement with different incremental surcharge loading

period has been implemented by different incremental surcharge loading of 40 kPa, 50 kPa, and 60 kPa, respectively. In Fig. 17.2, the comparison of fascia displacement with different incremental surcharge along the wall has been shown. The results show that the present model in ABAQUS [7] can predict accurate wall movement for any other loading conditions and soil properties.

Effect of Cohesion and Angle of Internal Friction

To study the effect of cohesion and angle of internal friction, the parametric study has been done as shown in Fig. 17.3. To understand the behavior of marginal soil, the cohesion of the reinforced backfill has been varied from 1, 5, and 10 kPa. The friction angle of the backfill has been varied from 30 to 26°. Fascia displacement has been considered as the measurement criteria of the GRS wall. Fascia displacement has been plotted from the top of the GRS wall to the leveling pad level. The critical movement layer has been observed as the third or fourth from the top of the wall for all the cases. The deviation from the standard backfill forces GRS wall to fail internally initially in terms of greater fascia movement. The friction angle value of 26° with minimal cohesion value can possess the overall global failure of the wall. The influence of the friction angle has been observed greater as compared to the cohesion. Since additional cohesion (more than 5 kPa) can restrict the movement of drainage and friction mobilization, the GRS wall has been experienced greater movement with poorly graded backfill material.

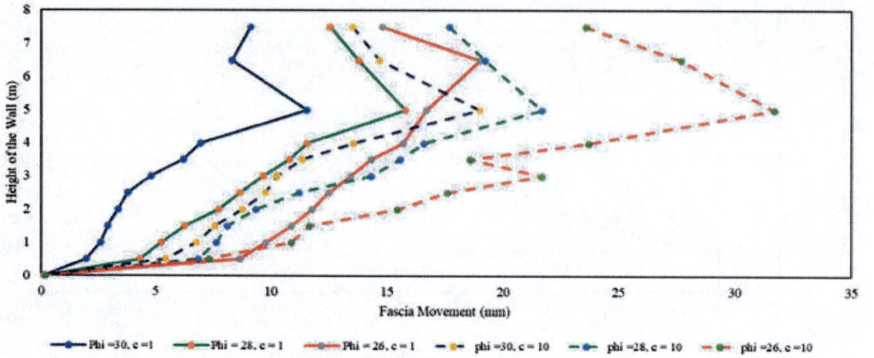

Fig. 17.3 Influence of the various angle of internal friction and cohesion on the fascia movement of the GRS wall

The strength of the geogrid for above analysis has been kept constant as 250 kN/m. The maximum facia displacement has been increased from 11.56 to 31.67 mm by using marginal backfill in GRS wall. The observed movement has been varied in between wall height. The criteria of safe vertical wall movement of the facia are 1 in 150 as per IRC SP 102. This criterion has been violated by using lower strength of the geogrid as reinforcement layer. For this purpose, geogrid strength has been considered as one of the parameters for restricting wall movement.

Effect of Geogrid Strength on the Fascia Movement

The geogrid strength has been varied from 100 to 750 kN/m, to understand the behavior of the marginally backfill soil in GRS wall. The reinforcement strength provides the effective frictional mobilization between the backfill and the geogrid area. The increase in geogrid strength can effectively reduce the fascia movement even using the marginally backfill soil. The worst combination of the c and ϕ has been used as 10 kPa and 26°. The fascia movement has been reduced from 17.65 to 3.45 mm by using 100 kN/m geogrid and 750 kN/m geogrid. In Fig. 17.4, the fascia movement of the different combination of shear strengths has been shown with the various geogrid strength. Overall, 47 to 80% reduction in fascia displacement has been observed in the horizontal direction. This establishes the utilization of the poorly graded backfill soil in GRS wall by using higher strength of reinforcement.

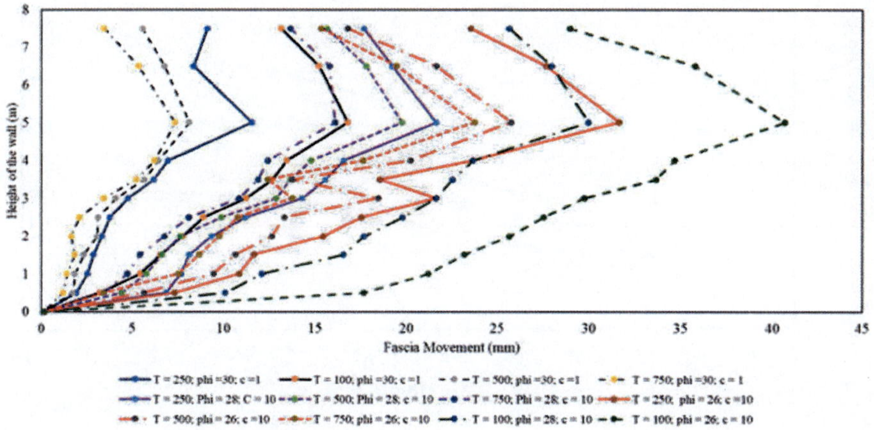

Fig. 17.4 Influence of geogrid strength on the fascia movement with different soil properties

Effect of GRS Wall Height

The height of the GRS wall is an important deciding factor for the implementation of marginally reinforced backfill material. In conventional GRS wall, the height up to 7.5 m has been given with standard reinforced backfill material. To understand the availability of the height of GRS wall with poorly backfill soil, the height of the GRS wall has been varied between 3.5, 5, 7.5, 10.5, 12.5 m. The five different heights of the GRS wall have been simuslated with same set of soil combinations in Section "Effect of Geogrid Strength on the Fascia Movement". The results have been presented in Fig. 17.5 with various heights of the GRS wall. The fascia movement at the critical layer has been the governing parameter of the permissible height of GRS wall. The observation from the maximum wall movement is beyond 7.5 m height of the GRS wall special measure that must be taken care in the field implementations.

Effect of Soil Marginality on the Connection Load

The connection load on the wall face can be quantified by the load of geogrid on the fascia block. The sum of the connection loads in each layer and toe load must be equal to the horizontal component of the lateral earth pressure to satisfy the horizontal equilibrium. The connection load has been measured along the height of the wall with different soil properties as shown in Fig. 17.6. The results show that the c = 10 kPa and $\phi = 26°$ give the maximum connection load of 9.89 kN/m which is significantly higher as compared to 3.34 kN/m with $c = 0$ and $\phi = 30°$. The increase in connection load can increase the wall movement and subsequently fascia can be bulged. Here the results have been shown with geogrid strength of 250 kN/m. The

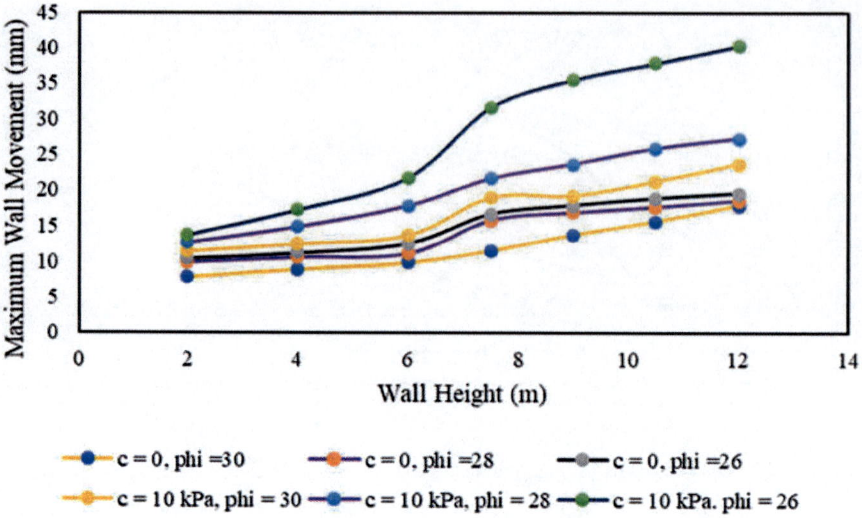

Fig. 17.5 Effect of GRS wall height on the wall movement for different soil properties

increase in geogrid strength can significantly reduce the connection load even with the poorly graded backfill as depicted in Section "Effect of Geogrid Strength on the Fascia Movement".

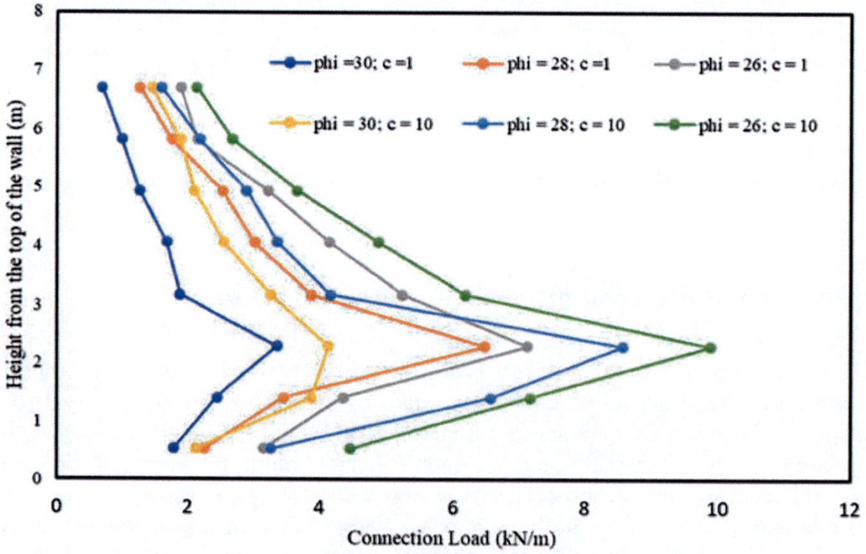

Fig. 17.6 Effect of soil marginality on the connection load of geogrid

Conclusions

In the present study, an attempt has been made to understand the behavior of GRS wall backfilled with poorly graded marginal material. A FE model has been developed in ABAQUS to investigate the problem under dry-state conditions. The major findings from the present study are the followings:

- The deviation from standard backfill can significantly affect the performance behavior of the GRS wall. The fascia movement has been observed to be increased significantly with marginal backfill material. The fascia movement can be increased up to 63% without using standard backfill.
- The geogrid strength can be improved to restrict the fascia movement even with the poorly graded backfill soil. The strength has been increased up to 750 kN/m to check the effective utilization of the poor backfill soil. The fascia movement has been significantly dropped down after using the high-strength geogrid.
- For using poorly graded soil, the GRS wall height must be restricted to 7.5 m. The increase in wall height can cause larger wall movement.
- The effect on connecting load of the GRS wall has been studied to check the utility of the marginal backfill soil. The connection load has been significantly increased due to the use of marginal backfill soil. The increase in geogrid strength can reduce the connection load.

Overall, marginal backfill soil can be utilized as the effective reinforced fill material with proper consideration of reinforcement and wall height.

References

1. Hatami K, Bathurst RJ (2006) Numerical model for reinforced soil segmental walls under surcharge loading. J Geotech Geoenvironmental Eng. https://doi.org/10.1061/(asce)1090-0241(2006)132:6(673)
2. Hatami K, Bathurst RJ (2006) Parametric analysis of reinforced soil walls with different backfill material properties. Natl Assoc Geosynth
3. Huang B, Bathurst RJ, Hatami K (2009) Numerical study of reinforced soil segmental walls using three different constitutive soil models. J Geotech Geoenviron Eng. https://doi.org/10.1061/(asce)gt.1943-5606.0000092
4. Hatami K, Bathurst RJ (2005) Development and verification of a numerical model for the analysis of geosynthethic-reinforced soil segmental walls under working stress conditions. Can Geotech J. https://doi.org/10.1139/t05-040
5. Berg R, Christopher B, Samtani N (2009) Design and construction of mechanically stabilized earth walls and reinforced soil slopes–volume I. Fed High W Adm. FHWA-NHI-10-024, FHWA-NHI-10–025
6. Jayanandan M, Viswanadham BVS (2020) Geogrid reinforced soil walls with marginal backfills subjected to rainfall: numerical Study. Indian Geotech J. https://doi.org/10.1007/s40098-019-00396-0
7. Smith M (2009) ABAQUS/standard user's manual. Version 6:9

8. Isola M, Zhao A, Scotto M (2018) Comparison of FHWA and BS8006 design methods for geosynthetic reinforcement over piles. In: 11th International conference on geosynthetics 2018, ICG 2018
9. Wilkins T, Worrall K, Smith R (2017) Briefing: BS 8006–1: a major amendment for 2016. Proc Inst Civ Eng Geotech Eng. https://doi.org/10.1680/jgeen.16.00217
10. Hsieh C, Chuang BF, Tseng YC (2010) Tensile creep behavior of a PVC coated polyester geogrid at different temperatures. In: 9th International conference on geosynthetics - geosynthetics: advanced solutions for a challenging World, ICG 2010

Chapter 18
Life Cycle Assessment for the Management of Municipal Solid Waste in Indian Cities

P. Devahi, R. Deendayal, and K. Muthukkumaran

Introduction

Industrialization, population growth, and rapid urbanization contribute to the enormous amount of municipal solid waste generation in every developing country [1]. Global solid waste generation has been estimated to reach more than 2.2 billion tons by 2025 [2]. In India, annual solid waste generation will be increasing at a rate of 5% and fetching 56 million tons every year [3]. The significant increase in solid waste generation directly impacts the environmental emissions, social, and economic issues in the country [4]. Municipal solid waste management has been identified as a well-known challenge for Indian cities. The traditional practice in most of the city region is the open dumping option [5]. It attracts severe environmental threats to the community and the atmosphere. However, sanitary landfills have been incorporated in many regions, considering them as a scientific approach. Progressive waste management approaches include material processing techniques, biomethanation, composting, and waste-to-energy techniques [6]. Irrespective of their benefits, every approach has an extent of the environmental debt. Life cycle assessment can be used as an evaluation tool to identify the significant impact categories associated with the integrated waste management scenarios.

The analyzing tools available for performing life cycle assessment include openLCA, Gabi, SimaPro, and Umberto. These tools have unique characteristics, which vary with associated databases, functionality, and data quality management [7]. In the last decade, many researchers [8, 9], and [10] have studied the possibility of choosing the best solid waste management strategies using LCA tools. Besides, there are numerous for managing the municipal solid waste, namely municipal solid waste management decision support tool (MSW-DST), waste and resources assessment tool for the environment (WRATE), environment assessment system for

P. Devahi (✉) · R. Deendayal · K. Muthukkumaran
National Institute of Technology, Tiruchirappalli, India
e-mail: devahi.civil@gmail.com

K. Muthukkumaran et al. (eds.), *Transportation and Environmental Geotechnics*,
Lecture Notes in Civil Engineering 298, https://doi.org/10.1007/978-981-19-6774-0_18

Table 18.1 Life cycle assessment management tool guidelines

Code	Description
ISO 14040: 2006	Environment Management—life cycle assessment—principles and framework
ISO 14044: 2006	Environment Management—life cycle assessment—Requirements and guidelines
ILCD Handbook	International reference life cycle data system handbook

environmental technologies (EASETECH), and solid waste optimization life cycle framework (SWOLF).

In the present study, LCA of municipal solid waste management for an Indian city with two different strategies has been performed using openLCA tool, and comparative analysis has been done using ReCiPe Midpoint (H) assessment method.

Methodology

The framework adopted for conducting life cycle assessment has been highlighted in Table 18.1.

LCA process was performed in four stages, namely goal and scope definition, life cycle inventory, impact assessment, and data interpretation (Refer Fig. 18.1). The functional unit adopted for the LCA study is 1 ton of MSW generated. The database used in the analysis is European Reference Life Cycle Database (ELCD). The assessment method adopted was ReCiPe Midpoint (H), and the impact categories focused on including human toxicity, global warming, ozone formation potential, and eutrophication.

LCA Study on Indian Cities

Life cycle assessment was performed by considering the study region as a Tiruchirappalli City. The system boundary emphasized was from the collection of municipal solid waste, transport, processing, and disposal of waste. In this study, two scenarios were taken into account. The landfill disposal option is labeled as a baseline condition. Eighty percent of the waste collected was disposed on landfills, and 20 percent of the waste material has been recycled. The recycling rate of 20% addresses the metals recovered, glass materials, paper, and plastics. The major objective of considering the recycling rate with ideal conditions helps the researchers to understand the importance and benefits of integrated waste management from the environmental perspective.

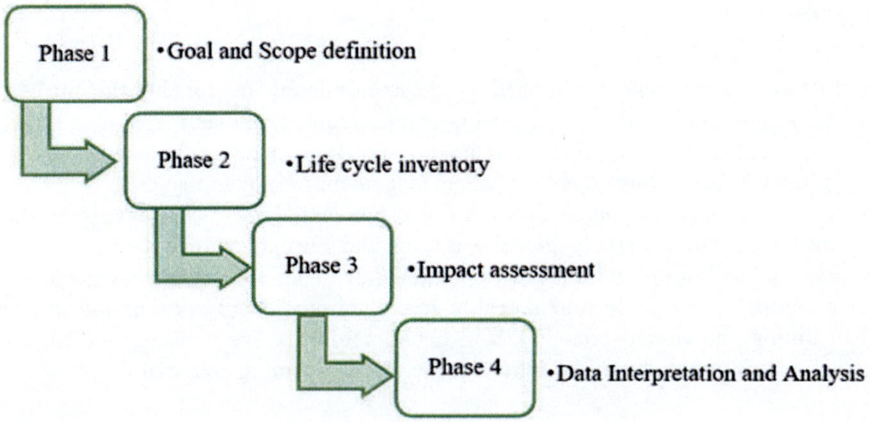

Fig. 18.1 Conceptual framework of LCA process

The inventory data of the present study was taken from the available database, ELCD. The inputs considered for two scenarios are the diesel requirement for trucks with reference distance (3 lit) and power requirement (0.045 kWh) for performing the waste management option [6]. Analysis was done using the ReCiPe Midpoint method, and the predicted impacts have been highlighted in Table 18.2.

From the environmental impacts potential, it is clearly visible that the emission and the exploitation rate has been reduced in scenario 2. Recycling practices have indirectly reduced the disposal load on the landfill, thereby preventing the emission corresponding to that particular disposal load. Eutrophication and acidification potential have been found to be minimum in both the scenario, whereas global warming potential demands attention for remedial measures. Effective utilization of waste-to-energy technologies might highly retard the global warming potential [11].

Table 18.2 Impact analysis by ReCiPe Midpoint (H) method

Environmental Impact	Scenario 1 (Baseline)	Scenario 2 (20% Recycling rate)
Human toxicity potential	340.18 kg 1, 4-DB eq t^{-1}	220.23 kg 1, 4-DB eq t^{-1}
Acidification potential	1.15 kg SO_2 eq t^{-1}	0.90 kg SO_2 eq t^{-1}
Eutrophication potential	0.7012 kg $PO_4{}^{3-}$ eq t^{-1}	0.5030 kg $PO_4{}^{3-}$ eq t^{-1}
Global warming potential	65.0 kg CO^2 eq t^{-1}	61.4 kg CO^2 eq t^{-1}

Conclusions

- Life cycle assessment can be used as a decision-making tool for choosing the best municipal solid waste management options and combinations.
- OpenLCA could be used as an analyzing tool for evaluating the environmental impacts in the municipal solid waste management for any Indian city.
- Incorporation of a 20 percent recycling rate has synthesized better environmental impacts in human toxicity, global warming, and eutrophication potential.
- Eutrophication and acidification potential fall under the acceptable range for scenarios 1 and 2. A high potential rate is observed only concerning global warming and climate change factor. Installment of waste of energy techniques and plants might drastically influence the global warming potential.

References

1. Laurent A, Bakas I, Clavreul J, Bernstad A, Niero M, Gentil E, Hauschild MZ, Christensen TH (2014) Review of LCA studies of solid waste management systems—Part I: lessons learned and perspectives. Waste Manage 34:573–588. https://doi.org/10.1016/j.wasman.2013.10.045
2. Tyagi VK, Fdez-Güelfo LA, Zhou Y, Álvarez-Gallego CJ, Garcia LIR, Ng WJ (2018) Anaerobic co-digestion of organic fraction of municipal solid waste (OFMSW): progress and challenges. Renew Sustain Energy Rev 93:380–399. https://doi.org/10.1016/j.rser.2018.05.051
3. Qu S, Guan D, Ma Z, Yi X (2019) A study on the optimal path of methane emissions reductions in a municipal solid waste landfill treatment based on the IPCC-SD model. J Clean Prod 222:252–266. https://doi.org/10.1016/j.jclepro.2019.03.059
4. Anshassi M, Townsend TG (2021) Reviewing the underlying assumptions in waste LCA models to identify impacts on waste management decision making. J Clean Prod 313:127913. https://doi.org/10.1016/j.jclepro.2021.127913
5. Khandelwal H, Thalla AK, Kumar S, Kumar R (2019) Life cycle assessment of municipal solid waste management options for India. Bioresour Technol 288:121515. https://doi.org/10.1016/j.biortech.2019.121515
6. Rana R, Ganguly R, Gupta AK (2019) Life-cycle assessment of municipal solid-waste management strategies in Tricity region of India. J Mater Cycles Waste Manage 21:606–623. https://doi.org/10.1007/s10163-018-00822-0
7. Lopes Silva DA, Nunes AO, Piekarski CM, da Silva Moris VA, de Souza LSM, Rodrigues TO (2019) Why using different Life cycle assessment software tools can generate different results for the same product system? A cause–effect analysis of the problem. Sustain Prod Consum 20:304–315. https://doi.org/10.1016/j.spc.2019.07.005
8. Christensen TH, Damgaard A, Levis J, Zhao Y, Björklund A, Arena U, Barlaz MA, Starostina V, Boldrin A, Astrup TF, Bisinella V (2020) Application of LCA modelling in integrated waste management. Waste Manage 118:313–322. https://doi.org/10.1016/j.wasman.2020.08.034
9. Llatas C, Bizcocho N, Soust-Verdaguer B, Montes MV, Quiñones R (2021) An LCA-based model for assessing prevention versus non-prevention of construction waste in buildings. Waste Manage 126:608–622. https://doi.org/10.1016/j.wasman.2021.03.047
10. Manjunatha M, Preethi S, Malingaraya MHG, Niveditha KN, Ravi (2021) Life cycle assessment (LCA) of concrete prepared with sustainable cement-based materials. Mater Today Proc. https://doi.org/10.1016/j.matpr.2021.01.248

11. Edun A, Hachem-Vermette C (2021) Energy and environmental impact of recycled end of life tires applied in building envelopes. J Build Eng 39:102242. https://doi.org/10.1016/j.jobe.2021.102242

Chapter 19
A Review on Removal of Heavy Metals from Contaminated Soils by Phytoremediation

Pratyasha Singh and Aparupa Pani

Introduction

Rapid urbanization and noteworthy industrial development associated with the mining and metal processing industry in our country is posing a big challenge. Mercury, cadmium, copper, zinc, cobalt, lead, arsenic, nickel, and other heavy metals are the most important trace elements in soil and groundwater pollution. Smelters and refinery plant facilities may be found all over the world including rural, agricultural, and natural locations. Large extent of heavy metals is found in the soil of such contaminated sites, posing a health danger to persons and the environment [1, 2]. To reform polluted soil, heavy metals must be utilized, but most of these procedures are neither sustainable nor low cost, and they have the ability to drastically alter the texture and properties of the soil mass. [3, 4]. As a result, [5] argued that new low-cost and ecologically friendly solutions are required. However, none of these solutions are cost-effective, exceedingly sophisticated, or commercially viable, not to mention environmentally damaging. As a result, phytoremediation has evolved as a green plant-based cleanup option capable of removing, metabolizing, and degrading a wide range of harmful soil HMs contaminants with minimal expense and without causing harm to the natural environment as per [6–9] experimented with plants to see whether they might remove organic contaminants and metals from the soil. Some of the limitations of phytoremediation are the lack of pollutants, the long treatment duration, and the depth of treatment to the roots zone, according to [10]. Various phytotechniques for the reduction of various pollutants, each with its own mechanism, depending on the application are shown in Fig. 19.1. The use of plants to remove and accumulate chemicals from the environment is known as phytoremediation. This includes the mitigation, transfer, stabilization, or degradation of pollutants in soil, sediments, and water. Indian lemongrass is reported to be a heavy metal

P. Singh · A. Pani (✉)
Department of Civil Engineering, KIIT University, Bhubaneswar-24, Odisha, India
e-mail: aparupa.panifce@kiit.ac.in

K. Muthukkumaran et al. (eds.), *Transportation and Environmental Geotechnics*,
Lecture Notes in Civil Engineering 298, https://doi.org/10.1007/978-981-19-6774-0_19

tolerant aromatic grass that can withstand any harsh environmental conditions. It has also been reported that Indian lemongrass and its essential oil have numerous medical and aesthetic properties. Hence, in the present research, a suitable plant species of Indian lemongrass (Cymbopogon citratus) will be used in phytoremediation to remove, transfer, stabilize, and/or eliminate contaminants in the soil and groundwater. Globally as every year 300–400 million tons of industrial wastes has been reported in India in the year 2017 in the year 2018 and its disposal is a big challenge. Apart from pollution, the current practice of depositing fly ash at landfills and disposal sites causes land use problems. Fly ash is a harmful pollutant that causes cancer, developmental delays, and behavioral challenges in children, as well as respiratory problems in adults [11]. There are numerous projects for the use of fly ash at the moment, but the most of them are still not commercially viable; so, better management and pollution reduction should be prioritized [12]. In this study, the plant growth and its morphological characteristics is evaluated under different conditions such as variation of industrial by-product that is fly ash and its response under different time intervals.

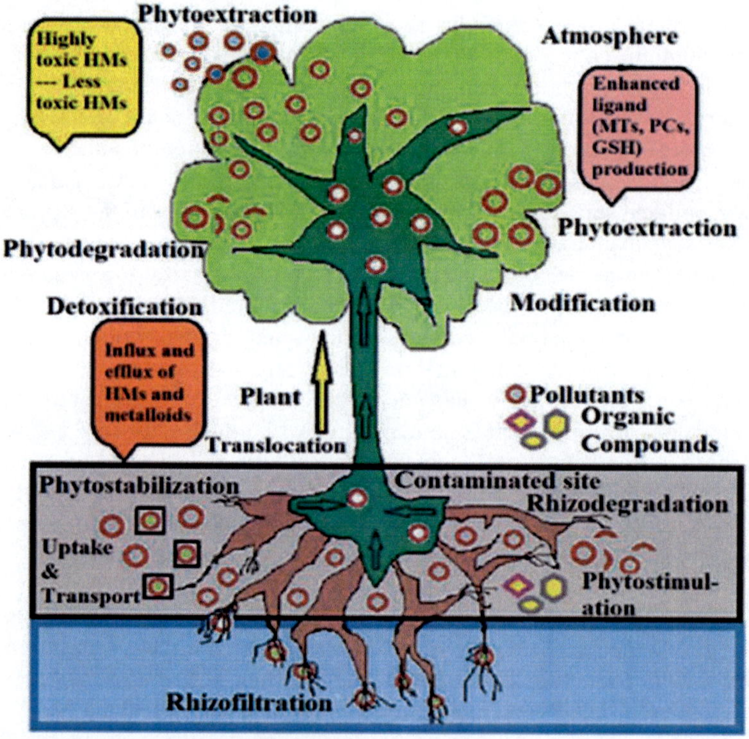

Fig. 19.1 Different phytoremediation techniques [13]

Table 19.1 Chemical properties of fly ash and soil with varying percentage [14]

Different parameters studied	Fly ash	Fly ash 25%	Fly ash 50%	Fly ash 75%	Garden soil
EC (μS/cm)	157 ± 12.0	98.0 ± 8.5	102.5 ± 3.5	110.1 ± 11.0	68.5 ± 11.0
pH	8.70 ± 0.10	6.78 ± 0.10	7.20 ± 0.04	7.50 ± 0.02	6.64 ± 0.11
Nitrogen (mg/kg)	0.10 ± 0.02	0.29 ± 0.03	0.16 ± 0.04	0.12 ± 0.02	0.25 ± 0.02
Organic Carbon (%)	0.08 ± 0.02	0.17 ± 0.02	0.16 ± 0.06	0.12 ± 0.03	0.16 ± 0.01
Potassium (mg/kg)	0.42 ± 0.03	0.81 ± 0.03	0.59 ± 0.07	0.39 ± 0.04	0.86 ± 0.05
Metals μg/g/dwt					
Fe	3150 ± 125	118 ± 21	1345 ± 42	1612 ± 56	108 ± 18
Cr	41.2 ± 2.5	18.4 ± 3.5	26.3 ± 5.2	33.5 ± 3.4	12.2 ± 1.3
Cd	46.3 ± 3.1	26.5 ± 2.4	34.3 ± 3.5	39.7 ± 3.8	21.1 ± 1.8
Zn	62.0 ± 4.3	78.4 ± 2.8	71.3 ± 7.3	66.4 ± 6.5	83.5 ± 2.4

Materials and Methods

Lemongrass species (Cymbopogon citratus) was taken from MS Swaminathan Research Foundation, Jeypore, India, for this study [14]. In three duplicates of a full randomize block design, healthy lemongrass shoot stumps of 10 cm each were immediately planted in plastic pots. For physicochemical stabilization and proper plantation, pots were left at the study site for 14 days pre to planting. Soil conditioning after treatment organic carbon, pH, total nitrogen, electrical conductivity (EC), and other chemical characteristics are all examined. Several proportions of fly ash and garden soil were created by combining garden soil and fly ash in various dry weight ratios, such as fly ash (100%), garden soil (100%), fly ash to garden soil (25: 75%), fly ash to garden soil (50: 50%), and fly ash to garden soil (75: 25%), as shown in Table 19.1. Photo-synthetic photon flux density, natural solar radiation, relative humidity, and air temperature all were exposed daily to the plants all hovering around 1460 ± 20 μ mol/m^2/s, 70–75%, and 33.6 ± 2 °C, respectively. Plant growth was measured using a variety of growth indicators, including dry and fresh weight of root and shoot after 90 days of treatment in each ectype model. After drying at 80 °C until a steady weight was recorded, total plant biomass was collected. The plant's metal tolerance ability was determined using the metal tolerance index formula (MTI).

MTI % = (Total plant biomass under treatment/Total plant biomass under control) × 100.

Metal Translocation and Accumulation

When estimating the metal intake level and mobility of the elements in lemongrass, the examination of the plant organs is critical. The metal content in the soil and the metal distribution in the plant tissues have a substantial impact on metal uptake and accumulation in different plant parts, and this is an indicator for evaluating the plants potential in a contaminated site remediation. The phytoextraction capacities of the plant were assessed using the bioconcentration factor (BCF) and the translocation factor (TF). By dividing the metal content in plant tissue by the metal content in the soil, the BCF was calculated. The metal concentration above the ground plant tissues was divided by the equivalent metal concentration in roots to calculate TF [15]. By dividing plant biomass in treatments by plant biomass in controls, the metal tolerance index (MTI) was calculated [16]. Growth metrics such as root, shoot, and total plant biomass were considerably boosted when 90-day-old lemongrass plants were exposed to a low amount of fly ash (25%) supplemented soil compared to plants maintained in garden soil.

Under modest levels of fly ash (25%) additions, lemongrass plant biomass rose by 11%, however, under 100 percent fly ash, shoot, and root biomass decreased by 70%, 73%, and 71%, respectively, compared to the control. Fly ash (25%) in soil amendments promotes the growth of lemongrass, according to the study. However, the high metal concentrations in fly ash can hinder plant growth [17, 18]. The fly ash has several physical and chemical qualities that could be effective in low-level soil amendments that are favorable to plant growth [19, 20]. The lemongrass demonstrated excellent tolerance to low levels of fly ash-amended soil and high potential to grow in metal-polluted soils [21]. The attributes and descriptions chemical and physical properties of fly ash and fly ash mixed with compost are shown in Table 19.2.

Results and Discussion

The use of fly ash for vegetation is limited due to the presence of non-essential and potentially harmful elements that may interfere with plant growth. Fly ash combined with a soil amendment may provide a sufficient mix to boost plant growth while reducing the danger of metal poisoning [22]. In comparison with garden soil, the electrical conductivity and accessible phosphorus concentration of fly ash were higher. In addition, fly ash has a higher concentration of aluminum, iron, chromium, manganese, copper, cadmium than garden soil. The concentrations of zinc and copper in garden soil were within the range of usual values for soil, and however, the concentrations of manganese were insufficient [23]. When a low level of fly ash (25%) was used in soil amendments, the pH, EC, and nitrogen levels improved significantly (*P* 0.05) when compared to other treatments. In comparison with garden soil, fly ash has a high phosphorus content. According to [20, 22] fly ash improves soil permeability, moisture retention capacity, acidity, and heavy metal availability, making it a potentially beneficial soil addition.

Table 19.2 Chemical and physical properties of heavy metals contents of fly ash, fly ash mixed with compost [14]

Properties	Fly ash	Compost	Fly ash + 10% compost
pH (1:2, Soil: water mixture)	8.8 ± 0.12	7.6 ± 0.08	8.2 ± 0.14
EC (1:2, Soil: water mixture, μS/cm)	2370 ± 185	890 ± 64	1984 ± 167
% Organic matter (Walkley—black acid digestor)	0.012 ± 0.008	42.28 ± 2.48	4.23 ± 0.02
Particle Size Distribution			
% Sand	21.05		21.00
% Silt	45.25		45.00
% Clay	33.70		34.00
Total Heavy Metals (mg/kg)			
As	10.08 ± 0.14	0.00 ± 0.00	9.06 ± 0.18
Cu	63.20 ± 2.64	32.40 ± 3.48	52.41 ± 3.47
Mn	781.71 ± 32.48	342.00 ± 24.82	718.5 ± 24.62
Ni	78.25 ± 3.45	2.12 ± 0.14	68.47 ± 6.57
Zn	25.31 ± 2.47	428.24 ± 38.43	65.48 ± 14.29
Cr	154.24 ± 5.84	3.28 ± 0.42	133.85 ± 6.42
Water soluble heavy metals (mg/kg)			
As	1.23 ± 0.04	0.00 ± 0.00	0.89 ± 0.02
Cu	2.07 ± 0.14	8.27 ± 0.20	5.43 ± 0.16
Mn	2.04 ± 0.08	32.48 ± 0.27	42.85 ± 0.63
Ni	0.51 ± 0.02	0.00 ± 0.00	7.62 ± 0.12
Zn	2.04 ± 0.21	48.42 ± 2.47	7.24 ± 0.27
Cr	4.21 ± 0.14	0.00 ± 0.00	2.94 ± 0.22

Conclusion

Fly ash is a waste product from coal-fired thermal power plants that could be exploited as a resource material in a variety of fields, allowing for more efficient disposal. According to the research, adding 25% fly ash to garden soil can help lemongrass grow faster and produce more photosynthesis in the leaves. As a result of the high MTI which was more than 100%, lemongrass can withstand up to 25% of fly ash. When lemongrass is exposed to high quantities of fly ash, various antioxidative enzymes are activated, indicating that the plant can tolerate metals. These findings imply that lemongrass is a plant species with the ability to grow. It can be used for phytorestoration of fly ash deposits and on fly ash deposits.

References

1. Chirakkara RA, Reddy KR, Cameselle C (2015) Electrokinetic amendment in phytoremediation of mixed contaminated soil. Electrochim Acta 181:179–191. https://doi.org/10.1016/j.electacta.2015.01.025
2. Reddy KR, Chirakkara RA, Martins Ribeiro LF (2020) Effects of elevated concentrations of co-existing heavy metals and PAHs in soil on phytoremediation. J Hazard Toxic Radioactive Waste 24(4):04020035
3. Marques AP, Rangel AO, Castro PM (2009) Remediation of heavy metal contaminated soils: phytoremediation as a potentially promising clean-up technology. Crit Rev Environ Sci Technol 39(8):622–654. https://doi.org/10.1080/10643380701798272
4. Jeelani N, Yang W, Xu L, Qiao Y, An S, Leng X (2017) Phytoremediation potential of Acorus calamus in soils co-contaminated with cadmium and polycyclic aromatic hydrocarbons. Sci Rep 7(1):1–9. https://doi.org/10.1038/s41598-017-07831-3
5. Ahmadpour P, Ahmadpour F, Sadeghi S, Tayefeh FH, Soleimani M, Abdu AB (2014) Evaluation of four plant species for phytoremediation of copper-contaminated soil. Soil Remediation and Plants: Prospects and Challenges 147
6. Cunningham SD, Berti WR, Huang JW (1995) Phytoremediation of contaminated soils. Trends Biotechnol 13(9):393–397
7. Flathman PE, Lanza GR (1998) Phytoremediation: current views on an emerging green technology. J Soil Contam 7(4):415–432
8. Ali H, Khan E, Sajad MA (2013) Phytoremediation of heavy metals—concepts and applications. Chemosphere 91(7):869–881
9. Zhang H, Dang Z, Zheng LC, Yi, XY (2009) Remediation of soil co-contaminated with pyrene and cadmium by growing maize (Zea mays L.). Int J Environ Sci Technol 6(2):249–258
10. Couto N, Guedes P, Ribeiro AB, Zhou DM (2015) Phytoremediation and the electrokinetic process: potential use for the phytoremediation of Antimony and Arsenic. In Phytoremediation. Springer, Cham, pp 199–209
11. Baba T, Makino Y, Yamada M, Fujiyama H (2016) Evaluation of the Cs-and Sr-absorption ability of plant species for phytoremediation. Eco-Eng 28(1):1–5
12. Ghosh D, Maiti SK (2021) Biochar assisted phytoremediation and biomass disposal in heavy metal contaminated mine soils: a review. Int J Phytorem 23(6):559–576
13. Saxena G, Purchase D, Mulla SI, Saratale GD, Bharagava RN (2019) Phytoremediation of heavy metal-contaminated sites: eco-environmental concerns, field studies, sustainability issues, and future prospects. Rev Environ a Contam Toxicol 249:71–131. https://doi.org/10.1007/398_2019_24
14. Panda D, Panda D, Padhan B, Biswas, M (2018) Growth and physiological response of lemongrass (Cymbopogon citratus (DC) Stapf.) under different levels of fly ash-amended soil. Int J Phytorem 20(6):538–544
15. Wu S, Zhou S, Li X (2011) Determining the anthropogenic contribution of heavy metal accumulations around a typical industrial town: Xushe China. J Geochem Explor 110(2):92–97
16. Wang J, Cui L, Gao W, Shi T, Chen Y, Gao Y (2014) Prediction of low heavy metal concentrations in agricultural soils using visible and near-infrared reflectance spectroscopy. Geoderma 216:1–9
17. Parveen R, Hisamuddin T, Azam T, Niyaz SS (2006) Effect of fly ash amended soil on the plant growth, yield, chlorophyll and oil content of Mentha citrata. In: Proceedings of the national symposium on issue and challenges for environmental management. Vision (Vol 2025, p 55)
18. Bisoi SS, Mishra SS, Barik J, Panda D (2017) Effects of different treatments of fly ash and mining soil on growth and antioxidant protection of Indian wild rice. Int J Phytorem 19(5):446–452
19. Verma SK, Singh K, Gupta AK, Pandey VC, Trivedi P, Verma RK, Patra DD (2014) Aromatic grasses for phytomanagement of coal fly ash hazards. Ecol Eng 73:425–428
20. Ram LC, Masto RE (2014) Fly ash for soil amelioration: a review on the influence of ash blending with inorganic and organic amendments. Earth-Sci Rev 128:52–74

21. Gautam M, Pandey D, Agrawal M (2017) Phytoremediation of metals using lemongrass (Cymbopogon citratus (DC) Stapf.) grown under different levels of red mud in soil amended with biowastes. Int J Phytorem 19(6):555–562
22. Pandey VC, Singh N (2010) Impact of fly ash incorporation in soil systems. Agric Ecosyst Environ 136(1–2):16–27
23. Kabata-Pendias A (2000) Trace elements in soils and plants. CRC Press

Chapter 20
Investigation of Soil Characteristics in and Around the Open Dump Site Near Sathyamangalam Town in Erode District

Geethamani Ramaswamy, Soundara Balu, and Kanmani Subramaniam

Introduction

The tremendous increase in population and township has led to the increase in the generation of solid waste among Indian cities. Due to improper management of solid waste, they are just dumped in open areas without any treatment. Dumping of solid waste in open dumpsites poses an important matter of concern, both to the surroundings and public health, causing air, water, and soil pollution resulting in various diseases to human beings. Municipal solid waste generated in the Indian cities is mostly dumped in the open dumpsites due to the absence of sanitary landfills. This improper method of solid waste management leads to soil and water pollution, both on the surface and subsurface. This in turn results in the impact on human health due to the use of polluted water and consumption of agricultural products cultivated from the field with polluted soil [1, 2]. The poor attitude of the public in handling solid waste also leads to environmental pollution. The authorities concerned should intensify their effort on the educational sector to create awareness among the public about the ill effects of improper dumping of solid waste and sensitize the public about the benefits of a sustainable environment [3]. However, the attitude varies for the age and educational qualifications of the people [1, 4]. A slow shift from open

G. Ramaswamy (✉)
Department of Civil Engineering, Bannari Amman Institute of Technology, Erode 638401, India
e-mail: geethamanir@bitsathy.ac.in

S. Balu
Department of Civil Engineering, SIMATS School of Engineering, Saveetha Institute of Medical and Technical Sciences, Chennai 602105, India
e-mail: soundarab.sse@saveetha.com

K. Subramaniam
Department of Civil Engineering, KPR Institute of Engineering and Technology, Coimbatore 641407, India
e-mail: kanmani.s@kpriet.ac.in

K. Muthukkumaran et al. (eds.), *Transportation and Environmental Geotechnics*, Lecture Notes in Civil Engineering 298, https://doi.org/10.1007/978-981-19-6774-0_20

dumpsites to sanitary landfills may reduce greenhouse gas emissions and reduce global warming [6, 8]. Heavy metal pollution is caused due to leachate penetration into the ground surface [5]. Municipal solid waste due to its rich nutrient content may be used to improve the fertility of soil thereby increasing the yield of crops.

The main objective of this research work is to identify the present characteristics of the soil around the open dumpsite and promote the development of sanitary landfilling, to reduce or eliminate environmental pollution due to open dumping of municipal solid waste.

Study Area Details

The open dumpsite is located in the municipal town of Sathyamangalam with a latitude and longitude of 11.5034° N, 77.2444° E, respectively, in Erode District, Tamil Nadu, India. River Bhavani, a tributary of Cauvery flows through the center of the town and forms the main source of water for the town. The area of the municipality is around 29 km. As per the 2011 census, the population of the town was 37,816. This place recorded a maximum temperature of 38.9 °C in the summer season and 29.1 °C in the winter season. The average annual precipitation is 730 mm.

The area of the open dumpsite is around 6.75 acres as shown in Fig. 20.1. It is protected by a compound wall and barbed wire fencing. It has accessibility roads inside for the garbage collection vehicles to deposit the waste. All the waste generated from the town is disposed of in this dumpsite. Table 20.1 shows the details of the type of area and the rate of waste generation in Sathyamangalam Town.

Materials and Methods

Soil Sample Collection

Soil samples were collected at six different locations. The soil sample sites were chosen based on a simple random sampling method [6]. Nine samples (S1–1, S1–2, S1–3, S2–1, S2–2, S2–3, S3–1, S3–2, S3–3, S4–1, S4–2, S4–3, S5–1, S5–2, S5–3, S6–1, S6–2, and S6–3) were collected inside and outside the dumpsite in three different locations. At each location samples were collected at depths 1 m, 2 m, and 3 m, respectively. The three sample locations outside the dumpsite were chosen based on the topography of the land. The three sample locations inside the dumpsite were chosen as old waste dumping location, new waste dumping location and in between the two.

The ground at the soil sampling locations was dug using a JCB, and the samples were collected in a ziplock polythene cover and were coded and labelled. The collected soil samples were taken to the laboratory. The lumps in the soil sample

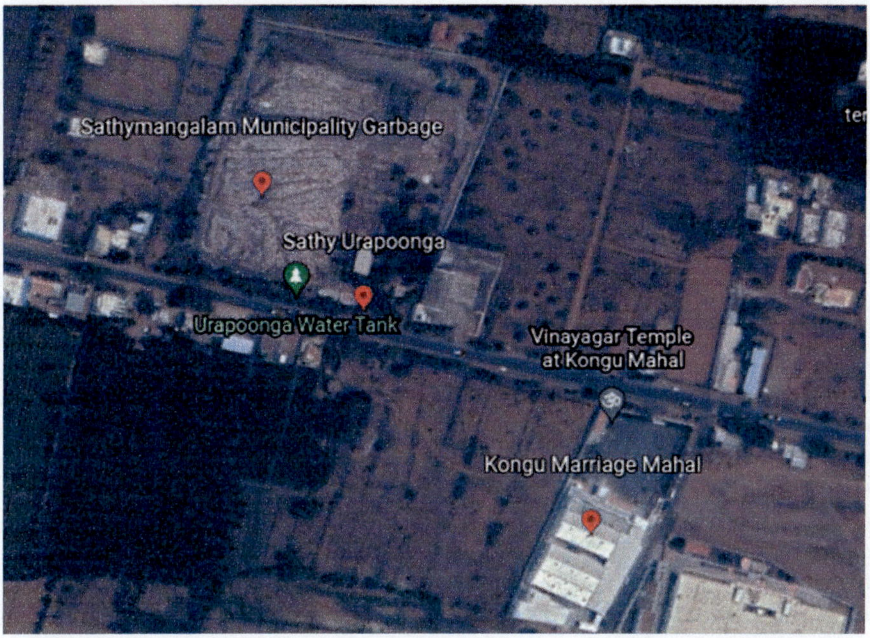

Fig. 20.1 Study area—Malaiyadipudur open dump site near Sathyamangalam. *Source: maps. Google.com*

Table 20.1 Type of area and waste generation rates

S. No	Type of area	Waste generation rate (Metric tons/day)
1	Residential	9.5
2	Markets	1.0
3	Commercial	5.0
4	Industrial	Nil
5	Silt	1.5
6	Construction and demolition waste	Nil
7	Total	16.0

[a]Source: Sanitary Division Commissioner Office–Sathyamangalam

were removed by breaking and dried inside the lab for three days. The soil lumps were broken and air-dried for 72 h. The air-dried soil samples passing through a 2 mm sieve were taken for further analysis [6].

The index properties of the soil such as grain size distribution and relative density were analyzed. The soil samples were water digested, and the parameters such as organic carbon [6], pH [7], electrical conductivity, total dissolved solids, chlorides, sulfates, and micronutrients [8] were analyzed. The soil samples were acid digested and analyzed for heavy metals such as copper (Cu), zinc (Zn), manganese (Mn), and iron (Fe). Nutrients such as nitrogen (N), phosphorus (P), potassium (K), organic carbon (OC), boron (B) [9] were also analyzed.

Analytical Methods

The grain size distribution of the collected samples was analyzed by sieve analysis and the specific gravity using a pycnometer as per Indian Standards 2720. The pH of the soil sample was analyzed using an electrode-type pH meter, the electrical conductivity of the soil was measured using an EC meter, by preparing a 1:2.5 soil-to-water extraction method [6]. The total organic carbon in the soil was measured using a TOC analyzer. The total dissolved solids, chloride concentration and sulfate concentration in the soil samples were analyzed by using APHA standards. The collected soil samples were dried in the laboratory naturally. The dried sample was sieved through 2.36 mm sieve. The passed soil content was acid digested using aqua regia (aqua regia is a combination of 1:3 hydrochloric acid: nitric acid) and was used for heavy metal analysis [10]. Heavy metals (copper, zinc, manganese, and iron) in the soil sample were analyzed using Atomic Absorption Spectrophotometer [10, 11]. To reduce the errors during the analysis of the soil sample, standards prescribed by the Indian Standard Organization or the American Public Health Association were followed. The samples analyzed using the standard methods are assumed to be devoid of more errors [12, 13].

Results

The coefficient of uniformity (Cu) and the coefficient of curvature (Cc) of the soil samples were represented in Fig. 20.2. Figure 20.3 represents the specific gravity of the soil samples. The physicochemical characteristics, heavy metal concentration, etc., of the collected soil samples were shown in Figs. 20.4, 20.5, 20.6, and 20.7.

Discussion

The coefficient of uniformity (Cu) and the coefficient of curvature (Cc) of the soil samples shown in Fig. 20.2 indicate the type of soil to be SP (poorly graded Sand). The specific gravity values in Fig. 20.3. Indicate that the soil samples collected inside the dumpsite showed lesser values when compared to the soil samples collected outside. The lower values of specific gravity indicate the presence of more organic content due to the dumping of solid wastes. The pH of the soil samples collected inside

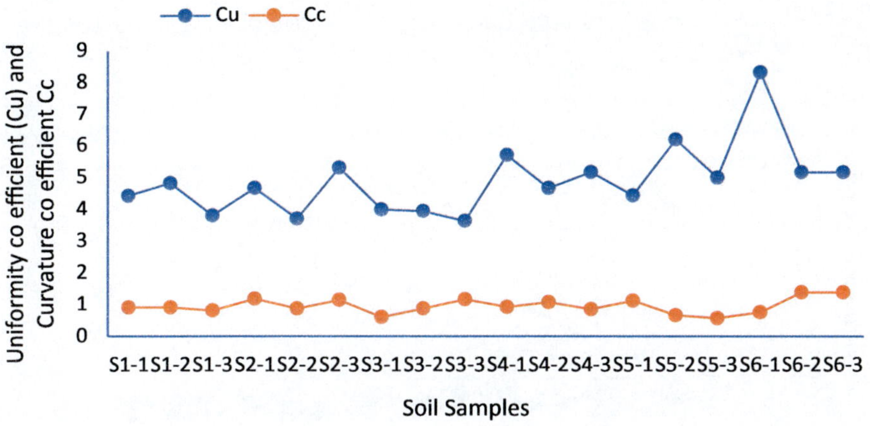

Fig. 20.2 Coefficient of uniformity (Cu) and coefficient of curvature (Cc) of the soil samples

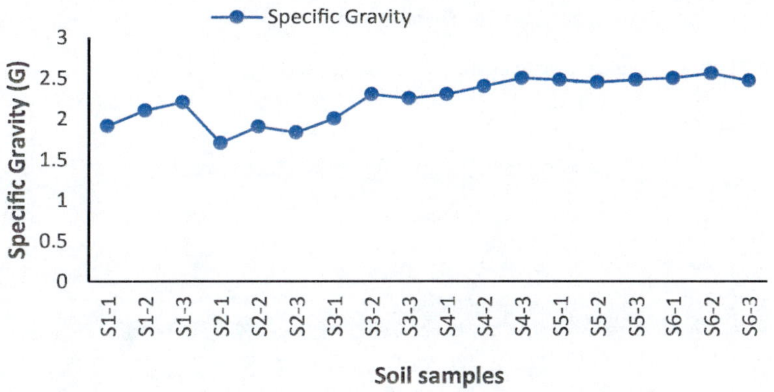

Fig. 20.3 Specific gravity of the soil samples

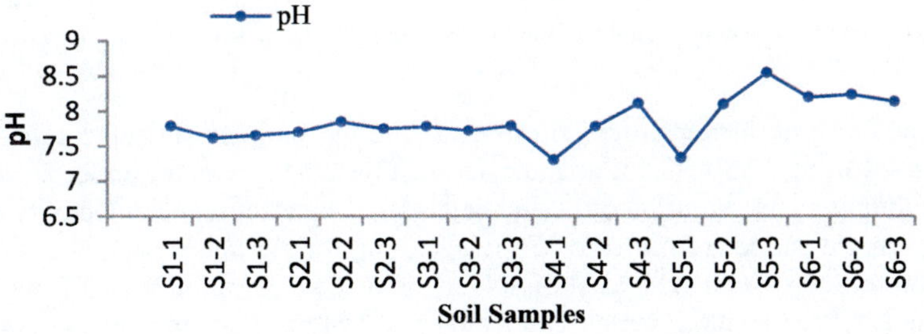

Fig. 20.4 pH for the soil samples collected in and out of the open dumpsite

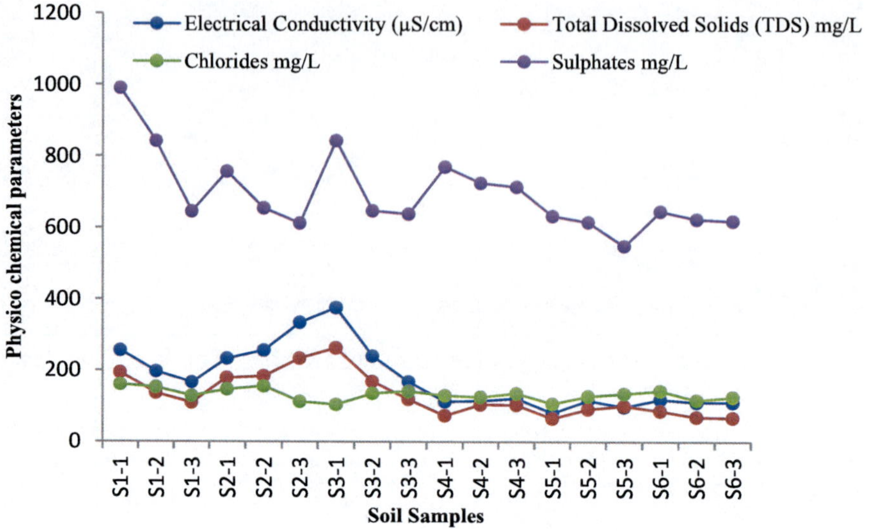

Fig. 20.5 Physicochemical characteristics of soil samples collected in and out of the open dump site

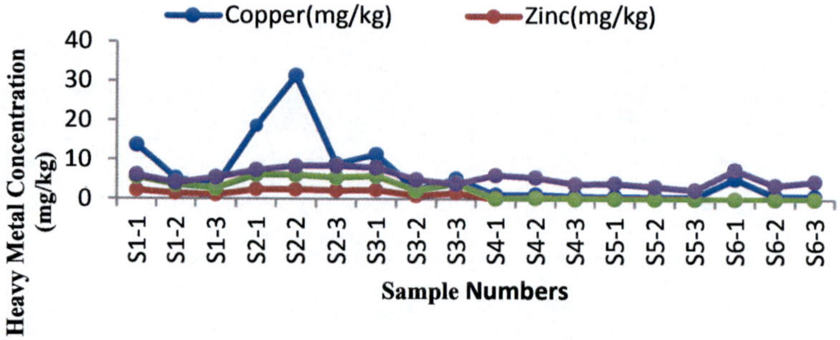

Fig. 20.6 Heavy metal concentration of the soil samples collected in and out of the open dump site

the open dumpsite varied from 7.30 to 8.11, and the samples collected outside the site varied from 7.33 to 8.55, as indicated in Fig. 20.4. pH is an important factor in deciding the concentration of heavy metals [8]. The electrical conductivity of the inside sample varied from 112 to 375 μS/cm, and for the outside sample, it varied from 80 to 256 μS/cm. The total dissolved solids concentration varied from 73 to 263 mg/L for the inside samples, and from 65 to 194 mg/L for the outside samples. The chloride ion concentration varied from 104 to 155 mg/L for the inside samples, and from 106 to 160 mg/L for the samples collected outside the site. The sulfate concentration varied from 612 to 842 mg/L for the inside samples, and from 615 to 990 mg/L for the outside samples, as shown in Fig. 20.5.

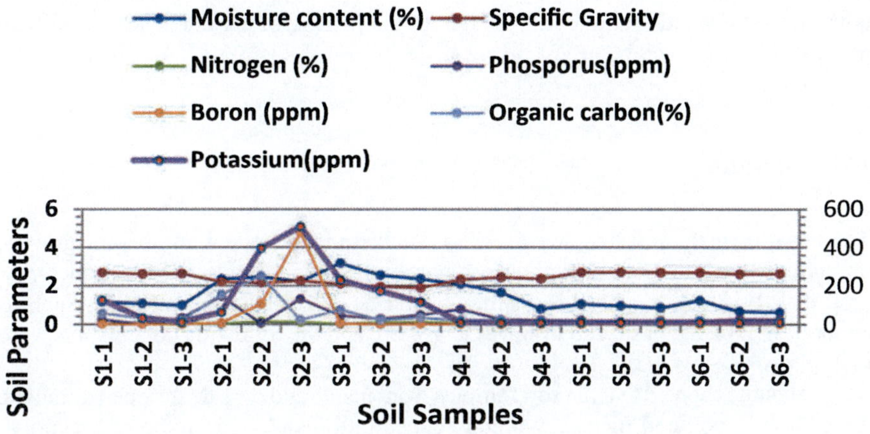

Fig. 20.7 Soil parameters for the collected soil sample in and out of the open dump site

The heavy metal concentration copper varied from 0.75 to 31.40 mg/kg for the inside samples, and it varied from 0.48 to 13.73 mg/kg. Zinc concentration varied from 0.00 to 2.43 mg/kg for the inside samples, and it varied from 0.00 to 2.22 mg/kg. Manganese concentration varied from 0.00 to 6.08 mg/kg for the inside samples, and it varied from 0.00 to 5.50 mg/kg. Iron concentration varied from 3.89 to 8.45 mg/kg for the inside samples, and it varied from 2.41 to 6.09 mg/kg, as shown in Fig. 20.6.

The moisture content varied from 0.79 to 3.17% for the inside samples, and the outside samples varied from 0.62 to 1.15%. The specific gravity varied from 1.88 to 2.43 for the inside samples, and for the outside samples, it varied from 2.60 to 2.70. The organic carbon concentration varied from 0.12 to 2.51% for the inside samples and from 0.10% to 0.56% for the outside samples. There is not much difference in the nitrogen and phosphorus values, they range from 0.03 to 0.13% and 0.04 to 1.45ppm. The potassium concentration for the inside soil sample showed the highest value of 509.2ppm, whereas for the outside sample it was only 124.8ppm, as depicted in Fig. 20.7.

There was not much difference in pH for the samples collected in and outside the open dumpsite. The maximum pH value was found for the sample collected outside the dumpsite. The electrical conductivity and total dissolved solids were more for the inside samples, but the chloride and sulfate concentration was found to be more for the outside soil samples. The highest concentration values of copper, zinc, manganese, and iron were found for the inside samples when compared to the outside samples. This may be due to the dumping of solid waste below the ground surface, inside the open dumpsite. The higher values of moisture content for the samples collected inside the dumpsite may be due to the decomposition of solid waste below the ground surface. The lower values of specific gravity inside the dumpsite may be due to the mixing of organic waste in the soil samples. The highest value of organic carbon content for the inside sample may be due to the mixing of organic solid waste with the soil.

Conclusions

The variation in the soil sample parameters such as moisture content, specific gravity, organic carbon, nitrogen, phosphorus, potassium, boron, pH, electrical conductivity, total dissolved solids, chlorides, sulfates, and heavy metal concentrations such as copper, zinc, manganese, and iron was analyzed, and the following conclusions were drawn from the results.

The results showed that the soil samples from inside and outside the open dumpsite were slightly polluted for heavy metal concentration and were in the order copper > iron > manganese > zinc. The higher concentration of heavy metals inside the dumpsite may be due to the contamination of the soil with the copper and iron-rich domestic solid waste disposal. The presence of manganese may be due to the mixing of sewage sludge with the soil. The presence of zinc may be due to the usage of fertilizers in the agricultural fields around the open dumpsite.

The highest concentration of moisture content for soil samples collected inside the dumpsite may be due to the dumping of organic wet waste. Dumping of organic solid waste leads to the emergence of leachate from the wet waste which in turn increases the wetness. The presence of wetness in the solid waste may lead to an increased moisture content of the soil. The lower value of specific gravity for the inside soil samples indicates that the soil is polluted with solid waste. Since the municipal solid waste will be rich in organic content, the weight of the soil is reduced due to the decomposition of organic matter in due course of time.

The above results were only concerning the soil samples collected at one particular time. However, there may be variations if the soil samples are analyzed for different seasons. Though an equal number of samples were collected in and outside the dumpsite, the samples were chosen only on a random basis. The present study reveals that the soil inside the dumpsite is polluted due to the dumping of municipal solid waste, and the soil outside the dumpsite is polluted due to the application of fertilizers for agricultural activities. Further study may be made by collecting soil samples at equal or specified intervals and for different seasons to analyze the variations in the result. To prevent the contamination of soil and groundwater, sanitary landfilling shall be adopted for better solid waste management and environmental pollution prevention.

References

1. Singh SK, Chokhandre P, Salve PS, Rajak R (2020) Open dumping site and health risks to proximate communities in Mumbai, India: a cross-sectional case-comparison study. Clin Epidemiol Global Health 9:34–40
2. Dos Muchangos LS, Tokai A (2020) Greenhouse gas emission analysis of upgrading from an open dump to a semi-aerobic landfill in Mozambique—the case of Hulene dumpsite. Sci Afr 10:e00638
3. Sankoh FP, Yan X (2013) Problems of solid waste management in developing urbancities: a case study of Freetown, Sierra Leone. Am J Environ Prot 2(5):113–120
4. Borjac J, El Joumaa M, Kawach R, Youssef L, Blake DA (2019) Heavy metals and organic compounds contamination in leachates collected from Deir Kanoun Ras El Ain dump and its adjacent canal in South Lebanon. Heliyon.5:e02212
5. Mekonnen B, Haddis A, Zeine W (2020) Assessment of the effect of solid waste dump site on surroundingsoil and river water quality in Tepi town, Southwest Ethiopia. J Environ Publ Health. https://doi.org/10.1155/2020/5157046
6. Bartoli G, Papa S, Sagnella E, Fioretto A (2012) Heavy metal content in sediments along the Calore river: Relationships with physicalechemical characteristics. J Environ Manage 95:S9–S14
7. Akanchise T, Boakye S, Borquaye LS, Dodd M, Darko G (2020) Distribution of heavy metals in soils from abandoned dump sites in Kumasi, Ghana. Sci Afr 10:e00614
8. Fernandez-Caliani JC, Giraldez MI, Waken WH, Del Río ZM, Cordoba F (2021) Soil quality changes in an Iberian pyrite mine site 15 years after land reclamation. CATENA 206:105538
9. Meena MD, Yadav RK, Narjary B, Yadav G, Jat HS, Sheoran P, Meena MK, Antil RS, Meena BL, Singh HV, Meena VS (2019) Municipal solid waste (MSW): strategies to improve salt affected soil sustainability: a review. Waste Manage 84:38–53
10. Ali SM, Pervaiz A, Afzal B, Hamid N, Yasmin A (2014) Open dumping of municipal solid waste and its hazardous impacts on soil and vegetation diversity at waste dumping sites of Islamabad city. J King Saud Univ Sci 26:59–65
11. Kaushal A, Sharma MP (2016) Methane emission from panki open dump site of Kanpur, India. Proc Environ Sci 35:337–347
12. Sharma A, Gupta AK, Ganguly R (2018) Impact of open dumping of municipal solid waste on soil properties in mountainous region. J Rock Mech Geotech Eng10(4):725–739
13. Kanmani R, Gandhimathi (2013) Assessment of heavy metal contamination in soil due to leachate migration from an open dumping site. Appl Water Sci 3:193–205. https://doi.org/10.1007/s13201-012-0072-z

Chapter 21
Experimental and Numerical Studies on Usage of Soil–Bentonite Mixture as Clay Liner in Engineered Landfill

S. Anjali, A. R. Ajitha, and P. Seethalakshmi

Introduction

One of the major threats faced by the society nowadays is the disposal of waste materials without emphasizing its safety and haphazard conditions. Conventional practice of disposing waste into a landfill lacks engineering design and criteria. Due to the increase in the industrialization, most of the municipal solid waste landfills are dumbed with industrial wastes which contain a number of heavy metals. These heavy metals can cause serious health-related consequences as well as harmful environmental impacts. Lead, cadmium, copper, mercury, nickel, chromium, iron, zinc, and selenium are the major heavy metals commonly found in the landfill leachate [4]. Even in traces these heavy metals are harmful to the organisms and the environment. The increase in industrialization has paved its way to heaping of toxic heavy metal on to the landfill site threatening human health to serious dimensions. The chief sources of these metals are from fertilizers, paints, textiles, electronic industries, batteries, iron and steel industries, chemical factories, etc. They are powerful enough to cause acute as well as chronic poisoning in man and other animals [12]. The major health issues are respiratory problems, skeletal damages, damages to the nervous system, pulmonary system, visual disorders, etc. Lead is an important heavy metal that threatens soil quality and human health at elevated levels. It causes damage to human health by disturbing the functions of the brain, kidney, and reproductive systems. There are several sources of lead in our environment including past uses in gasoline, lead battery recycling sites and firearms practice ranges, mineral resources exploitation. It reaches the soil by atmospheric deposition as well as by solid and liquid waste disposal as elemental lead, lead alloys, lead oxides, lead sulfate, lead

S. Anjali (✉) · P. Seethalakshmi
Department of Civil Engineering, National Institute of Technology, Kozhikode 673 601, India
e-mail: anjali_p200039ce@nitc.ac.in

A. R. Ajitha
Department of Civil Engineering, College of Engineering Trivandrum, Trivandrum 695 016, India

K. Muthukkumaran et al. (eds.), *Transportation and Environmental Geotechnics*,
Lecture Notes in Civil Engineering 298, https://doi.org/10.1007/978-981-19-6774-0_21

carbonate, etc. [1]. Their presence is more than the allowable concentration in most of the landfill sites. This heavy metal leaches through the soil and contaminates the groundwater underlying. So, it is necessary to construct an engineered landfill that could adsorb these heavy metals on to their surface and there by prevent the migration of these ions to the surrounding groundwater.

Adsorption is referred as a separation process in which adsorbate is concentrated from a bulk vapor or liquid phase on to the surface of a porous adsorbent. Usually, it is believed that adsorption is a monolayer process in which there is a requirement of a larger surface area for the material that is acting as the adsorbent [10]. The transport process involved in adsorption is analyzed by using batch adsorption, one-dimensional equilibrium adsorption on a column, and one-dimensional non-equilibrium adsorption. There are many empirical adsorption models, out of which Langmuir adsorption isotherm model is the commonly used model [5]. This model assumes that there is a specific amount of site that is available for adsorption and adsorbent molecules will adhere to that specific site. Several other models dealing with two, three, and multi-parameters are involved in the adsorption study [6]. The best fit of the experimental data with these models can give a suitable method for computing the adsorbed quantity and several other adsorption parameters.

Materials Used

Soil

Soil used for the study is collected from the Edathua region of Kuttanad, Alappuzha District. It is a dark brown medium-sensitive soil with high compressibility and low shear strength. The properties of the Kuttanad soil used for the study are shown in Table 21.1.

Bentonite

Bentonite used for the study is calcium montmorillonite. It is an adsorbent derived from the weathering of volcanic ashes. It is highly swelling clay which makes it suitable for the use of a liner material. Table 21.2 shows the properties of bentonite used for the study.

Table 21.1 Properties of Kuttanad soil

Property	Value
Natural water content (%)	62
Specific gravity	2.15
Liquid limit (%)	49
Plastic limit (%)	23
Plasticity index	26
Soil classification	CI
Shrinkage limit (%)	18
Percentage of clay (%)	23
Swell index (%)	0
Maximum dry density(g/cc)	1.56
Optimum moisture content (%)	27

Table 21.2 Properties of bentonite used in the study

Property	Value
Specific gravity	2.6
Liquid limit (%)	265
Plastic limit (%)	54
Percentage of clay (%)	66.0
Percentage of silt (%)	34
Swell index (%)	476.47

Landfill Leachate

Leachate sample for the study is collected from the Njeliyanparamba landfill, Kozhikode District. It is a municipal solid waste landfill that handles around 100 tons of waste per day. The properties of the leachate sample are listed in Table 21.3. The sample was tested according to central pollution control board ministry of environment and forest in order to identify the major heavy metals that are present in the leachate and then it is compared with the standards of ministry of environment and forests, Government of India, and it is shown in Table 21.4

Table 21.3 Properties of the leachate collected

Parameter	Value
pH	8.06
Chloride	4199 mg/L
Conductivity	28.25 mS/ppt
Turbidity	126.7NTU
Total dissolved solids	7.9 mg/L

Table 21.4 Chemical composition of the leachate

Leachate constituents	Mean value(mg/L)	Standard value
Fe	30	-
Cu	0.35	3
Zn	1.6	5
Cd	0.1	2
Ni	1.12	3
Pb	0.23	0.1

Methodology

In order to find the optimum bentonite percentage for a liner material, the hydraulic conductivity which is the important property of a liner material was studied by varying the bentonite content from 0 to 7%. The standard proctor test was done according to IS 2720(part 7) to obtain the maximum dry density and optimum moisture content. After fixing an optimum bentonite percentage, adsorption characteristics of the liner material were studied by conducting batch adsorption test and column studies. The adsorption data was analyzed using various numerical models.

Preparation of the Synthetic Leachate

The chemical composition of the leachate reveals that lead is a heavy metal that is more than the advisable quantity, and for the same reason, it is chosen as the heavy metal for the study. As it is dangerous to deal with the landfill leachate, synthetic leachate is prepared by using lead nitrate ($Pb(NO_3)_2$. A stock solution was prepared to get 1PPM (1000 mg/L) of solution for which 1.6 g of lead nitrate is mixed with a liter of distilled water. Further, it is diluted to 10, 20, 30, 40, 50 mg/L.

Batch Adsorption Test

Batch adsorption test was conducted as per ASTM D4646-16 standards for finding out the adsorption characteristics of the liner material. In a conical flask soil leachate mixture was taken and agitated in a mechanical shaker and then decanted. A small retention time was provided after which it is filtered through a Whatman filter paper. The filtrate is analyzed using an atomic adsorption spectrophotometer for finding out the equilibrium concentration of the liner material. Studies were conducted by varying the agitation time, concentration of the leachate solution, and pH. Equation 21.1 shows the relation between equilibrium concentration and percentage adsorbed, where q_e is the quantity adsorbed (mg/g), V is the volume of solution (ml), M is the mass of soil (g), C_i is the initial concentration (mg/L), and C_e is the equilibrium concentration (mg/L) [3, 6, 8].

$$q_e = (C_i - C_0)/M \tag{21.1}$$

$$\text{Percentage adsorbed} = ((C_i - C_e))/C_i * 100 \tag{21.2}$$

Numerical Modeling

Langmuir adsorption isotherm

Langmuir assumed that chemical adsorption does not occur beyond a single layer and hence layer adsorbed will only be one molecule thick. It is a theoretical model where the adsorption surface is plane without any corrugations and the molecules adsorbed don't interact with the adjacent sites. Equations 21.3, 21.4 show the general and linear form of the Langmuir adsorption isotherm model, where C_e represents the concentration of adsorbate at equilibrium is, q_e is the amount adsorbed at equilibrium, q_m is the maximum amount adsorbed, and b is the coefficient related to affinity [3, 4, 14]. One of the peculiarities of Langmuir adsorption isotherm is that it takes into account a special factor termed as separation factor (R_L) which defines the favorability of the adsorption. It is given by the Eq. 21.5 and if $0 < R_L < 1$ then Favorable adsorption occurs, $R_L > 1$ Unfavorable adsorption, $R_L = 1$ Linear adsorption and if $R_L = 0$ Irreversible adsorption [8].

$$C_e/q_e = 1/bq_m + C_e/q_m \tag{21.3}$$

$$q_e/q_m = (bC_e)/(1 + bC_e) \tag{21.4}$$

$$R_L = 1/(1 + (bC_o) \tag{21.5}$$

Freundlich Adsorption Isotherm Model

It is an empirical model that assumes adsorption occurs on to a rough heterogeneous surface. It is extended to multilayer adsorption where non-uniform heat of adsorption and adsorbate affinity is considered. This model lacks a strong thermodynamic mechanism. Eqs. 21.6 and 21.7 represent the general and linear form of Freundlich adsorption isotherm model respectively, where C_e is the concentration of adsorbate at equilibrium, q_e is the amount adsorbed at equilibrium, K_f is the intercept, and $1/n$ is the slope of the isotherm plotted [3, 4].

$$q_e = K_f C_e^{(1/n)} = K_f C_e \wedge (1/n) \tag{21.6}$$

$$\ln q_e = \ln K_f + 1/n \ln C_e \tag{21.7}$$

Temkin Adsorption Isotherm

One of the peculiarities of the Temkin adsorption isotherm is that it explicitly takes into account a factor that represents adsorbent–adsorbate interaction. It assumes a logarithmic relationship, and its derivation is put forward with an assumption of uniform distribution of binding energy. Equations 21.8 and 21.9 represent the general and linear form of Temkin adsorption isotherm where A_T is the Temkin isotherm equilibrium binding constant (L/g), b is the Temkin isotherm constant, R is the Universal gas constant (8.314 J/mol/K), T is the standard temperature (298 K), and RT/b is the constant related to heat of sorption [4].

$$q_e = \frac{RT}{b} \ln(A_T C_e) \tag{21.8}$$

$$q_e = \frac{\mathrm{RT}}{b} \ln A_T + \left(\frac{\mathrm{RT}}{b}\right) \ln C_e \tag{21.9}$$

$$\ln q_e = \ln K_f + \frac{1}{n} \ln C_e \tag{21.10}$$

Dubinin–Radushkevich Isotherm Model

Adsorption mechanism is expressed as Gaussian energy distribution over a heterogeneous surface. One of the main features of this model is that it is temperature-dependent. Equations 21.11 and 21.12 represent the general and linear form of D-R model, where q_s is theoretical isotherm saturation constant (mg/g), K_{ad} K_{ad} is the *D-R* isotherm constant ($\mathrm{mol_2/kJ^2}$), ϵ is the *D-R* isotherm constant, and ϵ is given by the equation [4]

$$q_e = q_s \exp -(K_{ad}\epsilon^2) \tag{21.11}$$

$$\ln q_e = \ln q_s - K_{\mathrm{ad}}\epsilon^2 \tag{21.12}$$

$$\epsilon = \mathrm{RT} \ln\left(1 + \frac{1}{C_e}\right) \tag{21.13}$$

Redlich-Peterson Adsorption Isotherm Model

It is an empirical equation that incorporates three parameters and features both Langmuir and Freundlich adsorption isotherms. The model is linearly dependent on the numerator and exponentially on to the denominator to express adsorption equilibrium over a wide range of concentrations. This model can be applied on to both homogeneous and heterogeneous surfaces. Equations 21.14, 21.15 represent the general and linear form of *R-P* isotherm model, respectively, where q_e is the percentage adsorbed, C_e is the equilibrium concentration, K_{RP} is the Redlich–Peterson model isotherm constant (L/g), α_{RP} is the Redlich–Peterson model constant (L/mg), and β_{RP} is the Redlich–Peterson model exponent [4, 12].

$$q_e = \frac{K_{RP}C_e}{1 + \alpha_{RP}C_e^{\beta_{RP}}} \tag{21.14}$$

$$\ln\left(K_{RP}\frac{C_e}{q_e} - 1\right) = \beta_{RP}\ln(C_e) + \ln\alpha_{RP} \tag{21.15}$$

Column Test

Laboratory column test is performed according to ASTM D4874:1995:R2001. A 30 cm long column is having a cross-sectional area of 10 cm^2, and it is made up of transparent fiber glass. Column is checked for leaks, and then soil is filled up to 10 cm height which is compacted in 5 layers to get the density similar to that at the field. Then it is saturated using distilled water, and later leachate solution is allowed to permeate maintaining constant head. The leached samples are collected periodically to allow continuous flow, and the collected leached samples are tested to obtain the equilibrium concentration. Test is repeated for different leachate concentration.

Results and Discussion

Determination of Optimum Bentonite Content

Compaction Curve for Various Bentonite Percentage

It is observed that with increase in the bentonite percentage from 0 to 7% there is a decrease in the maximum dry density and an increase in the optimum moisture content. This result may be attributed to the high activity of the bentonite which allows more water molecules to get adhere onto the clay surface, thereby reducing the maximum dry density. Compaction characteristics of the soil at different bentonite percentage are shown in Fig. 21.1.

Variation in The Hydraulic Conductivity

Hydraulic conductivity showed a decrease with the addition bentonite content. It is observed that when the bentonite content was 5% the material satisfies a liner criterion of $k < 10^{-7}$ cm/s, and for the same reason, it was chosen as the optimum bentonite content.

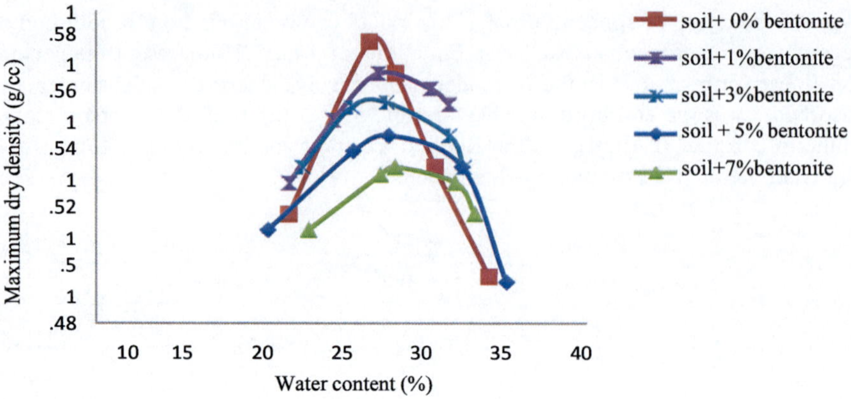

Fig. 21.1 Compaction curve for various bentonite percentage

Numerical Modeling

Batch adsorption test results are used for the identification of equilibrium concentration and percentage adsorbed using which different adsorption isotherm models are plotted. Regression analysis is carried out to obtain a best-fit model and also the maximum quantity that can be adsorbed by the liner is calculated.

Langmuir Adsorption Isotherm

When the experimental data was plotted according to Langmuir adsorption isotherm equation, it is observed that isotherm has got a distribution coefficient value (R^2) of 0.95. The maximum quantity that can be adsorbed (q_m) is obtained as 0.36 mg/g. Figure 21.1 shows the adsorption isotherm plotted. One of the other features of this Langmuir isotherm is a factor termed as separation factor (RL). It is seen from Table 21.5 that RL value for all the cases is below 1 showing that the adsorption is favorable.

Freundlich Adsorption Isotherm

The Freundlich adsorption isotherm plotted is shown in Fig. 21.2, from which it is clear that the R^2 value for this case is 0.965. Value of an adsorption parameter K_f is 0.63mg/g which gives an approximation of the maximum adsorbed quantity.

Table 21.5 Separation factor for different initial concentration

Initial concentration (C_0)	Value
10	0.0185
20	0.0094
30	0.0063
40	0.0047
50	0.0037

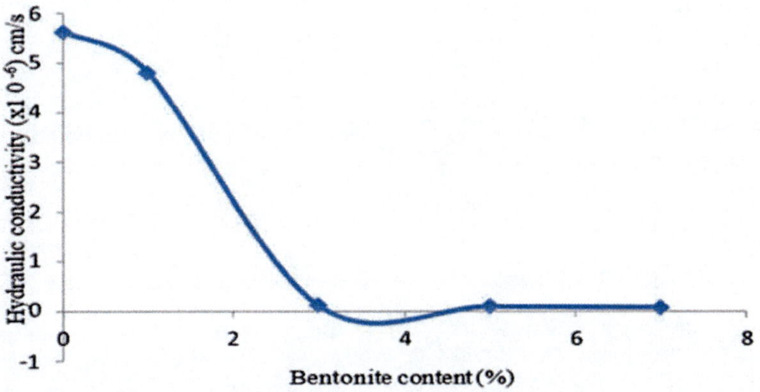

Fig. 21.2 Variation of hydraulic conductivity with different bentonite percentage

Temkin Adsorption Isotherm

The R^2 value obtained is 0.92 which is clear from the graph shown in Fig. 21.3.

Dubinin–Radushkevich Adsorption Isotherm

The R^2 value that is obtained is 0.77 showing a less fit. Figure 21.4 represents the *D-R* adsorption isotherm. The maximum quantity that can be adsorbed is obtained as 0.23mg/g (Figs. 21.5 and 21.6).

Out of the four, two parametric models considered Freundlich adsorption isotherm better adhere to the experimental data than the other three. The adsorption capacity for the liner material was observed as 0.63 mg/L.

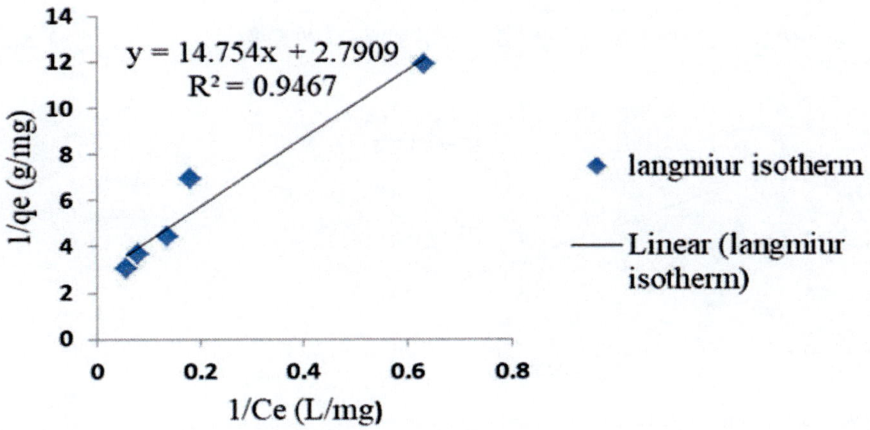

Fig. 21.3 Langmuir adsorption isotherm

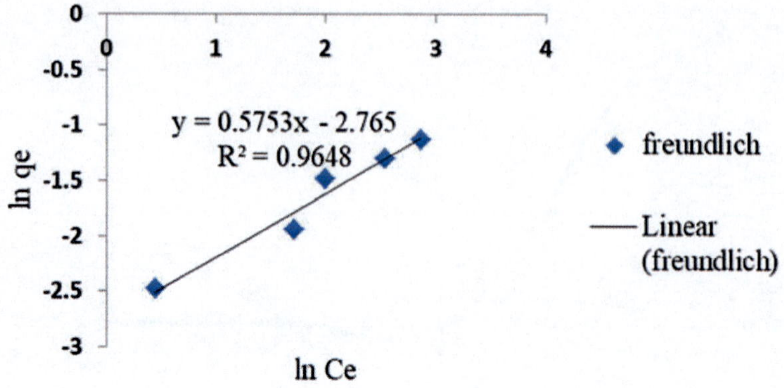

Fig. 21.4 Freundlich adsorption isotherm

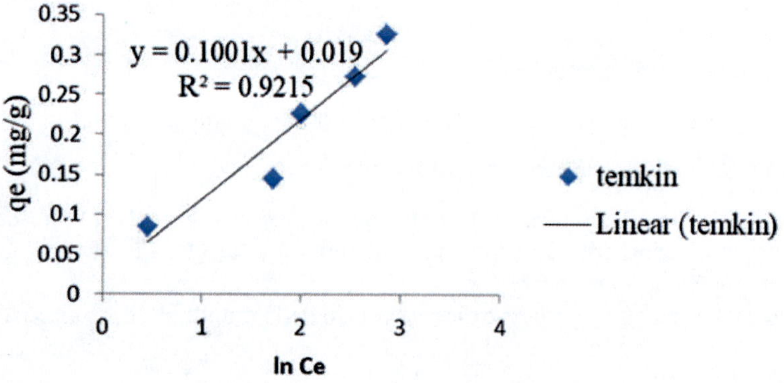

Fig. 21.5 Temkin adsorption isotherm

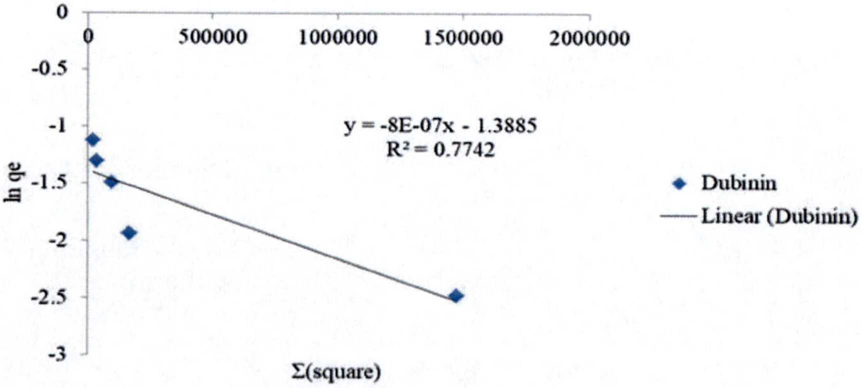

Fig. 21.6 Dubinin–Radushkevich adsorption isotherm

Redlich–Peterson Model

Redlich–Peterson model is three parametric models, both linear and nonlinear forms of the adsorption isotherms are plotted in the graphs as shown in Figs. 21.7 and 21.8 respectively. Linear *R-P* model has got an R^2 value of 0.94, and its nonlinear form has an R^2 value of 0.97 approximately. Table 6 shows the adsorption parameters for the nonlinear case. The maximum adsorption quantity obtained from the model is 0.64 mg/g.

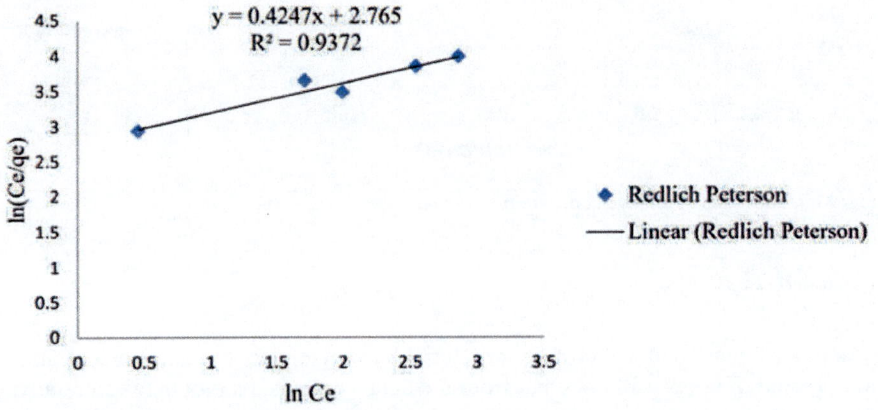

Fig. 21.7 Redlich–Peterson adsorption isotherm (linear form)

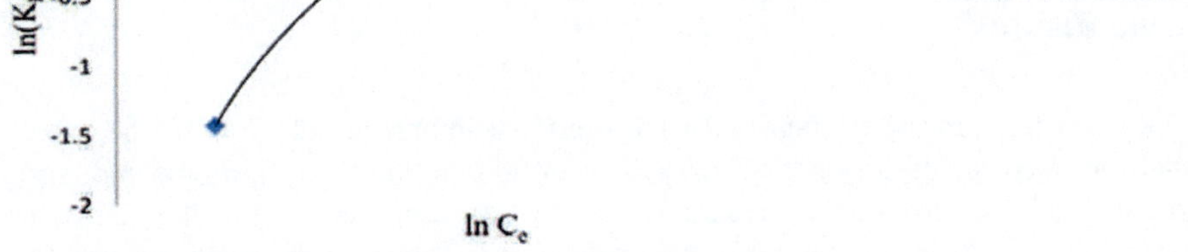

Fig. 21.8 Redlich–Peterson adsorption isotherm (nonlinear form)

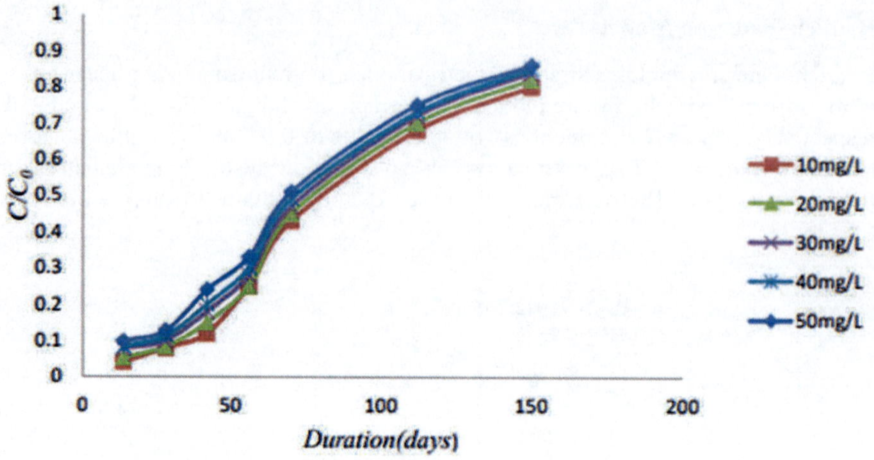

Fig. 21.9 Variation in relative concentration with duration

Column Test

Column test was conducted for liner material by varying the leachate concentration for a period of about 150 days and from the data's collected a plot between relative concentration and duration was plotted, which is shown in Fig. 21.9. It is observed that the relative concentration increases with duration, and it shows slight increase with concentration. From this, it is clear that the diffusion of leachate is not much dependent on concentration. Figure 21.10 shows the variation of retardation factor with the leachate concentration. It is observed that the retardation factor decreases with leachate concentration; it is attributed to the reduction in the diffuse double layer thickness which increases with concentration, thereby allowing less restriction in flow.

Conclusions

Optimum bentonite percentage that is required for a liner material is 5%. It is observed that the max dry density decreases with bentonite content and optimum moisture content showed an increase. This is due to the increased activity of the bentonite which allows a higher amount of water molecules to adhere on to its surface, thereby increasing the OMC and decreasing the maximum dry density. Numerical model studies revealed that out of the considered models Redlich–Peterson nonlinear model better fits the experimental data than any other model with an R^2 value greater than 0.96 and Freundlich adsorption isotherm also showed a similar fit with the R^2 value 0.96. Out of the two parametric models, Freundlich model showed a greater fit.

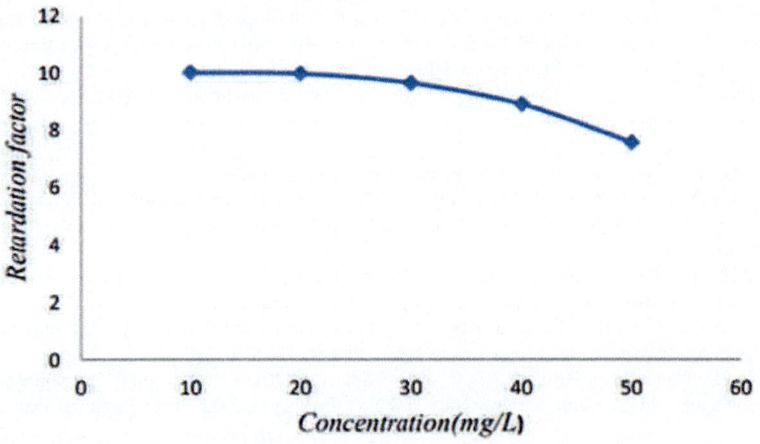

Fig. 21.10 Variation in the retardation factor with concentration

The maximum adsorbed quantity that is calculated is about 0.64 mg/g. The term separation factor in the Langmuir adsorption isotherm model reveals that adsorption is a favorable process in the considered material. Column test results show that the relative concentration of lead ions in the leachate increases with duration, and also, it is observed that the retardation factor showed a decrease with increase in concentration. Retardation factor gives the resistance to migration here since it is lower the resistance is less, it is because of the shrinkage in the diffuse double layer thickness which allows more void spaces and cracks in between and that leads to less resistance in flow.

References

1. Amadi AA, Osinubi KJ (2017) Transport parameters of lead(Pb) ions migrating through saturated lateritic soil bentonite column. Int J Geotech Eng 1–7
2. Chalermyanont T, Arrykul S, Charoenthaisong N (2008) Transport of heavy metals and chemical compactibility of hydraulic conductivity of a compacted sand bentonite mixture. Songklanakarin J Sci Technol 269–276
3. Desta MH (2013) Batch sorption experiments: lagmiur and freudlich isotherm studies for the adsorption of textile metal ions onto teff straw. J Thermodynamics 1–5
4. Foo KY, Hameed BH (2010) Insight into the modeling of adsorption isotherm systems. Elsevier 2–10
5. Li L, Lin C, Zhang Z (2016) Utilization of shale clay mixtures as a landfill liner material to retain heavy metals. Mater Des 1–10
6. Lu HJ, Laun MT, Zhang JL (2011) Study on transport of Cr (VI) through the landfill liner composed of two-layer soils. Elsevier 266:87–92
7. Lu H, Luan MT, Zhang JL (2010) Transport of Cr(VI) through clay liners containing activated carbon or acid-activated bentonite. Appl Clay Sci 99–105

8. Nethaji S, Sivasamy A, Mandal AB (2013) Adsorption isotherms, kinetics and mechanism for the adsorption of cationic and anionic dyes onto carbonaceous particles prepared from Junglans regia shell biomass. Int J Environ Sci Technol 231–242
9. Shan HY, Lai YJ (2002) Effects of hydrating liquid on the hydraulic properties of geosynthetic clay liners. Elsevier, pp 19–38
10. Sobti J, Singh SK (2017) Sorption behavior of heavy metals in sand bentonite coal ash mixes for use as a liner material in landfills. Int J Geotech Eng 1–14
11. Vijayaraghavan K, Padmesh TVN, Palanivelu K, Velan M (2006) Biosorption of nickel (II) ions onto Sargassum wightii: Application of two parameter and three parameter isotherm models. J Hazard Mater 304–308
12. Wu FC, Liu BL, Wu KT, Tseng RL (2010) A new linear form analysis of Redlich Peterson isotherm equation for the adsorption of dyes. Chem Eng J 21–27
13. Zaki AA, Ahmed KM, Rahman AE (2016) Sorption characteristic of a landfill clay soil as a retardation barrier of some heavy metals. Elsevier pp 1–18
14. Zito PF, Caravella A, Brunetti A, Drioli E, Barberi G (2015) Estimation of langmiur and sips models adsorption parameters for NaX and NaY FAU zeolites. J Chem Engineer Data

Chapter 22
Correlation of Rheological Properties with Slump and Undrained Shear Strength of Bauxite Residue

Shivesh Kumar and B. Hanumantha Rao

Introduction

With an increased number of tailings dams failures such as gold tailings dam in Russia in 2018 [1], Hindalco red mud dam in India in 2019 [2] and many more, across the world has resulted in enormously increased number of casualties across the world. Upon failure, these dams release a huge volume of debris, thus impacting several kilometers of cultivable land and causing huge economical loss to the country. The mobility of such debris can better be understood by studying the rheological properties such as yield stress and viscosity [3]. In general, two types of yield stresses are defined for any debris flow such as static and dynamic yield stress. The former is important for initiating the debris flow, while the latter represents the stopping of debris flow. These rheological properties are highly decisive in determining the flow characteristics of debris which is critical for estimating the frontal velocity, runout distance, impact pressure, etc., of the failed material and are influenced by numerous factors like soil type, moisture content, solids concentration, pH, mineralogy, shear rate history, temperature, and salinity [4]. Concerted efforts are made to understand the rheological properties of various materials. However, majority of the studies have focused mainly on assessing the dependency of rheological properties with moisture content [5], different types of soil mixtures [6], solids concentration [7], etc. Apart from this, limited research work is devoted on correlations between rheological properties and Atterberg's limit, as highlighted by few researchers [8].

In several circumstances, fixed yield stress parameter is defined during the movement, as it is the first conjecture to elucidate the post-failure behavior of the debris flow simulation. In this scenario, the yield stress determination is highly crucial.

S. Kumar · B. Hanumantha Rao (✉)
School of Infrastructure, IIT Bhubaneswar, Bhubaneswar, Odisha 752050, India
e-mail: bhrao@iitbbs.ac.in

S. Kumar
e-mail: sk80@iitbbs.ac.in

K. Muthukkumaran et al. (eds.), *Transportation and Environmental Geotechnics*,
Lecture Notes in Civil Engineering 298, https://doi.org/10.1007/978-981-19-6774-0_22

According to Jeong [9], the shear strength obtained from the vane or fall cone test would be more authentic for debris flow modeling. Therefore, in the present study, efforts are devoted toward finding the undrained shear strength and establishing its relevance with dynamic and static yield stress of the bauxite residue.

Materials and Methodology

Bauxite Residue (BR): Disturbed samples of bauxite residue collected from Vedanta Aluminium Limited, Kalahandi, Odisha, were used for detailed experimental investigations. The specific gravity, liquid limit, plastic limit, and shrinkage limit was measured as 3.10, 38, 29, and 25.5%, respectively. From particle size distribution, sand, silt, and clay size fractions are obtained as 18, 56, and 26%, respectively.

Measurement of Rheological Properties and Undrained Shear Strength: A 50 gm of oven-dry BR sample is mixed with water to attain a consistent slurry. The moisture content is varied from 38 to 48% with an increment of 2%, as can be seen in Table 22.1. Measurements of dynamic yield stress have been done with the help of dynamic shear rheometer (DSR) having parallel plate geometry configuration (diameter: 25 mm, gap: 0.75 mm), and the sample sandwiched between the plate was allowed to shear for 120 s. Moreover, static yield stress is also measured using DSR at different constant shear rates (0.001, 0.01, 0.1, 1, and 10 s^{-1}) allowed to shear for 300 s. The undrained shear strength of the soil is determined using vane shear apparatus for the moisture content varying from 38 to 48%. In all the tests, time lapse of 5 min as allowed between specimen mixing and beginning of test to ensure homogeneous mixing.

Measurement of Yield Stress from the Slump: The yield stress was also measured using cylindrical slump mold having aspect ratio equal to 1. In this test, 500 g of oven-dry BR sample was taken and mixed with water so as to maintain different moisture contents, as aforementioned. The same sample characteristics were used in this test as tabulated in Table 22.1. In order to avoid anomalies in using different cylinder heights, the results are presented in the form of dimensionless quantities.

Table 22.1 Sample characteristics of bauxite residue tailings

Red mud (g)	Moisture content (%)	Solids concentration (%)	Slurry density of tailings gradation (kg/m^3)
50	38	72.46	1935
	40	71.43	1910
	42	70.42	1886
	44	69.44	1863
	46	68.49	1842
	48	67.57	1821

The slump is measured from the top of the cylinder to the top of the slumped material with an accuracy of 0.5 mm.

Results and Discussion

Yield stress is determined by extrapolating the flow curve to the zero shear rate. Various rheological models have been used for this purpose such as the Bingham model and Herschel–Bulkley model. Due to the limitation in using the Bingham model in a low shear rate region as noticed by [10], Herschel–Bulkley model (shown in Eq. 22.1) is used in this study for determining the yield stress.

$$\tau = \tau_0 + K\gamma^n \tag{22.1}$$

where τ_0 is the yield stress, γ is the applied shear rate, K is the consistency index related to viscosity, and n is called as non-dimensional flow index. The above-mentioned equation (Eq. 22.1) is fitted to the experimental data according to the theory mentioned in [11]. After fitting the curve, it is found that the yield stress at 44% moisture content is 0.031 kPa and the value of flow index (n) is <1. This indicated the shear thinning behavior of the bauxite residue at all the concerned moisture contents. Also, static yield stress is denoted by the peak value of the curve obtained at different shear rates in the shearing stress versus shearing time graph as shown in Fig. 22.1. This indicates the complete cessation of the fluid structure. It can also be observed that at higher shear rate, the shearing time to attain the peak is very less in comparison with the low shear rate but the peak stress is higher than that obtained for low shear rate. Similar behavior was observed for other shear rates. The obtained result is in line with the [12] reported for other materials.

Figure 22.2 shows the variation of undrained shear strength, dynamic yield stress, and dimensionless slump height with the moisture content of BR. On comparison of the undrained strength with yield stress from Fig. 22.2, it is observed that the undrained shear strength is higher than the dynamic yield stress for a particular moisture content. For example, at a moisture content of 44%, the undrained shear strength and the yield stress are approximately 0.84 and 0.031 kPa. Whereas the undrained shear strength is comparable to the static yield stress (i.e., 0.783 kPa corresponding to shear rate of 10 s^{-1} obtained from Fig. 22.1). In the same line, [9, 13] reported an approximately tenfold increase of the static yield stress in comparison with undrained shear strength measured by fall cone test measured at 50% solids concentration. It is also reported that for fine-grained sediments, the yield strength is very close to the undrained shear strength. Further from Fig. 22.2, it can be seen that the undrained strength is inversely proportional to the slump value such as with an increase in the dimensionless slump height, the undrained strength decreased, and vice-versa.

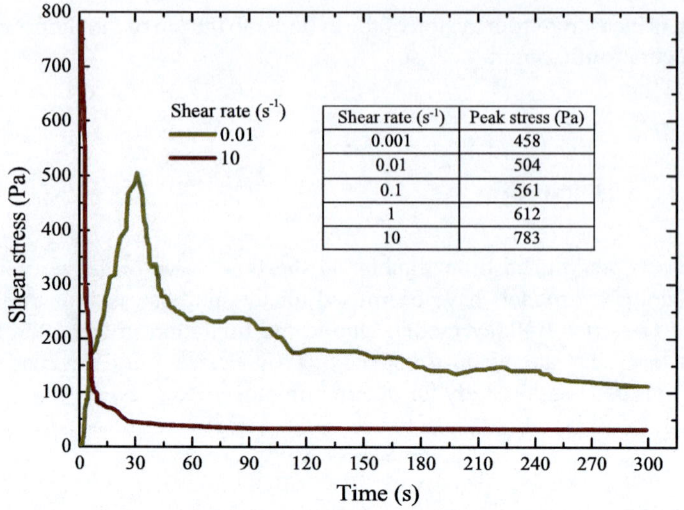

Shear rate (s^{-1})	Peak stress (Pa)
0.001	458
0.01	504
0.1	561
1	612
10	783

Fig. 22.1 Shear stress versus time response at 44% moisture content

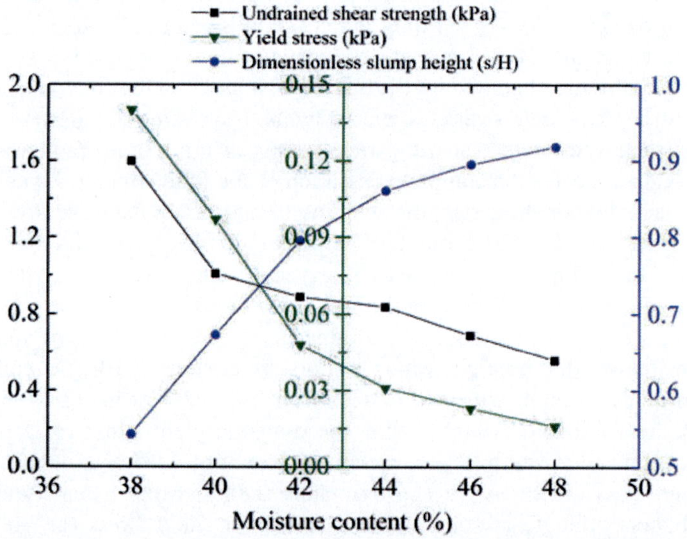

Fig. 22.2 Variation of undrained strength, yield stress, and dimensionless slump height with moisture content

Conclusions

In the present study, dynamic and static yield stress along with the undrained shear strength is measured using rheometer, slump, and vane shear tests. Based on the thorough experimental study, the following conclusions have been drawn.

- On increasing the moisture content, the yield stress of the bauxite residue is observed to be decreasing. The decrement rate is higher during the initial stage which gets reduced on further addition of moisture.
- With an increase in the moisture content, the dimensionless slump height is found to be increasing continuously.
- The undrained shear strength is higher than the dynamic yield stress at all the moisture contents, whereas it is approximately equal to the static yield stress measured at 44% moisture content.
- Moreover, the undrained shear strength is found to vary inversely with the dimensionless slump height.

The above-observed conclusions will help in the estimation of debris flow mobility. The decrease in yield stress on increasing the moisture content may be taken as indicative that the runout distance will be higher upon breaking of tailings dams. This provides an idea about the type of material to be stored in the dams; i.e., the material should have high yield stress (like that of highly concentrated material) which will enhance the storage and safety facility. The knowledge of undrained shear strength will help in the modeling of debris flow initiation and can be more reliable than yield stress for understanding post-failure characteristics.

References

1. Glotov VE, Chlachula J, Glotova LP, Little E (2018) Causes and environmental impact of the gold-tailings dam failure at Karamken, the Russian Far East. Eng Geol 245:236–247
2. Reddy PS, Reddy NG, Serjun VZ, Mohanty B, Das SK, Reddy KR, Rao BH (2020) Properties and assessment of applications of red mud (bauxite residue): current status and research needs. Waste Biomass Valorization 12:1185–1217
3. Wu H, He N, Zhang X (2015) Numerical model of viscous debris flows with depth-dependent yield strength. J GeoEng 10(1):1–10
4. Jeong SW, Locat J, Leroueil S, Malet JP (2010) Rheological properties of fine-grained sediment: the roles of texture and mineralogy. Can Geotech J 47(10):1085–1100
5. Coussot P, Laigle D, Arattano M, Deganutti A, Marchi L (1998) Direct determination of rheological characteristics of debris flow. J Hydraul Eng 124(8):865–868
6. Coussot P, Piau JM (1995) The effects of an addition of force-free particles on the rheological properties of fine suspensions. Can Geotech J 32(2):263–270
7. Nguyen VBQ, Kang HS, Kim YT (2018) Effect of clay fraction and water content on rheological properties of sand-clay mixtures. Environ Earth Sci 77(16):576
8. Cruden DM, Varnes DJ (1996) Landslide types and processes. Transportation Research Board, US National Academy of Sciences. Special report, vol 247, pp 36–75
9. Jeong SW (2019) Shear rate-dependent rheological properties of mine tailings: determination of dynamic and static yield stresses. Appl Sci 9(22):4744

10. Coussot P, Piau JM (1995) A large-scale field coaxial cylinder rheometer for the study of the rheology of natural coarse suspensions. J Rheol 39(1):105–124
11. Yang WY, Yu GL, WANG HK (2014) Rheological properties of dense natural cohesive sediments subject to shear loadings. Int J Sed Res 29(4):454–470
12. Sosio R Crosta GB (2009) Rheology of concentrated granular suspensions and possible implications for debris flow modeling. Water Resour Res 45(3)
13. Jeong SW (2015) Geotechnical and rheological characteristics of waste rock deposits influencing potential debris flow occurrence at the abandoned Imgi Mine, Korea. Environ Earth Sci 73(12):8299–8310

Chapter 23
Utilization of Red Mud (Bauxite Residue) in Environmental and Geotechnical Engineering Applications—An Overview

Aruri Vidya, Akella S. Poojitha, Narala Gangadhara Reddy, and Kundan Samal

Introduction

Bauxite residue, well known as red mud, a by-product of alumina industries, bauxite is used as a chief ore. Nearly 95% of the alumina is produced from Bayer's process, and only the remaining 5% is from sintering or combined process [1, 2]. Per every 1 ton of aluminum, around 1–1.5 tons of red mud is generated by consuming 2–3 tons of bauxite [2–4]. Table 23.1 shows the chemical properties of red mud and their percentage in various processes of production. The alkaline nature (high pH) of red mud and the presence of potentially toxic materials create serious environmental concern if disposed of without treatment [5–7]. Moreover, the red mud generated is mostly in a slurry-based form. It consumes huge land arising storage issues and, due to its high dispersive nature (when dried), leads to severe air pollution, negatively impacting terrestrial organisms [4, 8]. Recycling and neutralizing red mud is immensely important to avoid such environmental issues. This paper gives an overall view and possibilities of utilizing red mud in various environmental and geotechnical applications. The use of red mud in various applications through literature review and its application as an amendment, adsorbent, catalyst, and coagulant to remove pollutant/contaminants in soil or water are highlighted.

A. Vidya · A. S. Poojitha · N. G. Reddy (✉)
Department of Civil Engineering, Kakatiya Institute of Technology and Science, Warangal, India
e-mail: gn11@iitbbs.ac.in

K. Samal
School of Civil Engineering, KIIT Deemed to Be University, Bhubaneswar, India

K. Muthukkumaran et al. (eds.), *Transportation and Environmental Geotechnics*,
Lecture Notes in Civil Engineering 298, https://doi.org/10.1007/978-981-19-6774-0_23

Table 23.1 Chemical properties of red mud and their percentages in various processes of production [9–15]

Chemical constituents	Bayer's process (wt. %)	Sintering process (wt. %)	Bayer's sintering process (wt. %)
Fe_2O_3	28.41–42.34	8.50–10.97	10.97–11.4
Al_2O_3	16.26–31.39	7.28–7.68	7.68–10.66
SiO_2	6.97–19.29	22.67–24.02	21.06–22.67
Cao	0.15–21.35	40.78–44.00	40.62–40.78
Na_2O	1.79–3.83	0.69–2.93	1.49–2.93
TiO_2	4.27–7.36	3.26–4.17	3.26
K_2O	0.05–0.52	0.25–0.38	0.38–0.45
MgO	0.29–1.75	1.37–1.77	0.93–1.77
Loss	13.88	11.77–16.96	6.86–11.77

Utilization of Red Mud in Environmental Engineering Applications

Red Mud as an Adsorbent

Red mud can be transformed to utilize it as an adsorbent to remove the common pollutants found in wastewater like metal and non-metal ions (Table 23.2). Samal et al. [16] studies showed that red mud could be used to remove lead and chromium at different industrial sites. By using batch or column operations, the ionic concentrations of Cd^{2+} (1.78×10^5–1.78×10^3) and Zn^{+2} (3.06×10^5–3.06×10^3) with adsorbent dose as 5–20 g/L and pH of 1–6 are reduced, and around 100% of adsorption is done at lower concentrations and 60–65% is at higher concentrations (Table 23.3).

Few highly carcinogenic-phenolic chemicals like 4-chlorophenol and 2, 4 dichlorophenol are adsorbed up to 94–97% and 2-chlorophenol, phenol by 50–81% by red mud [8].

Another study by Li et al. [17] to remove pollutants (anionic micro and macro) like dichlorofenac (DCF) and phosphorus using RM-Ppy (RM with polypyrrole) were conducted successfully. Since Ppy is being non-toxic in nature and cost-efficient, it is made in contact with RM to generate positively charged functional groups like nitrogen in polymer matrix form so that anionic pollutants like DCF and phosphorous are adsorbed through electrostatic interaction.

Figure 23.1 shows that phosphorus has a lower adsorption capacity than DCF due to the existence of different anionic forms of phosphorous where ligand exchanges affect the adsorption process. Wang et al. [8] also came up with red mud as the best solution to remove dyes like congo red, procion orange, fast green, rhodamine from wastewater and also experimented to remove acid violet dye at various concentrations (10–40 mg/L), pH (2–11) and red mud as adsorbent dosages (0.05–1.50 g/50 ml).

Table 23.2 Removal of metallic and non-metallic ions by red mud

Ions	Type of red mud as adsorbent	Removal rate
Chromium (Cr)	RM activated by HCl at optimal pH RM activated by H_2O_2 at 500 °C	Chromate is 70% Cr^{6+} is 87.65%
Lead (Pb)	Mass ratio of RM, clay and coal powder is 80:20:5	20 °C—removal rate of Pb^{2+} is 60.4% 30 °C—removal rate of Pb^{2+} is 82.6% 40 °C—removal rate of Pb^{2+} is 60.4%
Cadmium (Cd)	RM activated by H_2O_2 in the pH range of 2–121	The removal efficiency of Cd^{2+} ions is 95.32%
Fluorine (F)	Activated RM by HCl	F removal rate is 98%
Arsenic (As)	Activated seawater neutralized RM	pH—4.5 removal efficiency of As^{5+} is 100% pH—8.5 As^{3+} can be removed efficiently
Phosphorous (P)	Activated RM	P removal rate is 74.1%

Table 23.3 Adsorption capacity of RM for various phenolic pollutants and dyes in aqueous solutions [8]

	Adsorbent	Pollutant	Amount adsorbed (mol/g)
Phenolic pollutants	RM	2,4 Dichlorophenol	0.80–0.85
	RM	4—Chlorophenol	0.78–0.82
	RM	2—Chlorophenol	0.72–0.79
	RM	Phenol	0.63–0.74
	Neutralized RM	Phenol	2.5×10^{-5}
	Acid RM	Phenol	2.98×10^{-5}
Dyes from water	RM	Acid violet	2.42×10^{-6}
	RM	Fast green	$(7.25–9.35) \times 10^{-6}$
	RM	Rhodamine B	$(1.01–1.16) \times 10^{-5}$
	Acid ARM	Congo red	5.81×10^{-6}
	RM	Methylene blue	$(4.35–5.23) \times 10^{-5}$

Red Mud as a Catalyst

Due to the high amount of ferric oxide (Fe_2O_3–39.89%), high specific surface area (40–70 m^2/g), and low capital cost, Wang et al. [8] considered red mud to be effectively used as a catalyst for many reactions so that pollutants get removed. Red mud can recover oil from PVC and fuel gas from waste oil mixtures.

Red mud can be used as a potential catalyst in process like hydro dechlorination, hydrocarbon oxidation and hydrogenation (Table 23.4).

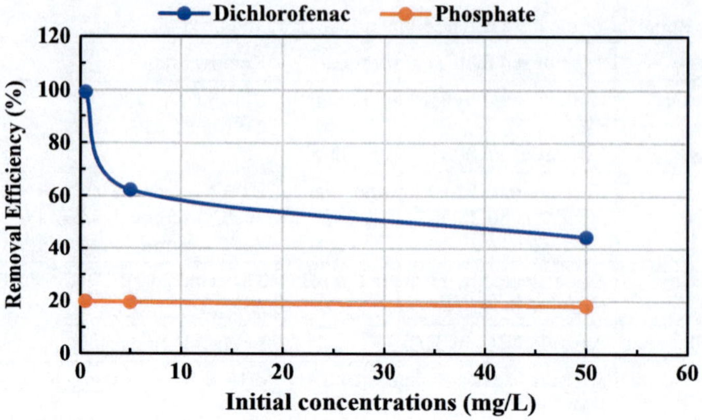

Fig. 23.1 Adsorption efficiencies of both DCF and phosphorus [17]

Table 23.4 Red mud as a catalyst for treating various reactions [16]

Application	Catalyst used
Coal hydrogenation	Untreated RM
Coal liquefaction	Sulfur promoted RM
Biomass liquefaction	Sulfur promoted RM
Naphthalene hydrogenation	Activated RM
Methane combustion	Activated RM
SO_2 reduction	Untreated RM
Nitric oxide oxidation	Cu impregnated RM

Red Mud as a Coagulant

Wang et al. [18] stated that one of the best ways to treat wastewater is coagulation using red mud due to high amounts of Al and Fe in it. To produce a coagulant, red mud is dissolved partially with H_2SO_4 (30 wt.%), and the liquid is separated by filtering and then neutralized with a base. 1 g of coagulant has a capacity to sorb 95.0 mg of Zn^{2+}, 98.7 mg of Pb^{2+}, 99.3 mg of Cu^{2+} when 100 mg of solution of the above elements is considered.

Use of Red Mud in Geotechnical Engineering Applications

Since various geotechnical properties of red mud are closely resemble that of conventional soils, specific gravity (2.5–3.2), and low amount clay size particle, red mud can be used in various geotechnical applications as potential resource material and

Table 23.5 Geotechnical properties of red mud [22, 23]

Liquid limit	Plastic limit	Plasticity index	Sand	Silt and clay size	Optimum moisture content
41	36	5	5–17	83–95	31

reduced disposal issues [19–21]. Red mud can be classified as silt of low plasticity index (Table 23.5).

Red Mud as an Amendment

Kalkan [24] investigated the use of red mud as an amendment and concluded that red mud alone and in combination with cement (cemented red mud) can be utilized in geotechnical applications as clay liners by examining properties like hydraulic conductivity, unconfined compressive strength (UCS), and swelling percentage (Table 23.6). Results revealed that compared to natural clay, samples (clay) amended with red mud and cemented red mud, decreased the hydraulic conductivity, and showed high UCS. Table 23.6 shows that addition of red mud and cemented red mud decreased the swelling compared to the natural clay sample to 18.2% and 5.4%, respectively. Sridevi et al. [22] performed experimental tests on expansive soils stabilized with red mud and lime mixes (50 wt.%) resulted an increase in UCS from 16.8 to 100.6 kPa upon 90 days of curing and cohesion reduced from 45 to 34 kPa increasing friction from 23° to 32°.

Table 23.6 Comparison of clay amended with red mud and cemented red mud on various geotechnical properties [24]

Properties	Increase or decrease	Red mud	Cemented red mud
Liquid limit (%)	Decrease	66–58	66–59
Plastic limit (%)	Increase	31–36	31–41
UCS (kPa)	Increase	120–207	120–355
Permeability (cm/s)	Decrease	1.96×10^{-7}–1.58×10^{-7}	1.96×10^{-7}–6.98×10^{-8}
Swelling percentage (%)	Decrease	23.6–18.2	23.6–5.4

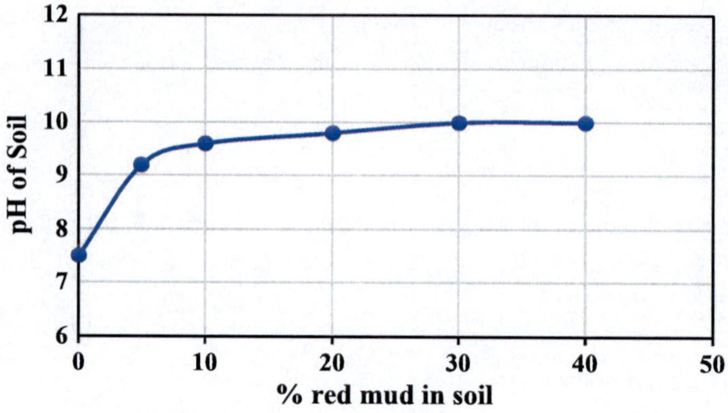

Fig. 23.2 Increase in pH as the % of red mud is increasing [25]

Red Mud as an Ameliorant

Ujaczki et al. [25] examined the properties of red mud to be reusable and recyclable material due to high pH (large sodium content). Based on the experiments, it is concluded that when 5% of red mud is mixed to the soil, the total amount of As, Cr, Pb, Ni, and Na content increases in the soil and when 20% is added, the water holding capacity of the soil is satisfactory with no toxic effect on subsoil layers. For chemically stabilizing the metal contaminated soils and acid sulfate soils, red mud can be applied effectively.

From Fig. 23.2, it can be observed that when different percentages of red mud is added to sandy soils, pH value increases till 20% red mud content. Coruh and Ergun [6] evaluated the leachability of heavy metals from zinc using Mine Waste Leaching Procedure (MWLP) and Toxicity Characteristic Leaching Procedure (TCLP) as 65–856 mg/l, and leaching can be reduced using red mud. The superalkalinity nature of red mud helps in leachate formation. Lima et al. [15] studied that phosphogypsum and NaCl solution helps in reducing both pH and zeta potential effectively, consequently enhancing the sedimentation rate of red mud. Alam et al. [3] analyzed that the leaching of Hg is controlled when red mud is stabilized with activated GGBS. A combination of iron, aluminum, and silicate present in red mud can immobilize toxic metals [26].

Red Mud as a Filler

Ujaczki et al. [25] reported that 10% of red mud can be used as a filler in bituminous mixtures and Zhang et al. [27] evaluated the radioactivity and leaching of red mud by using it as filler instead of natural mineral filler because the pore volume of this waste is 5 times higher than mineral filler. Moreover, the high specific gravity (30 m^2/g) in red mud, compared with natural mineral (<10 m^2/g), helps to increase the softening

Table 23.7 Comparison between virgin and optimum soil [28]

Experiments conducted	Virgin soil (clay)	Optimum soil (90% soil + 10% RM)
Maximum dry density (kN/m^3)	15.6	16
Optimum moisture content (%)	21.1	20.8
UCS (kPa)	286	347
Coefficient of compression	0.226	0.203

point, rutting parameter, thereby decreasing the non-recoverable creep compliance from 0.8 to 0.4. When sintered red mud is added in asphalt mastics, elastic recovery raised from 8 to 38%.

Red Mud as Geomaterial

Alam et al. [3] considered two Indian red mud samples from NALCO and HINDALCO plants and tried to improve red mud's geotechnical and durability properties by stabilizing with ground granulated blast furnace slag (GGBS) at various percentages. 25% GGBS and 1 M Na_2SiO_3 found to be optimum and observed a maximum UCS (8.2 MPa) and slake durability index (96.98–98.92%) for both the samples. These properties are comparable to the weathered basalt and better than sandstone. Based on these observations, it is clear that curing plays an important role in the improvement of UCS and durability. Parik and Patra [28] added oven dried red mud to the highly compressive clay soil considering proportions from 5 to 20%, and Table 23.7 shows the results of various tests conducted on both soils. Parik and Patra [28] studied that the water adsorption capacity and swelling potentials are reduced, thereby increasing soil strength due to the cation exchange process giving scope to use it as a stabilizing soil in pavements because of better shear modulus and low degradation index.

Conclusions

Due to smaller grain size and high alkaline content, disposal of red mud on land causes dust generation, and the caustic liquor contaminates the groundwater. So, instead of dumping the red mud, effective utilization in environmental and geotechnical engineering applications will reduce the environmental hazards. Red mud can be used as an adsorbent, catalyst, and coagulant to treat industrial wastewater, where it efficiently helps in removing metals ions like copper, chromium, lead, phosphorus, fluorine, phenols, and dyes that are mainly responsible for the pollution of water. Red mud can also be used to remove toxic gases like SO_2 and CO_2 due to possession of small particle size, high specific area, and chemical reactivity. Moreover, red mud has numerous geotechnical applications such as filler, amendment, ameliorant, and stabilizing agent. The utilization of red mud for various activities is still on research, and hence, the utilization is highly recommended to protect the environment.

References

1. Dentoni V, Grosso B, Massacci G (2014) Environmental sustainability of the alumina industry in Western Europe. Sustainability 6:9477–9493. https://doi.org/10.3390/su6129477
2. Mukizab E, Zhangb LL, Liua X, Zhang N (2019) Utilization of red mud in road base and subgrade materials: a review. Resour Conservation 141 187–199
3. Alam S, Das SK, Rao BH (2019) Strength and durability characteristic of alkali activated GGBS stabilised red mud as geo-material. Constr Build Mater 211:932–942
4. Reddy PS, Reddy NG, Serjun VZ, Mohanty B, Das SK, Reddy KR, Rao BH (2021) Properties and assessment of applications of red mud (bauxite residue): current status and research needs. Waste and Biomass Valorisation. 12:1185–1217. https://doi.org/10.1007/s12649-020-01089-z
5. Mishra MC, Reddy NG, Rao HB, Das SK (2020) A study on evaluating the usefulness and applicability of additives for neutralising extremely alkaline red mud waste. Lect Notes Civ Eng 89:139–149. https://doi.org/10.1007/978-3-030-51350-4_16
6. Coruh S, Ergun ON (2010) Use of fly ash, phosphogypsum and red mud as a liner material for the disposal of hazardous zinc leach residue waste. J Hazard Mater 173:468–473
7. Pappu A, Saxenaa M, Shyam R. Asolekar (2007) Solid wastes generation in India and their recycling potential in building materials. Build Environ 42:2311–2320
8. Wang L, Hu G, Lyu F, Yue T, Tang H, Han H, Yang Y, Liu R, Sun W (2019) Application of red mud in wastewater treatment. Minerals 9:281
9. Feng Y Yang C (2018) Analysis on physical and mechanical properties of red mud materials and stockpile stability after dilatation. Adv Mater Sci Eng, Article ID 8784232. https://doi.org/10.1155/2018/8784232
10. Ghorbani A, Nazarfakhari M, Pourasad Y, Abbasi SM (2014) Removal of Pb ion from water samples using red mud (bauxite ore processing waste). E3S Web of conferences. https://doi.org/10.1051/e3sconf/20130141019
11. Yildiz A, Oztas M, Karakaya N, Evrendilek F, Akyildiz.A, Agirgan OA, Tinmaz-Kose E, Sisman CB, Kaykioglu G, Ozden-Celik S (2015) Novel uses of red mud in textile wastewater treatment, dyeing and concrete production. Environ Eng Manag J 14(5):1171–1181
12. Alkan G, Schier C, Gronen L, Stopic S, Friedrich B (2019) A review on comprehensive utilisation of red mud and prospect analysis. Minerals 9(6):362. https://doi.org/10.3390/min9060362
13. Rai S, Wasewar KL, Mukhopadhyay J, Yoo CK, Uslu H (2012) Neutralisation and utilisation of red mud for its better waste management. Arch Environ Sci 6:13-33
14. Hua Y, Heal KV, Friesl-hanl W (2017) The use of red mud as an immobiliser for metal/metalloid-contaminated soil: a review. J Hazard Mater 325:17–30. https://doi.org/10.1016/j.jhazmat.2016.11.073
15. Lima MS S, Thives LP, Haritonovs V, Bajars K (2017) Red mud application in construction industry: review of benefits and possibilities. In: IOP conference series: materials science and engineering, vol. 251
16. Samal S, Roy AK, Bandhopadhyay A (2013) A proposal for resources, utilisation and processes of red mud in India. Int J Min Process 118:43–55

17. Li X, Ji M, Nghiem LD, Zhao Y, Liu D, Yang Y, Wang Q, Trinh QT, Vo DV, Pham VQ, Tran NH (2019). A novel red mud adsorbent for phosphorus and diclofenac removal from wastewater. J Mol Liq 303:112286
18. Wang S, Ang HM, Tade MO (2008) Novel applications of red mud as coagulant, absorbent and catalyst for environmentally benigh processes. Chemosphere 72:1621–1635
19. Reddy NG, Rao BH, Reddy KR (2018) Biopolymer amendment for mitigating dispersive characteristics of red mud waste. Géotech Lett 8(3):201–207. https://doi.org/10.1680/jgele.18.00033
20. Rubinos DA, Spagnoli, Barral MT (2016) Chemical and environmental compatibility of red mud liners for hazardous waste containment. Int J Environ Sci Technol 13:773–792. https://doi.org/10.1007/s13762-015-0917-8
21. Reddy NG, Nongmaithem RS, Basu D, Rao BH (2022) Application of biopolymers for improving the strength characteristics of red mud waste. Environ Geotech 9(6):340–359.https://doi.org/10.1680/jenge.19.00018
22. Sridevi G, Sahoo S, Sen S (2019) Stabilization of expansive soil with red mud and lime. In: Ground improvement techniques and geosynthetics Lecture notes in civil engineering, vol 14, https://doi.org/10.1007/978-981-13-0559-7_29
23. Reddy NG, Rao BH (2018) Characterisation of settled particles of the red mud waste exposed to different aqueous environmental conditions. Indian Geotech J 48(3):405–419
24. Kalkan E (2006) Utilisation of red mud as a stabilisation material for the preparation of clay liners. Eng Geol 87:220–229
25. Ujaczki E, Klebercz O, Feigl V, Molnár M, Magyar Á, Uzinger N (2015) Environmental toxicity assessment of the spilled ajka red mud in soil microcosms for its potential utilisation as soil ameliorant. Period Polytech Chem Eng 59(4):253–261
26. Gadepalle VP, Ouki SK, Van Herwijnen R, Hutchings T (2007) Immobilisation of heavy metals in soil using natural waste materials for vegetation establishment on contaminated sites. Soil Sed Contam 16:233–251
27. Zhang J, Liu S, Yao Z, Wu S, Jiang H, Liang M, Qiao Y (2018) Environmental aspects and pavement properties of red mud waste as the replacement of mineral filler in asphalt mixture. Constr Build Mater 180:605–613
28. Parik P, Patra NR (2020) Static and dynamic properties of expansive soil stabilised with industrial waste. https://doi.org/10.1061/9780784482780.048

Chapter 24
Assessment of Using Soil-Sized Material from Construction and Demolition Waste as an Earthfill

Apoorva Agarwal, Manoj Datta, G. V. Ramana, Narendra Kumar Soni, and Rajiv Satyakam

Introduction

The rapid increase in infrastructure development and thus construction activities lead to construction waste production due to these activities. Demolition of a building also produces waste in large quantities. The waste arising due to construction and demolition activities is termed construction and demolition waste (C&DW). As per the Building Material and Technology Promotion Council (BMPTC) assessment in 2017, India produces around 150 million tons of C&DW annually [1]. Due to the absence of legislation and awareness among the users, this waste mostly ends up in landfills. However, there is a potential to reuse C&DW in various applications like as an aggregate for concrete production [2], as aggregate in base or sub-base [3] for roads or use as structural fill for reinforced soil structures [4].

C&DW is primarily composed of concrete, bricks, unbound aggregates, and soil [5]. However, 35–40% of Indian C&DW is majorly composed of soil and a gravel-sized fraction [6], termed soil-sized material (SSM) (−4.75 mm). Reusing C&DW in value-added applications can only be feasible if C&DW is processed, which mainly includes size reduction and fines removal [7]. Recycling of C&DW in India is increasing, with more than 15 plants operational in different parts of the country [5]. The total capacity of these recycling plants is 6500 tons per day, whereas India produces approximately 0.4 million of C&D waste daily. Recycling accounts for only 1.6% of the total C&DW generated. Also, processing of C&DW in the processing plant requires massive investment, which results in cost addition to waste material otherwise. Thus, an economical, environmental, and sustainable way out is required to reutilize as received C&DW or, it will end up in landfill eventually.

A. Agarwal (✉) · M. Datta · G. V. Ramana
Indian Institute of Technology Delhi, New Delhi, India
e-mail: apoorvaagarwal6@gmail.com

N. K. Soni · R. Satyakam
NTPC-Netra, NTPC Limited, Greater Noida, India

K. Muthukkumaran et al. (eds.), *Transportation and Environmental Geotechnics*,
Lecture Notes in Civil Engineering 298, https://doi.org/10.1007/978-981-19-6774-0_24

The fines removal is necessary for value-added applications [8]. However, applications like filing embankments of small heights or low-lying areas do not require the removal of fines. SSM could be used at such places, which will be an economical and sustainable alternative and help reduce the amount of C&DW reaching landfills. However, no study is reported in the literature where the feasibility of SSM was assessed as fill for embankments and low-lying areas.

For the study, the sample of SSM was collected from an input unit of a C&DW processing plant. Extensive geotechnical testing was carried out to assess the feasibility of SSM for filling embankments of low heights or filling low-lying areas. Geotechnical properties like grain size distribution, plasticity, specific gravity, compaction, and shear strength of SSM were assessed. The results were compared with locally available fill material Delhi silt (DS), which is generally used to fill in an embankment and fill low-lying areas.

Materials and Testing Method

Material

The sample for testing comprised of soil-sized material (SSM) and was collected from the input unit of the C&D waste processing plant located in Ghaziabad, India. It was majorly comprised of silt and fragments of bricks, concrete, and unbound aggregates. Around 500 kg of representative sample was taken from a heap of material collected seven days by the quartering and coning method for testing purposes. Locally available silt, Delhi silt (DS), was also sampled as standard material, which is generally used as fill material for embankments.

Testing Methods

Extensive geotechnical testing was carried out to understand the physical and mechanical properties of SSM, and the results were compared with DS. The testing procedure was adopted as per Indian Standard (IS) codes. The geotechnical testing included grain size distribution, specific gravity, plasticity, compaction, permeability, and shear strength parameters using the direct shear test. All the tests were carried out at least thrice to ensure repeatability. The test results were also compared with the applicable codal provisions for the fill for an embankment.

Results and Discussion

Grain Size Distribution (GSD)

Grain size distribution of SSM and DS was carried out using Indian Standard (IS) 2720 Part IV [9]. The wet sieving method was adopted for obtaining GSD as fines were sticking to the surface of SSM. Figure 24.1 shows the GSD of SSM and DS. Table 24.1 gives the % of sand, silt, and clay in SSM and DS.

From Table 24.1, it can be seen that SSM was comprised of sand and silt, whereas DS comprised majorly of silt and some clay. Thus, SSM was coarser as compared to DS. The amount of silt (< 0.075 mm) was relatively high in SSM compared with the specifications laid for structural fill for reinforced soil structures. Thus, SSM is not suitable for use as value-added applications where there is a restriction on the percent of fines.

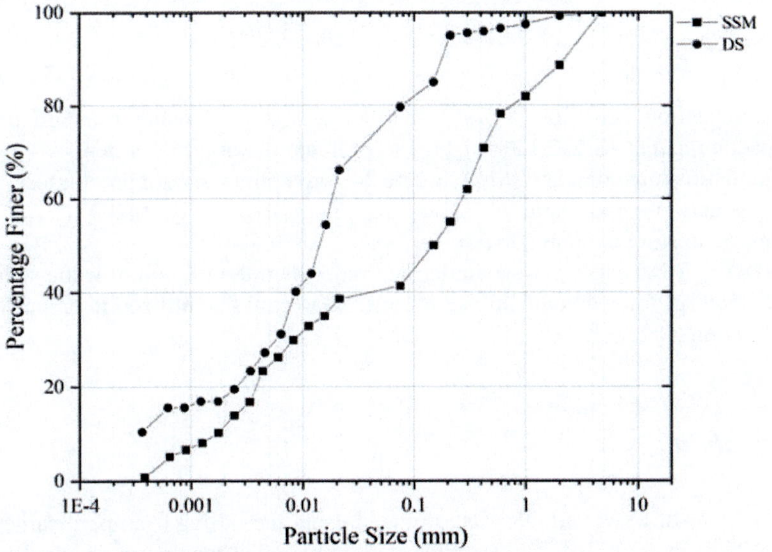

Fig. 24.1 Grain size distribution curves for SSM and DS

Table 24.1 Grain size distribution for SSM and DS

Property	Units	SSM	DS
Sand	%	59	20
Silt and clay	%	41	72
Clay	%	0	8

Table 24.2 Specific gravity and plasticity of materials

Property	Units	SSM	DS
Gs	–	2.63	2.66
PI	%	NP	7–8

Specific Gravity and Plasticity

As per IS 2720 Part III/Sec I [10], specific gravity was evaluated using a density bottle as the maximum particle size was less than 4.75 mm. Table 24.2 summarized the specific gravity of SSM and DS. Plasticity of materials was found using the procedure laid in IS 2720 Part 5 [11], and the results are tabulated in Table 24.2.

The specific gravity of SSM was less than DS due to the presence of fragments of bricks, concrete, and unbound aggregates apart from the soil. SSM contained only non-plastic fines, whereas DS contained 7–8% plastic fines.

Compaction

The compaction behavior of SSM and DS was studied using standard proctor compaction as per IS 2720 Part 7 [12] to evaluate maximum dry density (γ_{d}) and optimum moisture content (OMC). Figure 24.2 gives the standard proctor test curves for SSM and DS, and Table 24.3 gives maximum dry density (MDD) ($\gamma_{\text{d max}}$) and optimum moisture content (OMC).

MDD of SSM and DS was similar and more than the minimum value specified in IRC 36 [13] for road embankment. Thus, SSM could be utilized in embankment construction.

Permeability

Permeability of SSM and DS was estimated using the falling head permeameter as per IS 2720 Part 17 [14]. The samples were prepared at maximum dry density and optimum moisture content. The sample was kept for saturation for an extended period (around 5–7 days). The test results are reported in Table 24.4.

Permeability of SSM and DS is similar and is in the range of silt [15] and is not a free-draining material.

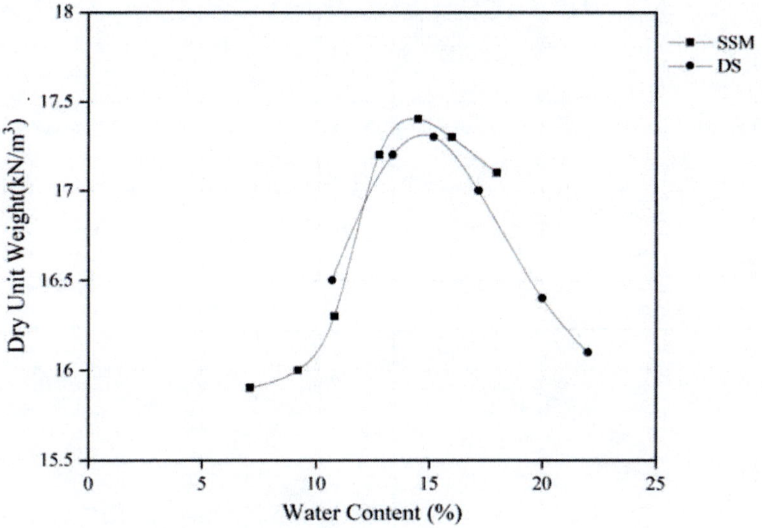

Fig. 24.2 Compaction behavior of SSM and DS

Table 24.3 Results from standard proctor compaction test

Property	Units	SSM	DS
$\gamma_{d\,max}$	kN/m^3	17.4	17.3
OMC	%	14.5	15.2

Table 24.4 Permeability of fill materials

Property	Units	SSM	DS
k (permeability)	cm/s	5×10^{-5}	7×10^{-5}

Shear Strength

The shear strength of fill materials was determined using a direct shear box as per IS 2720 Part 13 [16]. The test was carried out on a square box of a cross-sectional area of 3600 mm^2. The sample of SSM and DS was oven-dried. The sample was prepared by compacting into three layers and at 80% of maximum dry density. The height of the specimen was kept as 25 mm minimum. The test was carried out for three different normal stresses of 50, 100, and 200 kPa. The strain rate of 0.25 mm/min was adopted for testing, and the test was terminated at 15% strain.

Figure 24.3a and b gives the shear stress and volumetric change with respect to horizontal displacement for soil-sized material (SSM) and Delhi silt (DS), respectively. From shear stress and horizontal displacement graphs, it can be noticed that the initial slope and, thus, the elastic modulus of SSM was greater than DS. SSM showed strain-softening (gradual decrease in pullout resistance with increase in horizontal

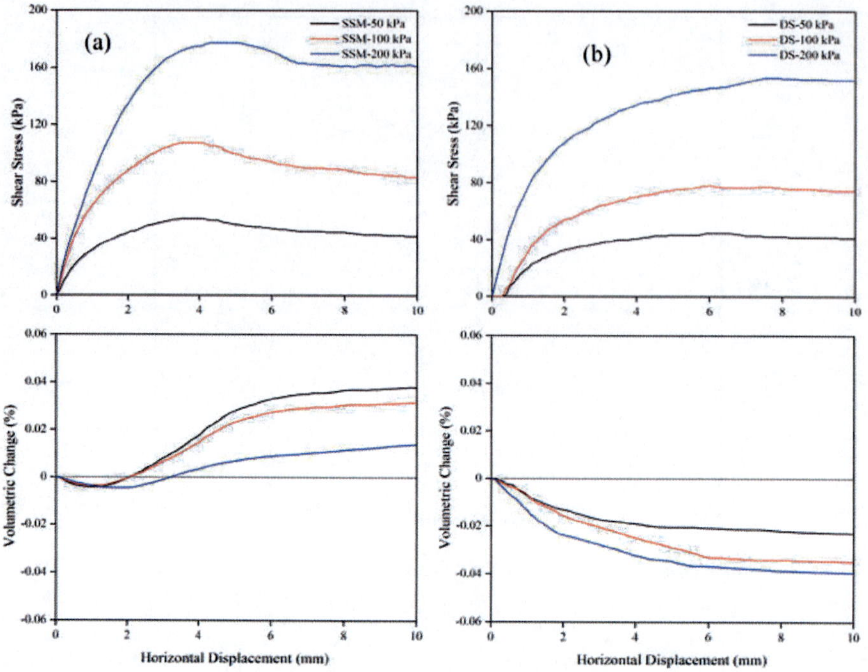

Fig. 24.3 Shear stress and volumetric change with respect to horizontal displacement for **a** soil-sized material (SSM) and **b** Delhi silt (DS)

displacement) behavior with an increase in strain, while DS showed progressive failure (no peak was observed). The horizontal displacement corresponding to peak shear stress also increased with an increase in normal stress. SSM exhibited compression initially but dilation with the increase in horizontal displacement, similar to dense soil. However, DS exhibited compression-only due to high fines.

Figure 24.4 gives the Mohr–Coulomb strength envelope for SSM and DS. From the strength envelopes, ϕ for SSM was in the order of 42–43°, whereas ϕ for DS was around 36–37°. Thus, SSM has the potential to be reused as fill material for embankments and low-lying areas.

Conclusion

From the preceding discussion on the geotechnical characteristics of SSM, the following conclusions can be drawn:

- SSM had 40–45% fines and was classified as silty sand. It was coarser as compared to locally available DS and had non-plastic fines.
- The specific gravity of SSM and DS was similar.

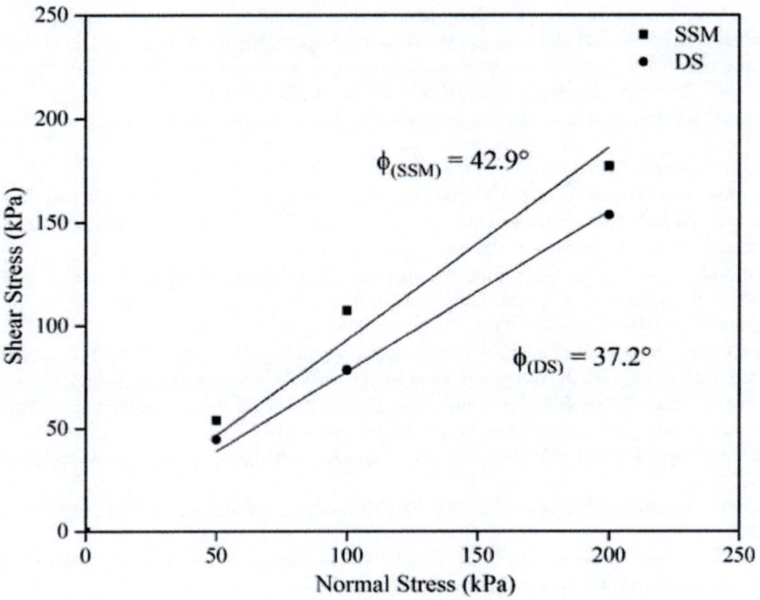

Fig. 24.4 Mohr–Coulomb strength envelop for SSM and DS

- The MDD of SSM and DS was similar and in the range of 17 kN/m^3. It was more than the minimum value specified for road embankment.
- SSM exhibited a strain-softening response, whereas DS exhibited progressive failure. Also, SSM exhibited dilation, whereas DS underwent compression.
- The shear strength of SSM was higher as compared to DS. The angle of shearing resistance for SSM was 42–43° and 36–37° for DS.

The study highlights the possible reuse of SSM of C&D waste as a replacement of natural sand in bulk geotechnical applications like earth fills. However, before using any waste material for geotechnical applications, possible environmental impact due to its usage needs to be assessed. Hence, more studies are required for assessing the leaching of soluble salts and heavy metals to ascertain the implications on the environment. With this study as the starting point, the policymakers can recommend the effective reuse of soil-sized material of C&D waste.

References

1. BMTPC (2018) Utilization of recycled produce of construction & demolition waste a ready reckoner. [Online]. Available: https://www.bmtpc.org/DataFiles/CMS/file/PDF_Files/C&D_Ready_Reckoner_BMTPC_web.pdf
2. Medina C, Zhu W, Howind T, Sánchez de Rojas MI, Frías M (2014) Influence of mixed recycled aggregate on the physical—mechanical properties of recycled concrete. J Clean Prod

68:216–225. https://doi.org/10.1016/j.jclepro.2014.01.002
3. Kumar R (2017) Influence of recycled coarse aggregate derived from construction and demolition waste (CDW) on abrasion resistance of pavement concrete. Constr Build Mater 142:248–255. https://doi.org/10.1016/j.conbuildmat.2017.03.077
4. Vieira CS, Pereira PM, Lopes ML (2016) Recycled construction and demolition wastes as filling material for geosynthetic reinforced structures. Interface properties. J Clean Prod 124:299–311. https://doi.org/10.1016/j.jclepro.2016.02.115
5. Somvanshi A, Verma A (2020) Another brick of wall: improving construction and demolition waste management in Indian Cities. Cent Sci Environ 1–148. [Online]. Available: https://www.cseindia.org/another-brick-off-the-wall-10325
6. Alsheyab MAT (2021) Recycling of construction and demolition waste and its impact on climate change and sustainable development. Int J Environ Sci Technol 0123456789. https://doi.org/10.1007/s13762-021-03217-1
7. Agarwal A, Datta M, Ramana GV, Soni NK, Satyakam R (2021) Geotechnical characterization of recycled aggregates (RA) comprising of mixed waste from construction & demolition (C&D) plants. In: Patel S, Solanki CH, Reddy KR, Shukla SK (eds) Indian geotechnical conference, IGC-2019, vol 134. Springer Singapore, Singapore, pp 185–194
8. I R Congress, IRC 121 guidelines for use of construction and demolition waste in road sector (2017)
9. Bureau of Indian Standards, IS 2720 (Part 4) methods of test for soil—grain size analysis (1995)
10. Bureau of Indian Standards, IS 2720 (Part III/Sec I) methods of test for soil—determination of specific gravity for fine grained soils (1997)
11. Bureau of Indian Standards, IS 2720 (Part 5) methods of test for soil—determination of liquid and plastic limit (1995)
12. Bureau of Indian Standards, IS 2720 (Part VII) methods of test for soils—determination of water content dry density relation using light compaction (2011)
13. I R Congress, IRC 36—recommended practice for construction of earth embankments and subgrade for road works (2010)
14. Bureau of Indian Standards, IS 2720 (Part 17) methods of test for soils—laboratory determination of permeability (1997)
15. Gulati S, Datta M (2008) Geotechnical engineering
16. Bureau of Indian Standards, IS 2720 (Part 13) methods of test for soil—direct shear test (1997)

Chapter 25
A Study on Pore Size Distribution of Compacted Expansive Soils

Sabari Ramesh and T. Thyagaraj

Introduction

Compacted expansive soils find its application in geoenvironmental engineering as landfill liners for the storage of municipal and hazardous wastes and also as buffer materials for nuclear wastes storage in deep geological repositories. For barrier applications, liner materials should exhibit very low hydraulic conductivity, adequate shear strength, and minimal volume changes. The compacted expansive soils have very low hydraulic conductivity and acceptable shear strength, but they may undergo large volume changes (swelling and shrinkage) due to changes in moisture content brought about by the seasonal fluctuations or variations in the surrounding environment or by the redistribution of moisture taking place in the soil structure with time. Compacted clay-sand mixtures, particularly sand-bentonite mixtures or natural expansive soil-sand mixtures are also provided in waste containment applications which give excellent mechanical stability and can also arrest the swelling and shrinkage deformations caused by expansive soils to a large extent [12, 14, 16, 17]. The hydraulic and volume change behavior of compacted soils is largely governed by the soil structure and the pore size characteristics.

The microstructural and pore size distribution studies on different compacted soils were carried out by many researchers using mercury intrusion porosimetry (MIP) and scanning electron microscopy (SEM) tests [2, 13, 20, 25–27]. Clays compacted on the dry side of optimum exhibited a double porosity structure with: (a) macropores or interaggregate pores and (b) micropores or intra-aggregate pores [1, 3]. Monroy et al. [13] and Tarantino [23] reported that, on saturation, the pore size distribution of compacted clays changes from bimodal to unimodal, but as per Thom et al. [24] the bimodal structure remains unchanged upon stress and suction changes. A study by Burton et al. [2] showed that an expansive clay compacted on the wet side of

S. Ramesh (✉) · T. Thyagaraj
Department of Civil Engineering, Indian Institute of Technology Madras, Chennai 600036, India
e-mail: sabarirameshjuly2@gmail.com

K. Muthukkumaran et al. (eds.), *Transportation and Environmental Geotechnics*, Lecture Notes in Civil Engineering 298, https://doi.org/10.1007/978-981-19-6774-0_25

optimum also exhibited the bimodal pore size distribution. But when subjected to higher stress, the size of the macropores reduced, and the structure became unimodal as the reduction in void ratio occurred first at the expense of larger pores. Wang et al. [27] observed the microstructural modification in compacted bentonite at a constant water content due to the redistribution of water in the interaggregate pores and water inside the aggregates to achieve equilibrium. Their observations revealed that the microstructural changes were slow for specimens with higher final dry density after saturation due to the smaller size of pores present in them.

In a compacted clay with a double structure, the macrovoid changes were dominant during loading at constant water content, whereas the microvoid changes were dominant with changes in water content [18]. An increase in compactive effort reduced the macropores, whereas it did not have any significant influence on the size of the micropores [11, 21]. The pore size distribution of compacted clays when subjected to wetting–drying cycles [20], wetting–loading paths [13], inundated with inorganic salt solutions [26], inundated with inorganic salt solutions and subjected to wet–dry cycles [25] were also studied by many researchers. X-ray computed tomography (X-ray CT), a non-destructive testing technique, was also used for studying the pore characteristics and crack formation in compacted soils [6, 7, 9, 10, 19, 22]. The data obtained from this test was based on the attenuation of X-rays when they pass through the specimen at different angles. The basic working principles of the X-ray CT technique were given in Kak and Slaney [8] and Herman [5].

A large number of studies were carried out on compacted expansive soils for understanding their pore size distribution by many researchers. The combined use of MIP and X-ray CT for studying the pore characteristics of compacted expansive soils is a recent development and needs further investigation. Also, the expansive soils differ among themselves by exhibiting different particle size distributions and dry densities and studies related to these are very scarce in the existing literature. Therefore in the present study, compacted soil specimens were prepared using four different expansive soils of different gradation and compaction characteristics. The X-ray CT and MIP tests were conducted on the prepared soil specimens, and the analysis of acquired data gave a better understanding of the pore size distribution of expansive soils considered in the study.

Materials and Method

Soils

Four expansive soils (designated as ES-1, ES-2, ES-3, and ES-4) from different areas of Tamil Nadu, India, were identified, procured, crushed, sieved, and used for the experimental study. The soils were collected from Kishkinta, Siruseri, Trichy, and Chemanchery. The three expansive soils (ES-1, ES-3, and ES-4) were sieved through a 2 mm IS sieve, whereas ES-2 was sieved through a 425-micron IS sieve. The

Table 25.1 Basic properties of different expansive soils taken for the study

Property	Expansive soil 1 (ES-1)	Expansive soil 2 (ES-2)	Expansive soil 3 (ES-3)	Expansive soil 4 (ES-4)
Plasticity index (%)	55	56	62	73
Specific gravity	2.71	2.72	2.70	2.70
Maximum dry density (Mg/m^3)	1.34	1.44	1.41	1.70
Optimum moisture content (%)	28	27	31	19
Maximum particle size (mm)	2.000	0.425	2.000	2.000
% Clay	82	65	60	41
% Silt	15	23	37	17
% Sand	3	12	3	42
Unified soil classification	CH	CH	CH	CH
Differential free swell index (%)	105	100	90	300

optimum moisture content and maximum dry density of all the soils were determined using standard Proctor compaction tests carried out in the laboratory. Table 25.1 shows the basic properties of all the expansive soils considered in the study. Soils were then mixed at water content corresponding to optimum moisture content and kept for moisture equilibration in desiccators for 48 h. Soil specimens of 75 mm diameter and 20 mm height were prepared in oedometer stainless steel rings by statically compacting the soil using a hydraulic press.

Methodology

The total void ratio of the compacted soil specimens was determined from height and diameter measurements carried out using Vernier calipers. X-ray CT tests were then performed on the compacted soil specimens. The soil specimens were placed inside the X-ray CT cabin on the specimen holder between the source and the detector. The optimal scan parameters were selected, and the test errors were minimized by following procedures as given in Du Plessis et al. [4]. X-rays were produced from the source, and the specimen was rotated 360° such that X-rays penetrate through the specimen at different angles. X-rays get attenuated when they pass through the material depending upon the material density, atomic composition, and energy of X-rays. This attenuation data was captured by the detector and converted to 2D images. Image reconstruction techniques were then applied on the attained 2D images, and the 3D image of the soil specimens was attained by stacking the 2D image slices.

After the X-ray CT test, the soil sample weighing less than 1 g was carefully extracted from the prepared compacted soil specimens and MIP tests were carried out. This test captures pore sizes ranging from 130 to 4 nm. The samples were freeze-dried before the MIP test to remove the moisture content without altering the structure. The MIP testing consists of two instruments with one instrument operating at low pressure (maximum pressure of 400 kPa) and the other instrument operating at high pressure (maximum pressure of 350 MPa). Mercury, a non-wetting fluid was injected into the specimen by applying pressure. The volume of intruded mercury corresponds to the pore volume in the sample ranging from 130 to 4 nm. The entrance pore diameters were calculated based on the pressure causing intrusion of mercury using the Washburn equation [28]. The pore size distribution and their volume for each compacted soil specimen were obtained from the MIP test results.

Results and Discussion

X-ray CT tests were conducted on all the four compacted expansive soil specimens considered for the study after calculating their total void ratios based on Vernier measurements. For each soil specimen, fifty reconstructed image slices were obtained. The thickness of each slice was 0.4 mm, and the minimum pixel size captured was 113 μm. Thus, this test could capture only the pores greater than 113 μm for the specimen size adopted in the study. The obtained X-ray CT images were then analyzed using an image analysis code written in MATLAB. The most probable threshold number for separating the solids and voids of the images was found out for each specimen. The image analysis procedure and methodology for finding out the most probable threshold number for segmenting X-ray CT images were adopted from Ramesh and Thyagaraj [15]. Figure 25.1 shows the grayscale image slices of the specimens obtained from X-ray CT tests at 5, 25, 50, 75, and 95% depths. The most probable threshold numbers obtained for ES-1, ES-2, ES-3, and ES-4 specimens were 120, 130, 121, and 101, respectively.

The obtained threshold number varied among the soil specimens prepared using different expansive soils. A higher threshold number of 130 was obtained for the ES-2 specimen owing to their smaller particle size and high clay content, whereas, for the ES-4 specimen with higher sand content and maximum dry density, a smaller threshold number of 101 was attained. The decrease in the volume of larger pores due to higher dry density resulted in the decrease of the threshold number. It was found out that the threshold number for image segmentation varied for compacted specimens of identical sizes prepared using soils of the same classification but different particle size distribution and therefore should be given importance. The grayscale images were thresholded using the most probable threshold numbers, and the segmented binary images are shown in Fig. 25.2.

The visual assessment of the binary images indicated that the distribution of large pores captured by X-ray CT varied for each soil specimen. The larger size pores were observed for the ES-2 specimen due to the clods formed during the specimen

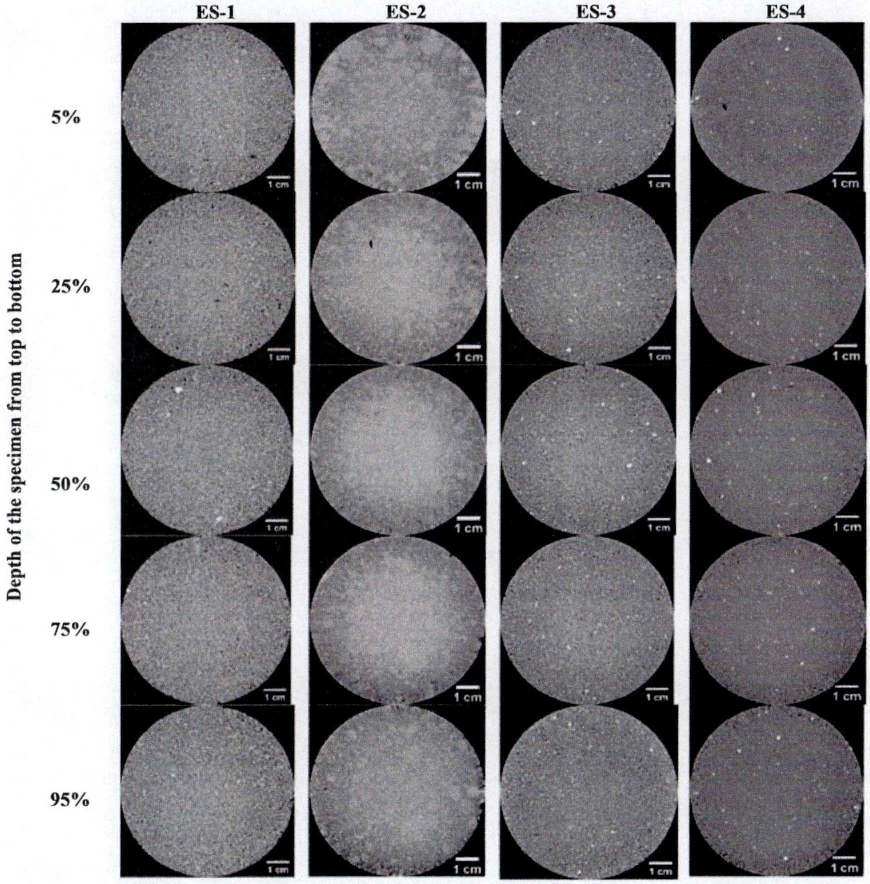

Fig. 25.1 X-ray CT image slices at 5, 25, 50, 75, and 95% depth from top to bottom surface for four different compacted expansive soils

preparation. The distribution of large pores in the case of ES-1, ES-3, and ES-4 was almost similar but a lesser percentage of pores was observed in the ES-4 specimen due to its higher dry density. For all the specimens, more pores were observed around the outer edges than at the center of the specimen. This might be due to the decrease in the effectiveness of compaction around the edges of the specimen due to the friction offered by the oedometer ring, as observed by Saba et al. [19].

The pore size distribution for all the specimens within the range of 130 μm and 4 nm was obtained from the MIP test. The pore entrance diameters obtained from MIP tests are presented in terms of cumulative mercury intruded void ratio and pore size density function in Fig. 25.3a and b, respectively. The total cumulative mercury intruded void ratio (e_{MIP}) was found to be much higher (0.555) for the ES-1 specimen compared to other specimens due to their higher total void ratio as a result of lower

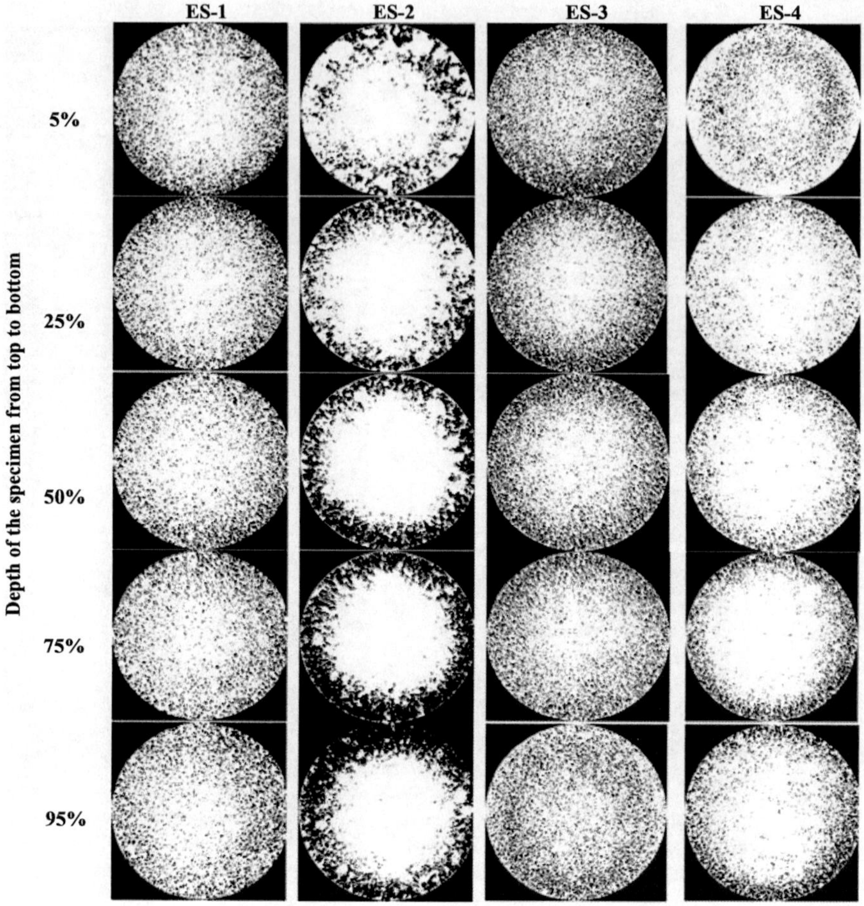

Fig. 25.2 X-ray CT image slices at 5, 25, 50, 75, and 95% depth from top to bottom surface for four different compacted expansive soils

dry density. The lowest e_{MIP} was obtained for the ES-4 specimen because of their higher maximum dry density. The e_{MIP} obtained for all the specimens was much lower than the calculated total void ratio indicating that a large volume of pores remained uncaptured by this technique. The pore sizes larger than 130 μm could be obtained from X-ray CT test but those smaller than 4 nm stayed undetected.

As observed from Fig. 25.3b, the mean pore entrance diameters of macropores are 65 μm for ES-1 and ES-2 specimens, 30 μm for ES-3 specimens, and 80 μm for ES-4 specimens. A slightly higher value of the mean pore entrance diameter was obtained for the ES-4 specimen due to the higher sand content, whereas it exhibited the lowest macropore volume as a result of its higher dry density. The micropore volume distribution and mean pore entrance diameters obtained for ES-2, ES-3, and

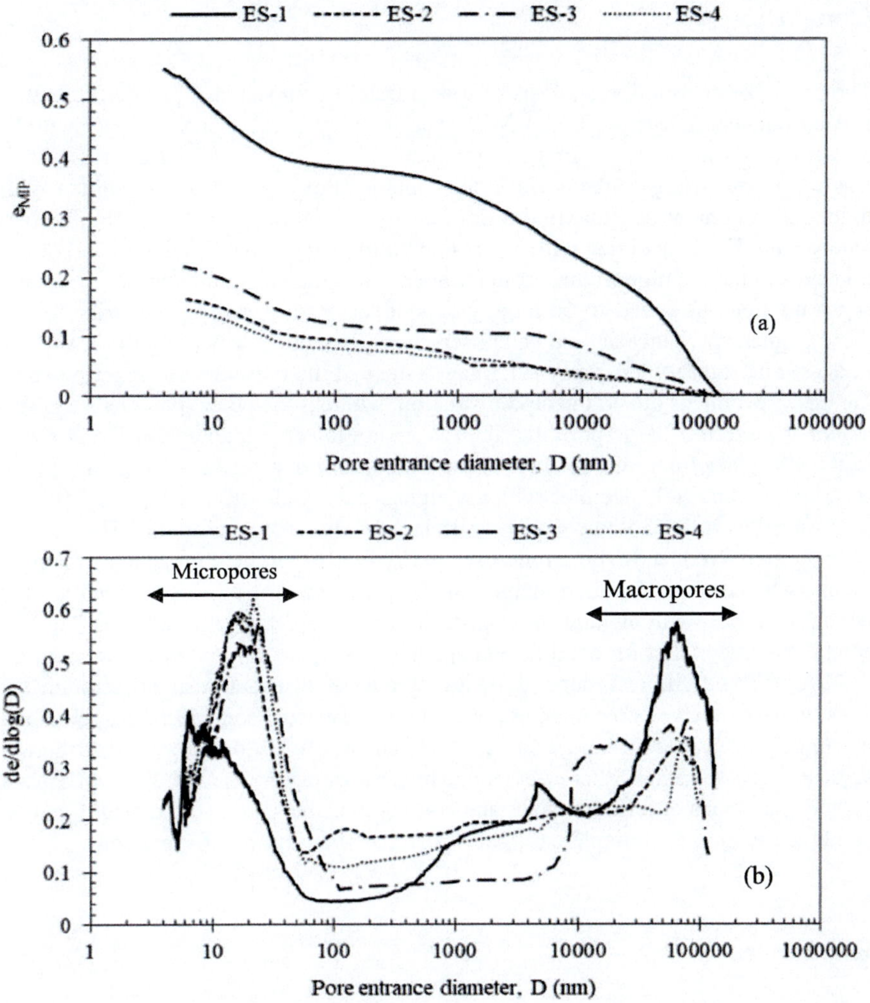

Fig. 25.3 Pore size distribution curves in terms of **a** cumulative mercury intruded void ratio versus pore entrance diameter **b** pore size density function

ES-4 specimens were found to be identical (18 nm), whereas a mean pore entrance diameter of 9 nm found out for the ES-1 specimen was slightly lower due to its higher clay content of 82%. The macropore volume obtained was higher than the micropore volume for the ES-1 specimen due to the lower compaction dry density. From Fig. 25.3b, it can be inferred that the volume of macropores reduced as specimen dry density increased, whereas micropores remain almost unaltered [11, 21].

Conclusions

The pore size distribution analysis of four different compacted expansive soils was carried out using X-ray CT and MIP tests in the present work. The most probable threshold number for image segmentation of each compacted expansive soil was found out using a newly developed methodology. The ES-4 specimen with higher sand content and maximum dry density gave the lowest threshold number, whereas the ES-2 specimen with lower maximum particle size (425 μm) and larger pores exhibited the highest threshold number. Thus, the threshold number for image segmentation was found to be largely dependent on the grain size distribution of the specimen and hence should be chosen carefully while segmenting the X-ray CT images. The thresholded X-ray CT images showed the presence of bigger pores in the ES-2 specimens due to the clod formation, whereas the other specimens showed a similar pattern of large pore distribution with a lower percentage of larger pores in the ES-4 specimen due to their higher maximum dry density. The MIP results showed that the ES-1 specimen with the highest clay content and lowest dry density gave a smaller volume of micropores and a larger volume of macropores. The volume of macropores reduced with an increase in dry density of the compacted soil specimens, whereas micropores remained unaltered except for the ES-1 specimen for which micropore volume and mean pore entrance diameter were reduced. Thus, it can be concluded that for compacted expansive soils, the volume of pores captured by X-ray CT and MIP tests depends on their grain size distribution and maximum dry density. Also, a large percentage of pores in the compacted soil specimens remained uncaptured even after utilizing both X-ray CT and MIP tests for pore size distribution studies. The study can further be extended by performing the X-ray CT and MIP tests on soil specimens of different sizes and compaction energies, and the results can be combined to obtain a complete range of pore size distribution of specimens.

References

1. Alonso EE, Gens A, High DW (1987) Special problem soils. In: General report. Proceedings of 9th European conference on soil mechanics and foundation engineering, Dublin, vol 3, pp 1087–1146
2. Burton GJ, Sheng D, Campbell C (2014) Bimodal pore size distribution of a high plasticity compacted clay. Géotech Lett 4:88–93
3. Diamond S (1971) Microstructure and pore structure of impact-compacted clays. Clays Clay Miner 19:239–249
4. du Plessis A, Broeckhoven C, Guelpa A, Roux SG (2017) Laboratory X-ray micro-computed tomography: a user guideline for biological samples. GigaScience 6:1–11
5. Herman GT (2009) Fundamentals of computerized tomography: image reconstruction from projections, 2nd edn. Springer, Dordrecht Heidelberg London, New York
6. Julina M, Thyagaraj T (2020) Combined effects of wet-dry cycles and interacting fluid on desiccation cracks and hydraulic conductivity of compacted clay. Eng Geol 267:105505
7. Julina M, Thyagaraj T (2019) Quantification of desiccation cracks using X-ray tomography for tracing shrinkage path of compacted expansive soil. Acta Geotech 14:35–56

8. Kak AC, Slaney M (1988) Principles of computerized tomographic imaging. IEEE Press, New York, p 329
9. Kawaragi C, Yoneda T, Sato T, Kaneko K (2009) Microstructure of saturated bentonites characterized by X-ray CT observations. Eng Geol 106:51–57
10. Kozaki T, Suzuki S, Kozai N, Sato S, Ohashi H (2001) Observation of microstructures of compacted bentonite by microfocus X-ray computerized tomography. Nucl Sci Technol 38:697–699
11. Lloret A, Villar MV, Sanchez M, Gens A, Pimtado X, Alonso EE (2003) Mechanical behaviour of heavily compacted bentonite under high suction changes. Géotechnique 53(1):27–40
12. Louafi B, Bahar R (2012) SAND: an additive for stabilization of swelling clay soils. Int J Geosci 3:719–725
13. Monroy R, Zdravkovic L, Ridley A (2010) Evolution of microstructure in compacted London clay during wetting and loading. Géotechnique 60(2):105–119
14. Omidi GH, Prasad TV, Thomas JC, Brown KW (1996) The influence of amendments on the volumetric shrinkage and integrity of compacted clay soils used in landfill liners. Water Air Soil Pollut 86:263–274
15. Ramesh S, Thyagaraj T (2021) Segmentation of X-ray tomography images of compacted soils (Submitted for publication)
16. Ramesh S, Thyagaraj T (2020) Effect of sand content on cyclic swell-shrink behavior of compacted expansive soil. Geotech Spec Publ 319:141–150
17. Rawat A, Baille W, Tripathy S (2019) Swelling behavior of compacted bentonite-sand mixture during water infiltration. Eng Geol 257:105–141
18. Romero E (2013) A microstructural insight into compacted clayey soils and their hydraulic properties. Eng Geol 165:3–19
19. Saba S, Delage P, Lenoir N, Cui YJ, Tang AM, Barnichon JD (2014) A further insight into the microstructure of compacted bentonite-sand mixture. Eng Geol 168:141–148
20. Seiphoori A, Ferrari A, Laloui L (2014) Water retention behaviour and microstructural evolution of MX-80 bentonite during wetting and drying cycles. Géotechnique 64(9):721–734
21. Sivakumar V, Tan WC, Murray EJ, Mckinley JD (2006) Wetting, drying and compression characteristics of compacted clay. Géotechnique 56(1):57–62
22. Tang CS, Zhu C, Leng T et al (2019) Three-dimensional characterization of desiccation cracking behavior of compacted clayey soil using X-ray computed tomography. Eng Geol 255:1–10
23. Tarantino A (2011) Compacted versus reconstituted states. In: Alonso EE, Gens A (eds) Unsaturated soils. CRC Press, BocaRaton, FL, pp 113–136
24. Thom R, Sivakumar R, Sivakumar V, Murrary EJ, Mackinnon P (2007) Pore size distribution of unsaturated compacted kaolin: the initial states and final states following saturation. Géotechnique 57(5):469–474
25. Thyagaraj T, Julina M (2019) Effect of interacting fluid and wet-dry cycles on microstructure and hydraulic conductivity of compacted clay soil. Géotechnique Letters 9(3):1–26
26. Thyagaraj T, Salini U (2015) Effect of pore fluid osmotic suction on matric and total suctions of compacted clay. Géotechnique 65(11):952–960
27. Wang Q, Cui YJ, Tang AM, Li XL, Ye WM (2014) Time- and density-dependent microstructure features of compacted bentonite. Soils Found 53(2):232–245
28. Washburn EW (1921) The dynamics of capillary flow. Phys Rev 17(3):273–283

Chapter 26
Stabilized Soil as a Sustainable Construction Material Using the Rammed Earth Technique

Kola Vikas, V. Ramana Murthy, and G. V. Ramana

Introduction

The natural soil has been serving as a sustainable construction material in the human history until the development of new construction materials. However, with the enormous increase in the world population and industrialization, several alternative materials have been developed for construction which is generally energy-intensive and lead to environmental concerns. Parallelly, a group of researchers has been continuously exploring the possibility of using upgraded natural soils in place of energy-intensive materials. Stabilized soil blocks using cement, lime, and other suitable admixtures were investigated for their use in construction. Alternatively, rammed earth as a construction technique has been studied by several researchers using both the natural earth and stabilized earth. For any soil to be favorable for rammed earth construction, the UCS must be over 2 MPa [1].

Stabilized rammed earth is a technique for constructing foundations, floors, roads, and walls utilizing locally available soils that are mixed with the optimum dosage of stabilizers at OMC and compacted in desired molds to form structural elements. It offers social, economic, and environmental benefits [2]. The usage of nearby earth materials reduces transportation costs, carbon emissions, and the associated energy consumption [3]. It expands the accessibility of quality housing and enhances the local economy [4]. Rammed earth (RE) was a building method that had already been in existence for thousands of years before the first documentation in about 77 AD by the Roman Pliny the Elder circa in natural history [5]. But people today are hesitant to

K. Vikas (✉) · V. Ramana Murthy · G. V. Ramana
Department of Civil Engineering, National Institute of Technology, Warangal 504006, India
e-mail: kvikas1@student.nitw.ac.in

V. Ramana Murthy
e-mail: vramana@nitw.ac.in.com

G. V. Ramana
e-mail: gvramana@nitw.ac.in.com

K. Muthukkumaran et al. (eds.), *Transportation and Environmental Geotechnics*,
Lecture Notes in Civil Engineering 298, https://doi.org/10.1007/978-981-19-6774-0_26

adopt this technique as there is a stigma that it uses dirt as a construction material and is categorized for use in agricultural families and economically distressed families. RE provides the solution to the problem of expensive housing in remote and rural areas [4]. The effect of lime in imparting long-term strength in earth blocks was studied and indicated that to attain higher strength, and lime should be added in the optimal quantity combined with cement [6]. Most of the previous studies were confined to compressive strength.

In the present work, an attempt is made to study the compressive strength and tensile strength of soil using stabilization techniques along with fiber reinforcement. The polypropylene fibers were used for the purpose with the available understanding of the long fibers having dispersion problems and tend to not mix well with the dry soil and higher amounts tend to accumulate closely and disperse poorly in the soil [7, 8]. Lower amounts result in pockets of soil that has very little fiber reinforcement. So, it is necessary to add the fibers in proper proportion. The fibers can add shear strength and flexural strength to rammed earth material [7, 8].

Materials and Methodology

The materials used and the methodology adopted in the present work are presented below.

Materials Used

In this study, three different soils having the following properties (Table 26.1) were used. The properties were determined as per IS: 2720-1985. The optimum moisture content and maximum dry density were determined as per IS: 2720-1985 (Part-8): standard proctor test, and these values are plotted in Fig. 26.1.

Cement. Ordinary Portland cement (Gr 53) is used as a stabilizer. Cement-stabilized soils rammed at optimum moisture content forms a strong and durable building material with reduced energy requirement.

Lime. Commercial lime is used in this study, and a sample of lime used is shown in Fig. 26.2. Lime is used for the pre-stabilization of black cotton soil to modify its plasticity before cement stabilization. Though the B.C. soil has become non-plastic at 6% lime, only 3% lime was used for pre-stabilization as the plasticity index is reduced considerably from 45 to 12% with this lime content.

Polypropylene Fiber. The fibers used are 6–12 mm long as shown in Fig. 26.3. The tensile strength is 450 MPa, and density is 0.91 g/cc.

Table 26.1 Three different soils index properties

S. No.	Properties	Red soil	Black cotton soil	Moorum soil
1	*Grain size distribution*			
	(a) Gravel %	1	2	7
	(b) Sand %	37	22	73
	(c) Silt + clay (Fines %)	62	76	20
2	Specific gravity	2.66	2.70	2.67
3	*Atterberg limits*			
	(a) Liquid limit %	43	67	33
	(b) Plastic limit %	24	22	17
	(c) Plasticity index	19	45	16
4	*Compaction properties*			
	(a) OMC %	20	25	15
	(b) MDD KN/m^3	17.7	15.5	18.3
5	Indian Standard Soil Classification	CI	CH	SC

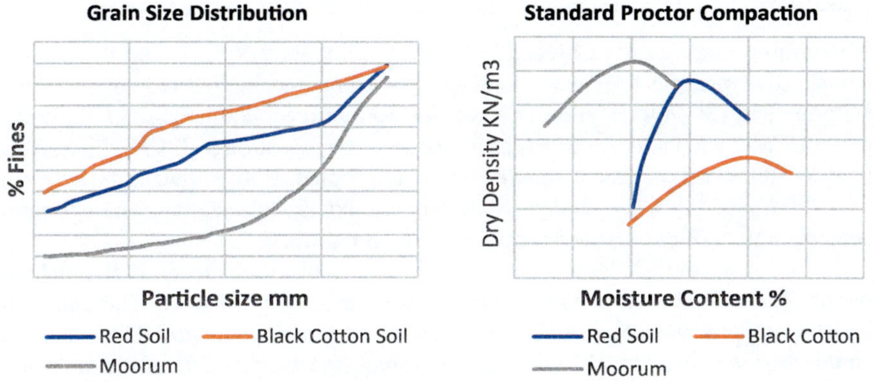

Fig. 26.1 Gradation and compaction curves of three soils

Fig. 26.2 Lime

Fig. 26.3 Polypropylene fiber

Sample Preparations and Test Procedure

The influence of admixture stabilization and fiber reinforcement on the properties of rammed earth was investigated. In the present work, unconfined compressive strength tests, Brazilian tests, and durability tests were conducted.

Unconfined Compressive Strength Test (IS 2720 Part-X [9]**).** The unconfined compressive strength tests were conducted on cement-stabilized red soil (CI) and Moorum soil (SC) mixes in order to get the optimum cement content to attain the target UCS of 2 MPa for rammed earth. In case of black cotton soil (CH), lime stabilization followed by cement stabilization was adopted. Polypropylene fiber (PPF) of 0.4, 0.6, and 0.8% was added along with admixture stabilization. The optimum cement content of 8% was arrived at based on initial trials.

The samples for UCS tests were prepared using 75 mm diameter and 150 mm long molds. The soil admixture combinations were prepared in dry state. The optimum moisture content was added, and the test specimens corresponding MDD of modified compaction were prepared by compacting the required weight of sample in the mold. These samples were tested at 14 days and 28 days curing period.

In case of fiber-reinforced samples, the soil–fiber–additive mixes were prepared in dry state and the required water content was added before molding the test samples. The samples were tested at a displacement rate of 1 mm/min until failure. The proving ring and dial gauge readings were taken to get the stress–strain patterns. Tests were also conducted on samples soaked for 24 h. The post-test moisture contents were also determined (Fig. 26.4).

Brazilian Test (IS 10082 [10]**).** The test specimens of 50 mm diameter and 50 mm height were prepared using the mold by following a similar procedure as for the UCS test. Figure 26.5 shows the samples and test setup. The specimens were tested at 14 days and 28 days of the curing period. These tests were performed as per IS: 10,082 (1981). The cylindrical samples were tested by placing the samples horizontally between the two steel bearing plates at a displacement rate of 0.5 mm/min until failure. Testing was conducted in both soaked and dry states.

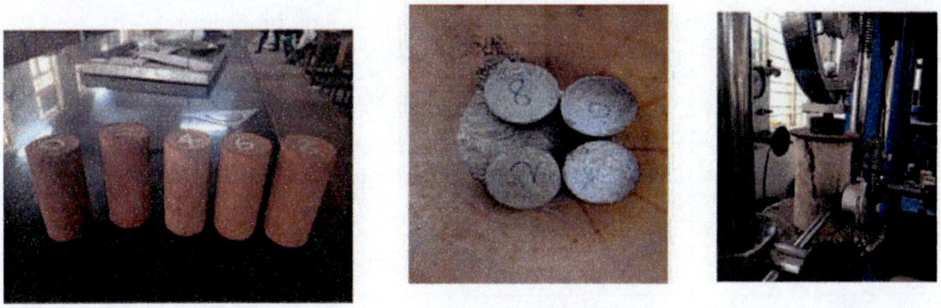

Cured Specimens Soaked Specimens Test Setup

Fig. 26.4 UCS samples and test setup

Sample Specimens Test Setup

Fig. 26.5 Brazilian test samples and test setup

To study the degree of durability of stabilized rammed earth, the stabilized test specimens were subjected to alternate wetting and drying cycles as per IS: 4332-1969. The percentage volume change and weight loss were found after 12 cycles of wetting and drying.

Discussion on Test Results

Unconfined Compressive Strength Properties of Stabilized Soil

Keeping in view the required target strength of 2 MPa for rammed earth, initial trials were made on soil admixture combinations. The UCS test results are presented in Table 26.2, and it can be seen from this table that the unsoaked compressive strength of all the stabilized soils at 28 days curing period is more than the required target value of 2 MPa. However, upon soaking, the CH soil could not retain the required compressive strength. Instead of adding more cement content, the addition of fiber was attempted to satisfy the compressive strength while imparting tensile strength also.

Table 26.2 UCS of different soils treated with additives at 28 days curing period

Cement content (%)	Unconfined compressive strength (MPa)					
	Red soil (CI)		Black cotton soil (CH)		Moorum soil (SC)	
	Dry	Soaked	Dry	Soaked	Dry	Soaked
0	0.63	–	0.7	–	0.66	–
2	0.86	–	1.12	0.19	0.81	0.26
4	2.02	0.49	2.17	0.77	1.87	1
6	2.85	1.34	2.59	1.23	3.34	1.65
8	4.34	2.48	2.93	1.56	3.89	2.2

Influence of Fiber Content on Compressive Strength. The influence of PPF content on the compressive strength of stabilized soil samples is presented in Table 26.3. It can be observed that the required compressive strength of 2 MPa could be attained by the addition of fiber content for all the soils even at 14 days curing period. In the case of CI and CH soils, the strength is increasing up to 0.6% PPF content, and in the case of SC soil, the optimum fiber content is obtained as 0.4% beyond which the strength reduction is observed. The increase in strength could be due to the development of interlocking friction between fiber and soil–cement mixture. Further, it can be noted that the required strength of 2 MPa can be achieved at 14 days curing period when the fiber is added for all the soils for the optimum cement content of 8%. The variation in optimum PPF content across the soils could be due to the variation of finer fractions in the respective soils (Fig. 26.6).

Tensile Strength Properties of Stabilized Soils

Brazilian tests were conducted on binder and fiber-treated soil samples after 14 and 28 days of curing on both dry and soaked specimens and peak strength values are obtained and presented in Table 26.4.

From Table 26.4, it can be observed that the trend of tensile strength variation of stabilized soils with cement content and fiber content is similar to that observed with compressive strength. The optimum content of polypropylene fiber for cement-stabilized rammed earth is found to be 0.4% for SC soil and 0.6% for CI and CH soils.

Durability

A durability study is carried out on three soils stabilized with their optimum mix found earlier by wetting and drying cycles as per IS 4332 Part IV [11]. Percentage

Table 26.3 Influence of fiber content on cement-stabilized soils at different curing periods

Fiber content (%)	Unconfined compressive strength (MPa)											
	Red soil (CI)				Black cotton soil (CH)				Moorum soil (SC)			
	Days (14)		Days (28)		Days (14)		Days (28)		Days (14)		Days (28)	
	Dry	Soaked	Dry	Soaked	Dry	Soaked	Dry	Soaked	Dry	Soaked	Dry	Soaked
0.4	4.68	2.42	4.89	2.65	4.3	2.62	5.27	2.82	5.39	4.21	6.19	4.67
0.6	5.19	2.75	5.89	3.03	4.59	2.66	5.44	3.19	5.04	3.71	6.1	4.42
0.8	4.74	2.56	4.78	2.69	4.05	2.21	4.29	2.82	4.77	3.7	5.94	4.1

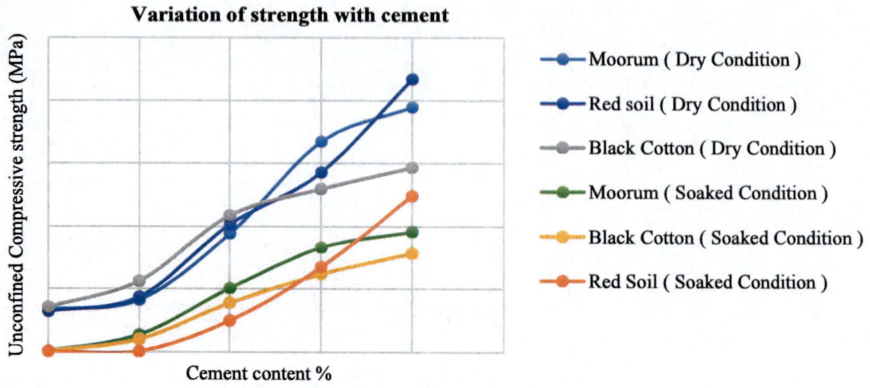

Fig. 26.6 Influence of variation in cement on UCS of different soils after 28 days curing

volume change and weight loss are found with number of cycles of wetting and drying up to 12 cycles.

- Volume change of specimens is calculated based on the difference between the initial volumes of specimens and the post-test volumes as a percentage of the initial volume. The soil stabilizer loss of specimens shall be calculated as a percentage of the initial oven-dry weight of the specimen as follows:

$$\text{Soil} - \text{Stabilizer loss, percent} = (A/B) * 100$$

where A = original oven-dry weight—oven-dry weight at end of the cycle, and
B = original oven-dry weight.

From Figs. 26.7 and 26.8, it is found that the Moorum (SC) soil has less effect of weathering compared to red (CI) and black cotton (CH) soils. The effect was predominantly observed for CH soil followed by CI and then SC soil; this is due to unbound soil particles present in soils with higher fines content. But still, the samples are found stable enough to take the stress of 2 MPa and the performance was found to be in permissible limits according to ASTM-D559M.

Conclusions

From the present study, the following conclusions are drawn.

- The locally available soils taken for the study do not satisfy the required strength criteria for rammed earth construction in their natural condition and stabilization becomes inevitable.
- The target UCS of 2 MPa can be obtained with 8% cement content at 28 days curing period for all the soils. However, for BC soil, 3% lime stabilization before cement

Table 26.4 Variation in tensile strength of different soils with variation in PPF content

Fiber content (%)	Tensile strength (MPa)											
	Red soil (CI)				Black cotton soil (CH)				Moorum soil (SC)			
	14 Days		28 Days		14 Days		28 Days		14 Days		28 Days	
	Dry	Soaked	Dry	Soaked	Dry	Soaked	Dry	Soaked	Dry	Soaked	Dry	Soaked
0.4	7.61	3.35	9.79	3.40	4.84	2.99	7.30	3.21	8.11	5.17	10.48	4.90
0.6	9.28	5.11	10.33	5.62	5.71	3.13	8.63	3.54	6.56	4.37	10.24	4.49
0.8	6.52	3.15	9.90	3.26	4.03	2.55	6.69	3.26	6.40	2.67	9.94	4.22

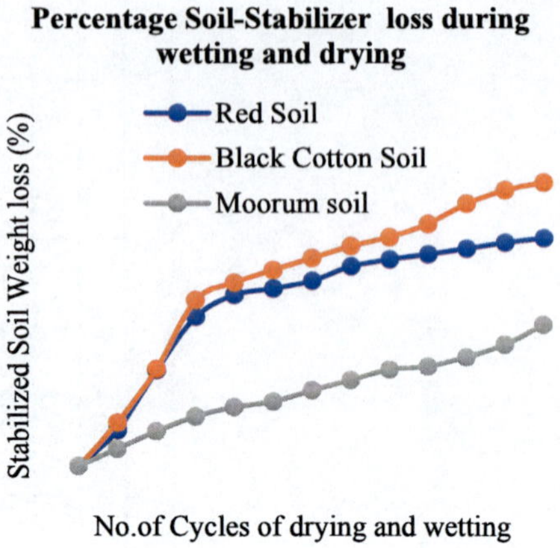

Fig. 26.7 % Soil stabilizer loss with no of cycles of drying and wetting

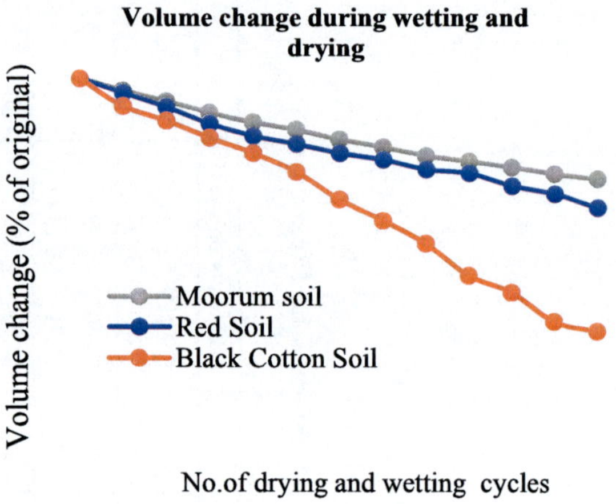

Fig. 26.8 % Volume change with no. of cycles of drying and wetting

stabilization is required. When fiber is used along with admixture stabilization, the target strength can be obtained right from 14 days curing period with 0.4% fiber content for the soils.

- The addition of fiber increased both the compressive strength and tensile strength of the stabilized soils.
- For all the stabilized soils, the durability criteria in terms of volume reduction are within the permissible limits.

The study revealed that the locally available soils can be adopted for sustainable rammed earth construction by stabilizing them to satisfy the suggested compressive strength of 2 MPa.

References

1. Burroughs S (2008) Soil property criteria for rammed earth stabilization. J Mater Civ Eng 20(3):264
2. Hallal MM et al (2018) Evaluation of engineering characteristics of stabilized rammed-earth material sourced from natural fines-rich soil. J Build Appraisal 30(11)
3. Ciancio D, Jaquin P, Walker P (2013) Advances on the assessment of soil suitability for rammed earth. J Constr Build Mater 42(3):40–47
4. Lee A (2009) Economic benefits of contemporary earth construction in low-cost urban housing—state-of-the-art review. J Build Appraisal 5:259–271
5. Gramlich AN (2013) A concise history of the use of the rammed earth building technique including information on methods of preservation, repair, and maintenance
6. Nagaraj HB, Sravan MV, Arun TG, Jagadish KS (2014) Role of lime with cement in the long-term strength of compressed stabilized earth blocks. Int J Sustain Built Environ 3(1):54–61
7. C. Tang, Shi B, Gao W, Chen F, Cai Y (2007) Strength and mechanical behavior of short polypropylene fiber reinforced and cement stabilized clayey soil. Geotext geomembr 25(3):194–202
8. Araki H, Koseki J, Sato T (2016) Tensile strength of compacted rammed earth materials. Soils Found 56(2):1–16
9. IS 2720-Part 10: Methods of tests for soils—determination of unconfined compressive strength of soil. Bureau of Indian Standards, New Delhi (1985)
10. IS 10082: Methods of tests for soils—determination of tensile strength by indirect tests. Bureau of Indian Standards, New Delhi (1981)
11. IS 4332-Part 4: Methods of tests for soils—wetting and drying and freezing and thawing tests for compacted soil-cement mixtures. Bureau of Indian Standards, New Delhi (1968)
12. IS 2720-Part 4: Methods of tests for soils—grain size analysis. Bureau of Indian Standards, New Delhi (1985)
13. IS 2720-Part 5: Methods of tests for soils—determination of Atterberg limits. Bureau of Indian Standards, New Delhi (1985)
14. IS 2720-3-1: Methods of tests for soils—determination of specific gravity. Bureau of Indian Standards, New Delhi (1980)
15. IS 2720-Part-8: Methods of tests for soils—standard proctor test. Bureau of Indian Standards, New Delhi (1983)

Chapter 27
Assessment of Adsorptive Removal of Naphthalene Through Constructed Wetland Using Physical and Numerical Modeling

Avishek Adhikary, Pradyumna Konar, Tapabrata Chakraborty, and Supriya Pal

Introduction

Various hydrocarbons are carried into surrounding bodies of water by industrial effluents, which eventually leak and contaminate the groundwater [1]. When fossil fuels and other organic materials are partially burned, poly-aromatic hydrocarbons (PAHs) are chemical substances containing two or more aromatic rings. According to the International Agency for Research on Cancer [IARC (1987)] and Bayazit et al. [2], aromatic hydrocarbon compounds are highly toxic and cause mutagenic and carcinogenic effects.

Naphthalene, one of the most widely utilized PAHs, can discharge up to 2 mg/L of Naphthalene into the water via industrial leachate. According to the Encyclopedia of Toxicology [3], the permitted limit of Naphthalene in drinking water is 20 μg/L. The maximum limit of Naphthalene for marine life protection is 0.6 mg/L, according to the US Environmental Protection Agency (USEPA) [4]. The presence of large quantities of Naphthalene in drinking water may pose serious health risks to humans [5].

Naphthalene removal from effluents is commonly accomplished using oxidation processes, adsorption, biodegradation, and membrane separation [6, 7]. Nevertheless,

A. Adhikary (✉) · P. Konar · S. Pal
Department of Civil Engineering, National Institute of Technology, Durgapur, West Bengal 713209, India
e-mail: avishek.adhikary31@gmail.com

P. Konar
e-mail: pk.19p10090@mtech.nitdgp.ac.in

T. Chakraborty
Department of Earth and Environmental Studies, National Institute of Technology, Durgapur, West Bengal 713209, India
e-mail: tc.19p10021@mtech.nitdgp.ac.in

K. Muthukkumaran et al. (eds.), *Transportation and Environmental Geotechnics*, Lecture Notes in Civil Engineering 298, https://doi.org/10.1007/978-981-19-6774-0_27

due to low operating and maintenance costs, the recyclability of adsorbents, the increased design life of waste containment structures, and high pollutant removal effectiveness, adsorption of solid adsorbents have a distinct advantage over alternative procedures [8, 9]. As a result, adsorption has established itself as one of the first approaches for eliminating organic pollutants from wastewater.

Clay minerals, zeolites, and agricultural wastes have all been acknowledged as promising adsorbents for removing Naphthalene [10, 11]. Researchers have recently been pondering the use of naturally existing adsorbents with a high abundance, large specific surface area, and remarkable chemical stability [12].

Therefore, batch scale adsorption investigations, one-dimensional conventional solute transport tests, and vertical flow laboratory-scale built wetland were utilized in this study to assess the efficacy and compatibility of locally available silty-sandy soil to adsorb Naphthalene.

The physicochemical parameters of the soil were thoroughly described using various approaches. Following that, the impact of different essential parameters such as solution pH, contact time, adsorbent dosage, and Naphthalene concentration on the removal efficacy of the employed soil was evaluated. A study of the kinetics as well as isotherms of Naphthalene adsorption by sandy soil was also carried out. Finally, the results from all of the experiments were verified using the HYDRUS-2D software's standard solute transport module and CW2D module.

Materials and Methods

Materials

In accordance with USEPA-625/1-88/022, 1998 [13], while choosing a wetland location, the underlying soil permeability must be taken into account. The ideal soil permeability ranges from 10–6 to 10–7 m/s.

Keeping this in mind, the adsorbent chosen in this investigation was silty-sandy soil utilized for cultivation of tubers like potato. The soil was obtained from agricultural fields around Durgapur Barrage (23.4792° N–87.3089° E) in India, as shown in Fig. 27.1. M/s Bengal Chemicals in India provided the analytical grade Naphthalene used in the inquiry. The remaining solvents and chemicals were all of analytical purity. Deionized water was used to make all of the Naphthalene aqueous solutions.

Physicochemical Properties

Under IS 2720:1983, soil physicochemical parameters were investigated in the laboratory at standard temperature and pressure (STP).

Fig. 27.1 Soil collection area

pHzpc Test

The pHzpc experiment was used for determination of the soil's point of zero surface charge. A BOD shaker was used to continuously shake 150 mg of soil in a weak potassium chloride (KCl) solution with a pre-set pH for 24 h. The pH was initially adjusted to five distinct levels, ranging from 2 to 10. Following the conclusion of the shaking, the filtration of solution was done by the use of vacuum filter and the pH was checked once again. The zero-point charge pH was obtained by plotting the initial pH on the horizontal axis against the final pH on the vertical axis.

Batch Adsorption Test and Response Surface Methodology

Stat-Ease Inc's Design-Expert statistical software tool is used to optimize the parameters utilized in batch adsorption studies to explore the adsorption phenomena of the selected soil to remove PAHs from wastewater. Comparative testing, optimization, screening, characterization, mixture designs, resilient parameter design, and integrated designs are all available through Design-Expert [14]. The input factors considered in this investigation were soil dosage (g/L), PAH concentration as input (mg/L), and time of contact (minutes). The dose was kept between 10 and 40 g/L, the time was varied between 60 and 120 min, and the initial PAH content was varied between 2 and 12 mg/L [4]. The temperature was kept constant at 25 °C, and the shaker speed was set to 150 rpm. The sets were removed from the shaker once the appropriate time intervals had passed, and the contents were centrifuged at 4000 rpm for 5 min before being filtered using Whatman no. 2 filter paper. After that, the solutions were examined in a UV–visible spectrophotometer to determine the PAH output concentration. Determination of removal efficiency is done by the following equations:

$$\text{Adsorption Capacity of soil } (q_e) = \frac{(C_0 - C_e)}{m} * V \tag{27.1}$$

$$\text{Naphthalene removed } (\%) = \frac{(C_0 - C_e)}{C_0} * 100 \tag{27.2}$$

where C_0 is the initial Naphthalene concentration (mg/L), C_e is the equilibrium concentration (mg/L), V is the solution volume (L), m is the adsorbent mass (g), and q_e is the amount of Naphthalene adsorbed per unit mass of adsorbent (mg/g). The percentage of Naphthalene removed is represented by R percent.

Vertical Column Test

The laboratory-scale column test method was used to replicate dynamic conditions. The overall height of the column was 250 mm, and the diameter was 60 mm (Fig. 27.2). To avoid clogging throughout the experiment, the soil was put into the column in distinct layers up to 40 mm depth, having a 90% relative compaction. The initial Naphthalene concentration was fixed at 12 mg/L. A UV–visible spectrophotometer with a 220 nm wavelength was used to measure the exit concentrations of Naphthalene.

Constructed Wetland

The study used a laboratory-scale artificial wetland with 450 mm × 300 mm × 350 mm dimensions, as illustrated in Fig. 27.3. In the wetland, silty-sandy soil acquired for the study was employed as an adsorbent. Up to 300 mm of soil sample was provided, with 50 mm of freeboard. A constant supply of 12 mg/L Naphthalene solution after every 4 h for 15 min was done for observation of the efficacy of the soil in contaminant removal.

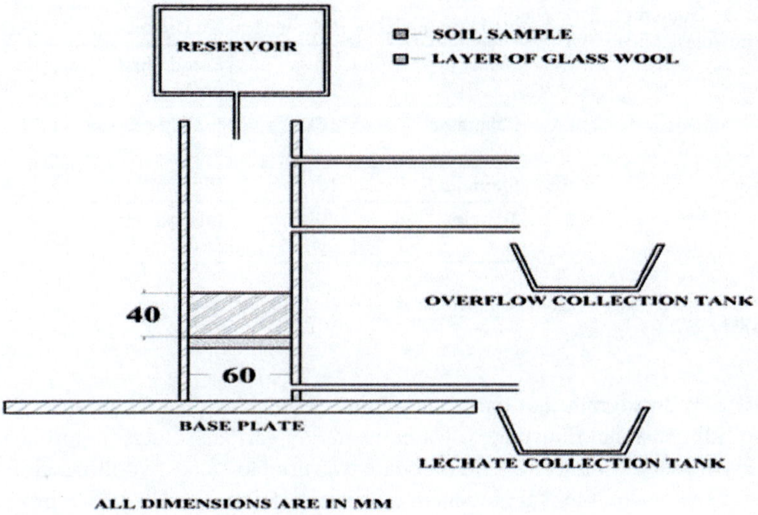

Fig. 27.2 Vertical column schematic diagram

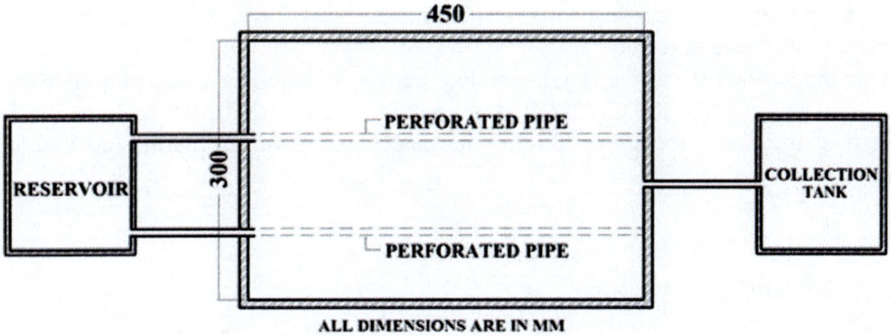

Fig. 27.3 Constructed wetland schematic diagram

Isotherm and Kinetic Studies

Four different isotherm and kinetic models were applied in this study, as mentioned in Table 27.1. The following equation was used to calculate the sorption capacity of soil at time *t*, q_t (mg/g).

$$q_t = \frac{(C_0 - C_e)}{m} * V \tag{27.3}$$

Table 27.1 Isotherm and kinetic models applied in the study

Isotherm model	Reference	Kinetic model	Reference
Linear	–	Pseudo-first order	[15]
Langmuir	[16]	Pseudo-second order	[17]
Freundlich	[18]	Elovich	[19]
Temkin	[20]	Intraparticle diffusion	[21]

HYDRUS Modeling

HYDRUS is a program that calculates finite element models for modeling single and multidimensional transport of water, heat, and various solutes in porous media with varying saturation. HYDRUS software was used to create two-dimensional and three-dimensional numerical models in this investigation. The HYDRUS program is used to solve Richard's equation for saturated–unsaturated water flow and the Fickian advection–dispersion equation for solute transport. Advective–dispersive transport is considered in the liquid phase, while diffusion is considered in the gaseous phase. The governing flow and transport equations are numerically solved using Galerkin-type linear finite element techniques.

In this present study, the standard solute transport module is used to simulate the column setup, while the CW2D module is used to simulate the lab-scale wetland. Both of the above models employed the batch adsorption, isotherm, and kinetic studies as input parameters.

Results and Discussions

Characteristics of Soil

The physicochemical characterization of the soil was conducted in the laboratory according to IS: 2720-1983, and the results are shown in Table 27.2.

Wetland demands a sandy type of soil that can let the liquid pass through while absorbing the contaminants mixed with it. Also, the hydraulic conductivity falls within the range as prescribed by USEPA.

Table 27.2 Physicochemical properties of soil

Parameter	Output
Permeability	$K = 1.042 \times 10^{-5}$ cm/s
Specific gravity	$G = 2.64$
pH of soil	7.2
MDD[a]	1.732 gm/cc
OMC[b]	9.75%
Particle size distribution (%)	Sand = 49
	Silt = 39
	Clay = 12

[a] Maximum dry density; [b] Optimum moisture content

pHzpc Test

The pH_{ZPC} results are given in Fig. 27.4. The pH_{ZPC} is the pH at which the total surface charge is zero, whereas at pH > pH_{ZPC} and < pH_{ZPC}, respectively, the surface possessed negative and positive charges [22]. A plot of initial pH against final pH was used to calculate the pH_{ZPC}. The pH_{ZPC} value achieved from the experiment was 6.3, as shown in Fig. 27.4.

To eliminate the effect of adsorbent surface charge on the adsorption process, all of the studies were repeated with the pH set to the pHzpc value.

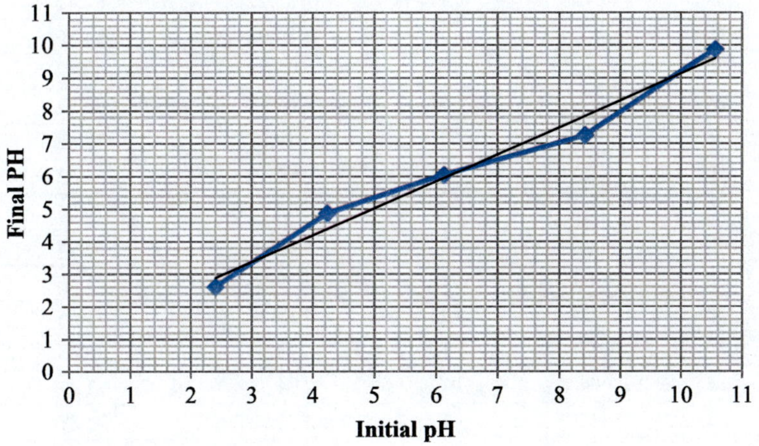

Fig. 27.4 pHzpc test result

Response Surface Methodology

The Box–Behnken approach yielded 17 trial combinations with varying soil dosage, initial concentrations, and time of contact. The following removal efficiencies were discovered in laboratory batch experiments under those conditions, as reported in Table 27.3.

The removal results acquired from the batch study (Table 27.3) were used to design Design-Expert, which kept contact time and soil dose to a minimum, starting concentration to a maximum, and maximum removal efficiency. The quadratic model was the best fitting model by Design-Expert, and the analysis of variance (ANOVA) is listed in Table 27.4. The model's F-value of 34.06 shows that it is significant. There is just a 0.01% chance that an F-value of greater magnitude would occur due to noise. Significant model terms have p-values less than 0.0500. In this circumstance, the model terms A and C are crucial. If the p-value is more than 0.1000, the model terms are not significant. The lack of fit has an F-value of 0.21, indicating that it is not significant compared to the pure error. A lack of fit higher F-value has an 88.18% chance of occurring due to noise. As a result, a non-significant lack of fit suggests that the model has a good prediction potential.

Table 27.3 Response surface values for various experimental settings

S. No.	Soil dosage (g/L)	Contact time (min)	Initial concentration (mg/L)	Predicted removal efficiency (%)	Actual removal efficiency (%)
1	25	120	12	85.22	87.41
2	10	60	7	86.28	84.55
3	25	120	2	55.38	60.22
4	25	60	2	66.18	65.21
5	25	90	7	79.87	77.65
6	10	90	2	69.19	67.63
7	25	90	7	78.28	70.66
8	40	60	7	66.43	65.98
9	25	60	12	78.55	81.96
10	40	90	2	53.29	51.65
11	40	90	12	75.16	73.85
12	25	90	7	74.98	72.22
13	25	90	7	73.72	72.87
14	10	90	12	85.28	87.27
15	25	90	7	76.36	75.21
16	10	120	7	85.55	83.19
17	40	120	7	68.16	66.63

Table 27.4 ANOVA for the quadratic model in response surface methodology

Source	Sum of squares	df	Mean square	*F*-value	*p*-value	
Model	1525.245457	9	169.4717	34.06217	5.82E-05	Significant
A-soil dosage	520.5151125	1	520.5151	104.6185	1.84E-05	
B-contact Time	0.0078125	1	0.007813	0.00157	0.969498	
C-initial Conc	919.77605	1	919.7761	184.866	2.74E-06	
AB	1.010025	1	1.010025	0.203005	0.66593	
AC	1.6384	1	1.6384	0.329302	0.584029	
BC	27.2484	1	27.2484	5.476664	0.051828	
A^2	5.255779211	1	5.255779	1.056361	0.338246	
B^2	25.95388447	1	25.95388	5.216478	0.056307	
C^2	26.41588447	1	26.41588	5.309336	0.054652	
Residual	34.827555	7	4.975365			Not significant
Lack of fit	4.826475	3	1.608825	0.214502	0.881751	
Pure error	30.00108	4	7.50027			
Cor total	1560.073012	16				

The removal efficiency is obtained in terms of the coded factor with the help of Eq. 27.4.

$$\begin{aligned}\text{Removal efficiency}(\%) = {} & 73.72 - 8.07 * A - 0.0312 * B + 10.72 * C \\ & + 0.5025 * \text{AB} + 0.64 * \text{AC} + 2.61 * \text{BC} \\ & - 1.12 * A^2 + 2.48 * B^2 - 2.50 * C^2 \end{aligned} \tag{27.4}$$

where *A* stands for the adsorbent dose, *B* for the time, and *C* for the Naphthalene concentration. The equation in terms of coded factors can be used to predict the reaction for specified values of each ingredient. High levels of the components are expressed as +1 by default, whereas low levels are recorded as −1.

The most desirable combination, according to Design-Expert, was an initial concentration of 12 mg/L, a soil dosage of 10 mg/L, and a contact time of 60 min. 88.65% was discovered to be the best feasible removal efficiency. Figures 27.5 and 27.6 illustrate the ramp plot and contour diagrams, respectively.

The obtained values, which corresponded to 60 min of contact time, 10 g/L of soil dosage, and 12 mg/L of Naphthalene concentration, were cross-checked in the laboratory and found to provide 85.78% removal efficiency, compared to 88.6562% removal efficiency in the Design-Expert predicted result.

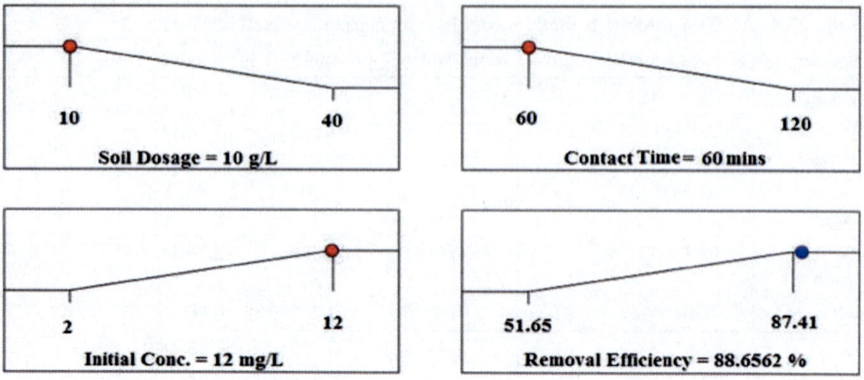

Fig. 27.5 Ramp plot of Naphthalene

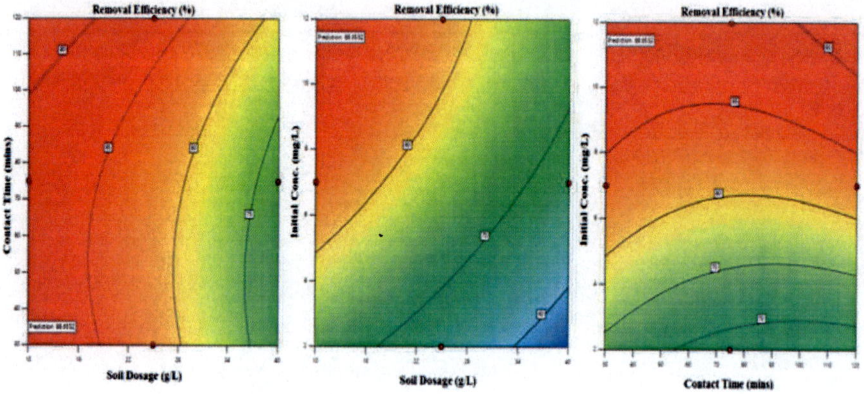

Fig. 27.6 Contour plot of Naphthalene

Isotherm and Kinetic Modeling

In this investigation, the Langmuir model suited the data well for Naphthalene, which states single layer adsorption on the adsorbent's surface, with the adsorption system's energy assumed to be constant and all adsorbed molecules are in direct contact with the surface layer. The Langmuir adsorption isotherm was obtained by plotting $1/Q_e$ versus $1/C_e$ (Fig. 27.7). Q_o signifies metal ion absorption per unit mass of adsorbent (mg/g), and c is the Langmuir constant.

Adsorption kinetics is crucial for determining the solute uptake rate and the adsorption time. The pseudo-second-order model (Fig. 27.8) showed to be the most suited in this situation. The quadratic driving force model is used to derive the pseudo-second-order model. In the pseudo-second-order model involving chemisorption, surface adsorption is the rate-limiting phase. The physicochemical interaction

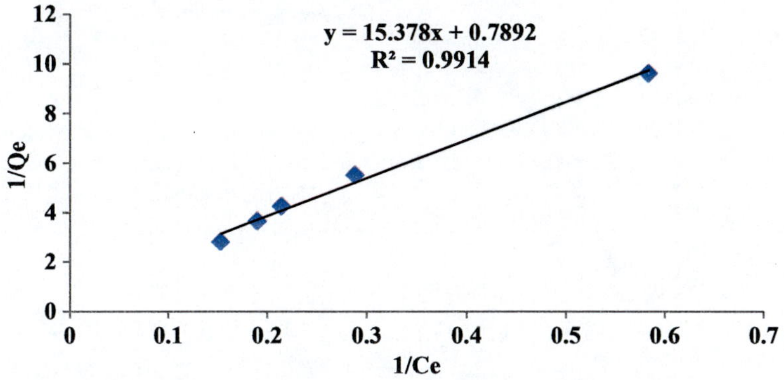

Fig. 27.7 Langmuir adsorption isotherm plot of Naphthalene

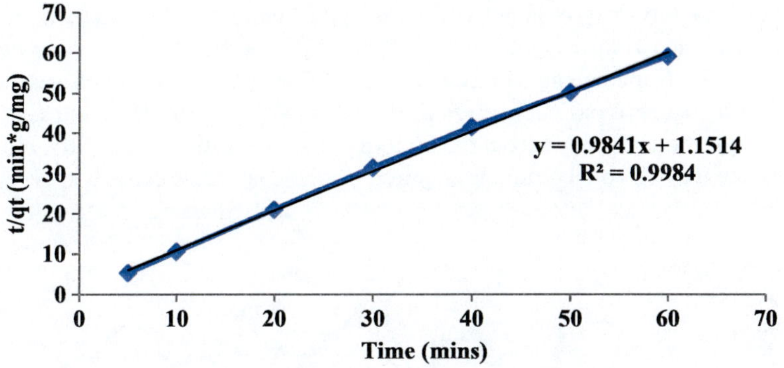

Fig. 27.8 Pseudo-second-order adsorption kinetic plot of Naphthalene

between the two phases is responsible for eliminating contaminants from a solution [23].

HYDRUS Modeling of Column and Wetland

The initial Naphthalene concentrations were 12 mg/L in a laboratory-scale column test with a soil depth of 40 mm. The exit concentrations of Naphthalene from the column were measured at 3, 6, 9, 12, 24, 36, 48, 60, 72, 84, and 96 h. The end time was kept at four days in both the lab research and the HYDRUS model, and 90% soil saturation was obtained around 1.5 days in the numerical model, which validates the lab result, as shown in Fig. 27.9.

A two-dimensional CW2D wetland module was used to model the constructed wetland. The built wetland was charged with 12 mg/L of Naphthalene solution for

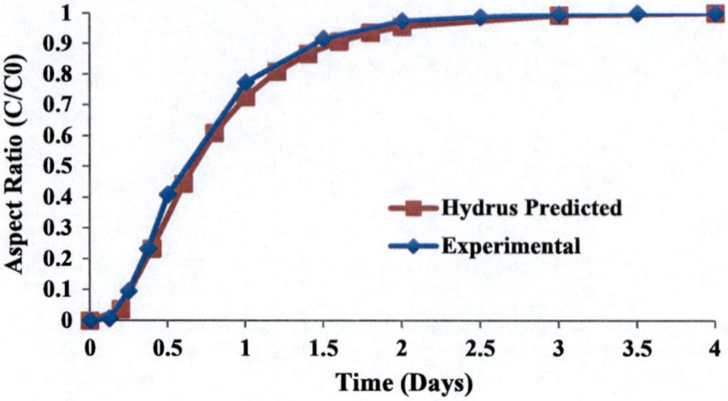

Fig. 27.9 Comparison of lab-scale column test result and HYDRUS predicted result

15 min at 4-h intervals throughout a 15-h run in the laboratory. After that, the effluent was collected at 1-h intervals from the bottom of the wetland for 15 h and utilized to measure the Naphthalene exit concentration. The HYDRUS model was run for 24 h with the same input parameters as the lab model. Figure 27.10 compares the exit concentrations obtained from the bottom most observation node of the modeled wetland and the exit concentrations acquired from the lab-scale constructed wetland.

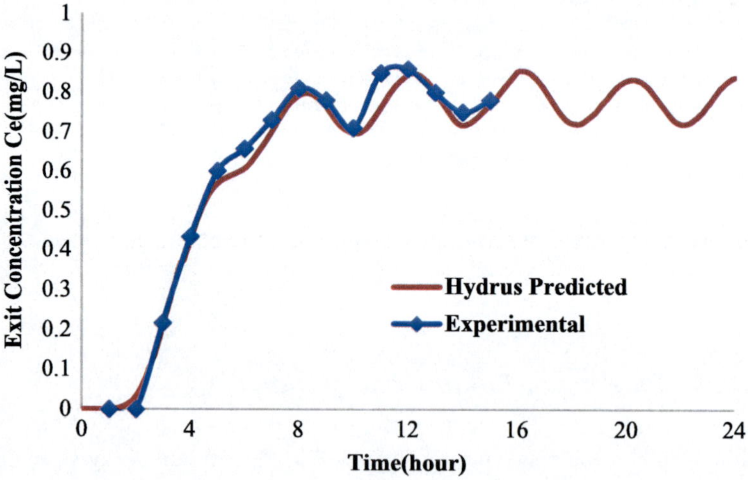

Fig. 27.10 Comparison of lab-scale wetland and HYDRUS predicted result

Conclusions

In this investigation, adsorbents made from locally accessible soil were utilized to extract PAHs from contaminated wastewater. A silty sand soil was taken from the Damodar River's bank in Durgapur, West Bengal, as an adsorbent. To thoroughly analyze the potential of this soil as an adsorbent of PAHs and also the design life of this soil to be used as an adsorbent, a detailed study was conducted that included isotherm studies, kinetic studies, column test, and lab-scale constructed wetland test, as well as numerical modeling of the same with HYDRUS-2D.

The Langmuir isotherm was the best-fitted isotherm with an R^2 value of 0.97 in soil isotherm research. Pseudo-second-order Kinetics was the best fitting kinetics in the Kinetic research, with an R^2 value of 0.9996. On the selected soil, vertical column research was done to observe PAH movement through the soil. HYDRUS-3D was used to model the same column. It was discovered that 90% of the initial PAH concentration reached the outlet after 1.6 days in a column with a height of 40 mm and an internal diameter of 60 mm. This made it easier to figure out how long the 40 mm dirt column would last if the contamination could migrate vertically.

A laboratory-scale constructed wetland was built, and experiments were conducted to see how effective the selected soil was at removing PAH. HYDRUS-2D with a CW2D module was used to model the same. To avoid any media clogging and achieve maximum removal effectiveness, the PAH loading was done at a rate of 15 min at a 4-h interval. The test lasted 15 h, with a maximal exit concentration of 0.86 mg/L obtained during the 12 h. As a result, removal efficiency of 90.25% was achieved. The HYDRUS model projected a maximum exit concentration of 0.8495 mg/L, indicating a clearance efficiency of 91.79%. As a result, the results from the lab-scale wetland, and the HYDRUS projected models are comparable.

References

1. Mirzaei M, Sabbaghi S, Zerafat MM (2018) Photo-catalytic degradation of formaldehyde using nitrogen-doped TiO_2 nano-photocatalyst: statistical design with response surface methodology (RSM). Can J Chem Eng 96:2544–2552. https://doi.org/10.1002/cjce.23192
2. Bayazit ŞS, Yildiz M, Aşçi YS, Şahin M, Bener M, Eğlence S, Salam MA (2017) Rapid adsorptive removal of naphthalene from water using graphene nanoplatelet/MIL-101 (Cr) nanocomposite. J Alloy Compd 701:740–749. https://doi.org/10.1016/j.jallcom.2017.01.111
3. Naphthalene. In: Encyclopaedia of toxicology, 2nd edn. pp 185–188 (2005)
4. Arizavi A, Mirbagheri NS, Hosseini Z, Chen P, Sabbaghi S (2020) Efficient removal of naphthalene from aqueous solutions using a nanoporous kaolin/Fe_3O_4 composite. Int J Environ Sci Technol 17:1991–2002. https://doi.org/10.1007/s13762-019-02521-1
5. Ashour I, Abu Al-Rub F, Sheikha D, Volesky B (2008) Biosorption of naphthalene from refinery simulated wastewater on blank alginate beads and immobilized dead algal cells. Sep Sci Technol 43:2208–2224. https://doi.org/10.1080/01496390801887351
6. Jiménez S, Andreozzi M, Micó MM, Álvarez MG, Contreras S (2019) Produced water treatment by advanced oxidation processes. Sci Total Environ 666:12–21. https://doi.org/10.1016/j.scitotenv.2019.02.128

7. Srivastava P, Sreekrishnan TR, Nema AK (2017) Degradation of low molecular-weight PAHs: naphthalene, acenaphthylene, phenanthrene, and fluorine. J Hazard Toxic Radioact Waste 21:04017008. https://doi.org/10.1061/(ASCE)HZ.2153-5515.0000360
8. Da Silva CMF, Rocha QdC, Rocha PCS, Louvisse AMT, Lucas EF (2015) Removal of naphthalene from aqueous systems by poly (divinylbenzene) and poly (methyl methacrylate-divinylbenzene) resins. J Environ Manage 157:205–212. https://doi.org/10.1016/j.jenvman.2015.04.025
9. Karri RR, Sahu JN (2018) Modeling and optimization by particle swarm embedded neural network for adsorption of zinc (II) by palm kernel shell-based activated carbon from the aqueous environment. J Environ Manage 206:178–191. https://doi.org/10.1016/j.jenvman.2017.10.026
10. Kim J, Hyun S (2018) Sorption of ionic and nonionic organic solutes onto giant Miscanthus-derived biochar from methanol-water mixtures. Sci Total Environ 615:805–813. https://doi.org/10.1016/j.scitotenv.2017.09.296
11. Younker JM, Walsh ME (2015) Impact of salinity and dispersed oil on adsorption of dissolved aromatic hydrocarbons by activated carbon and organoclay. J Hazard Mater 299:562–569. https://doi.org/10.1016/j.jhazmat.2015.07.063
12. Shabeer TPA, Saha A, Gajbhiye VT, Gupta S, Manjaiah KM, Varghese E (2014) Removal of polyaromatic hydrocarbons (PAHs) from water: effect of nano and modified nano-clays as a flocculation aid and adsorbent in coagulation-flocculation process. Polycycl Aromat Comp 34:452–467. https://doi.org/10.1080/10406638.2014.895949
13. Design Manual: constructed wetlands and aquatic plant systems for municipal water treatment. United States Environmental Protection Agency Office of Research and Development, EPA/625/1-88/022, September 1998
14. Rezic´ I (2011) Prediction of the surface tension of surfactant mixtures for detergent formulation using design-expert software. Monatsh Chem 142:1219–1225. https://doi.org/10.1007/s00706-011-0554-y
15. Lagergren S (1898) About the theory of so-called adsorption of soluble substances. Kungliga Suensk Vetenskapsakademiens Handlingar 241:1–39
16. Langmuir I (1918) Adsorption of gases on plain surfaces of glass mica platinum. J Am Chem Soc 40:1361–1403
17. Ho YS, McKay G (1999) Pseudo-second order model for sorption processes. Process Biochem 34:451–465
18. Freundlich H, Heller W (1939) The adsorption of cis- and transazobenzene. J Am Chem Soc 61:2228–2230
19. Low MD (1960) Kinetics of chemisorption of gases on solids. Chem Rev 60:267–312
20. Tempkin MJ, Pyzhev V (1940) Kinetics of ammonia synthesis on promoted iron catalysts. Acta Physicochim URSS 12:217–222
21. Weber WJ, Morris JC (1963) Kinetics of adsorption on carbon from solutions. J Sanit Eng Div 89:31–60
22. Pezzotti G, McEntire BJ, Bock R, Zhu W, Boschetto F, Rondinella A (2016) In situ spectroscopic screening of osteosarcoma living cells on stoichiometry-modulated silicon nitride bioceramic surfaces. ACS Biomater Sci Eng 2:1121–1134. https://doi.org/10.1021/acsbiomaterials.6b00126
23. Ray SS, Gusain R, Kumar N (2020) Chapter 5—Adsorption equilibrium isotherms, kinetics and thermodynamics. In: Carbon nanomaterial-based adsorbents for water purification-fundamentals and applications, micro and nano technologies, pp 101–118. https://doi.org/10.1016/B978-0-12-821959-1.00005-2

Chapter 28
Study of Soil Erosion Process in Laboratory Flume Experiments

Charu Chauhan, Tarun Semwal, and K. V. Uday

Introduction

Soil erosion is a complex phenomenon that involves the removal of topsoil from a region by natural forces such as excessive rainfall, wind, and human-caused activities majorly construction and deforestation. Soil erosion by water is one of the main global environmental problems which occur by the removal of topsoil particles by raindrops, transportation and deposition [1]. The three processes of detachment, transportation, and deposition of particles occur simultaneously, and one will be the major process at a certain point according to the hydrodynamics characteristics and rill bed resistance to erosion [2].

Various forms of erosion were observed comprising splash-induced sheet erosion or inter-rill erosion and rill erosion. On the basis of morphological advancement of rills, the progress of hillslope erosion involves three phases; the first phase is the sheet erosion phase, the second is the rill development phase, and the third is the rill maturity phase [3–5]. Over recent times, various studies have been done to interpret the erosion process, and it has been reported that erosion is directly related to runoff and soil; other factors like rainfall intensity, slope gradient, rainfall duration indirectly influence the erosion by changing the influence of direct factors [6, 7]. Also, related researches revealed that the extent of soil erosion was assigned to erosivity of rainfall and high rainfall intensity results in the formation of rills with great depth [8–10]. During erosion, the transportation of sediment by runoff and deposition can be governed by its size distribution and surface morphology, slope, and surface cover [11]. The particle size distribution of sediment can provide information on the different erosion processes as sediment size depends upon the

C. Chauhan · T. Semwal · K. V. Uday (✉)
School of Engineering, Indian Institute of Technology Mandi, Kamand Campus, Mandi, Himachal Pradesh 175005, India
e-mail: uday@iitmandi.ac.in

K. Muthukkumaran et al. (eds.), *Transportation and Environmental Geotechnics*, Lecture Notes in Civil Engineering 298, https://doi.org/10.1007/978-981-19-6774-0_28

dominance of erosion phases and particles eroded in inter-rill erosion are finer as compared to the particles eroded from the rill erosion process [5, 8, 12].

The study of the development of inter-rills and rills is crucial to understanding the erosion process which will help to improve erosion control, mitigation, and to restore degraded landmasses. Though the topography initiated by erosion is often complicated and its survey is difficult and unfeasible in the field [9]. The use of laboratory flume is advantageous for mimicking the actual field conditions and experimental conditions such as temperature, flow rate, gradient, slope, and rainfall can be controlled [12]. However, many differences exist between laboratory flume and natural conditions as experiments are conducted under simulated conditions. The soil erosion phenomenon cannot be anticipated by a single general procedure as it depends upon various factors.

With this in view, the present study aims to (1) study the different erosion processes occurring in soil having different particles size and depict the dominant erosion process. (2) Study of soil loss rate during different erosion processes. (3) Compare the particle size distribution, characteristics diameter (D10, D30, D60, and D90) of natural soil and sediment particle size distribution.

Materials and Methodology

Materials

The soils used in current study were collected from Mandi, Himachal Pradesh. The soils are classified as per USCS (ASTM D2487-17), and the physical properties of soils are presented in Table 28.1. The experimental runs were carried out on unsieved soil containing gravel and sieved soil passing from 2.36 mm sieve. The particle distribution curve of both sieved and unsieved soils was given in Fig. 28.1. For the implementation of the experiment, the laboratory apparatus consists of a box, a rainfall simulator, and a camera setup.

Experimental Setup and Soil Box Preparation

A plastic box (48 × 39 × 13) cm with an adjustable gradient was used for the experiments as shown in Fig. 28.2. The box used in the current study was smaller to measure soil erosion as it is more feasible to conduct experiments [17]. For the current study, the box was rested at a slope of 30°. Considering the adverse condition for erosion, slope angle was taken as 30° because a high slope angle reduces percolation of water and initiated more surface runoff and corresponding erosion [13]. The setup of the test was as shown in Fig. 28.2. For the collection of runoff and sediments during erosion, a plastic container was used and at the exit of the flume box. At the

Table 28.1 Physical properties of the soil

Soil properties	Sieved soil	Unsieved soil
Specific gravity	2.66	2.66
Gravel (%)	0	52
Sand (%)	32	45
Silt (%)	61	2
Clay (%)	6	1
Liquid limit (%)	21	21
Plastic limit (%)	NP	NP
Plasticity index	NP	NP
USCS	ML	GP
OMC (%)	15.27	15.27
MDD (g/cc)	1.85	1.85

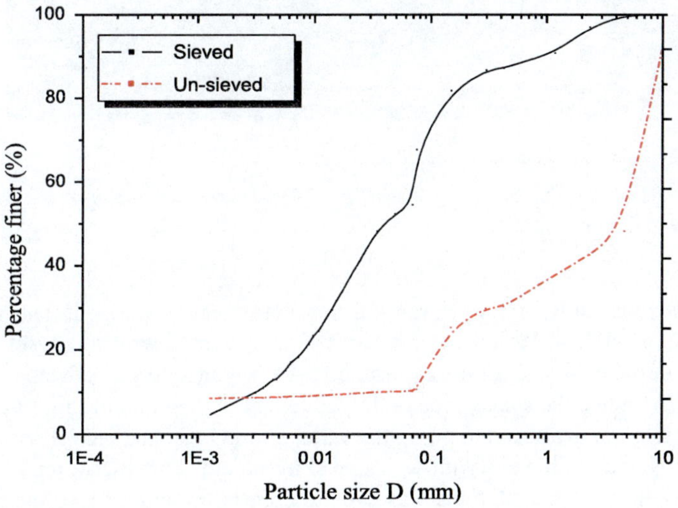

Fig. 28.1 Particle size distribution of sieved and unsieved soil

bottom side of the box, holes were made for the drainage of water. To prepare the soil specimen, oven-dried soil was mixed with water to achieve optimum water content and then compacted in a flume box at a bulk density of 90% of MDD. The bottom of the flume box was packed with sand up to a height of 3 cm for the purpose of drainage and the remaining 10 cm were filled with soil in tree layers. The prepared soil box was kept for 24 h for moisture homogenization.

Fig. 28.2 Experimental setup and rainfall simulator

Rainfall Simulator

For the current study, a simple rainfall simulator was developed. The simulator consists of a total of 16 no. of sprinklers attached on a wire mesh set at 1.5 m above the flume box used for the experiment. A spray nozzle was selected instead of a dropper to reduce the splash erosion by reducing the diameter of raindrops [14]. A submersible pump was used to pump the water stored in a tank and released through sprinkles. To maintain a constant water level in the tank running water was continuously supplied. A control valve was used to control the intensity of the precipitation. For the measurement of rainfall intensity, non-recording type rain gauges were used which consist of a collector funnel of diameter 105 mm, storage bottles, and a measuring cylinder. The storage bottles with funnel were set at nine points including four corners of the box. The amount of rainfall falling into the box was estimated by taking the average of the water in nine cylinders. The average rainfall intensity calculated was 90 mm/h; chosen intensity is typical of intense rainfall. The error for simulating rainfall was less than 5%.

Test Procedure

Before experimenting, the box was set to a slope of 30° and camera calibration and orientation were done before the experiment. The rainfall duration for each test was kept at 10 h considering the adverse condition. Images were acquired throughout the experiment at 1-min intervals. Throughout the experiment, surface runoff and sediments were collected in the vessel provided at the outlet of the flume box. To collect the infiltrated water, container was provided at the bottom of the box but the amount of infiltrated water was negligible owing to the slope provided. The collected runoff containing sediments was dried in an oven at 105 °C, and the weight of the remaining soil was measured as sediment yield. For the comparison of grain size distribution and characteristics diameter (D10, D30, D60, and D90) of sediment and original soil, particle size distribution curve was obtained for sieved soil.

Image Analysis

The photogrammetric equipment consists of a CMOS 800TVL HD camera. For the study of soil surface morphology and different erosion processes, image analysis was done using ImageJ software. ImageJ software is relatively easy to use, and eroded surfaces were traced based on which eroded area was calculated. Before tracing the surface, a scale for pixel size (pixel per cm) was set by taking the average of all known distances.

Results and Discussion

Soil Erosion Process

At the initial stage, the runoff rate was lower because in the initial stages raindrop splash erosion was dominant. From the soil surface shown in Fig. 28.3 for sieved soil, both sheet erosion and rill erosion were observed, as soil surface was eroded unvaryingly by a splash of a raindrop and small discontinues shallow flow occur. In the case of the sheet erosion phase, the sediment eroded along with runoff is fine particles. The depth and length of rills increased with time. When the length of rills on the hill slope becomes equal to 20 cm, the sheet erosion phase changes into the rill advancement phase [4]. In the case of the rill advancement phase when the rill channel develops, no selectivity occurs in sediment transport, and there was enrichment of fine as well as coarse particles. Whereas in the case of unsieved soil, only sheet erosion was observed as shown in Fig. 28.4, and eroded sediments were majorly fine particles. Similar results were also reported by various researches which revealed that inter-rill erosion favorably transports finer particles [11, 15].

Fig. 28.3 Surface morphology for sieved soil after 10-h rainfall event

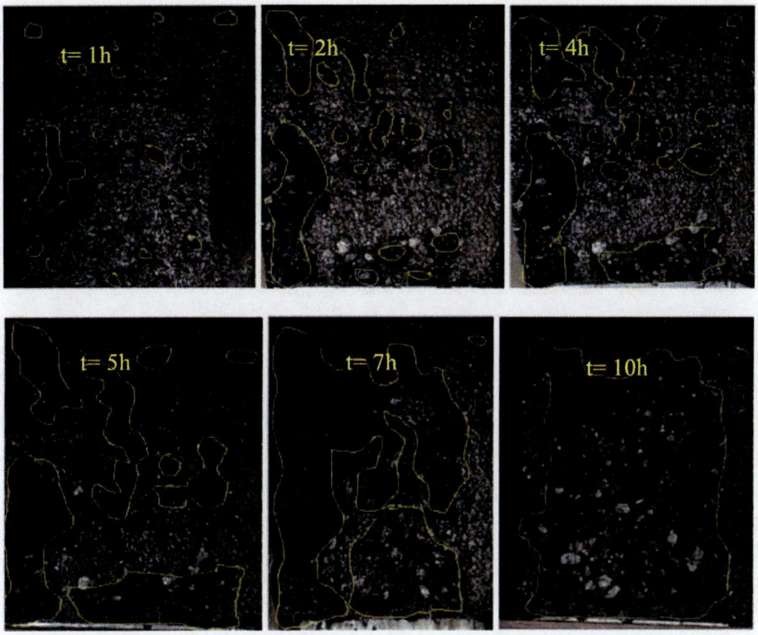

Fig. 28.4 Surface morphology for unsieved soil after 10-h rainfall event

To study the surface morphology of eroded surface, the plot between eroded area and time were obtained using Image J software as shown in Fig. 28.5, and it has been observed that in the case of sieved soil after 3 h of rainfall eroded area becomes constant. From past studies, it has been observed that a larger portion of eroded soil approximately 80% is transported through rills [16]. In the case of unsieved soil, the eroded area was continuously increased with time and soil erosion.

Soil Loss Rate

For the study of hillslope soil detachment and transport process sediment concentration and sediment, the loss rate is an essential factor. For a total rainfall duration of 10-h occurring at an intensity of 90 mm h^{-1}, the total amount of sediment was measured, and soil loss rates were calculated as shown in Table 28.2. The soil loss rate was defined as the amount of sediment lost per unit area per unit time [5, 7, 17].

$$E = \frac{M}{At}$$

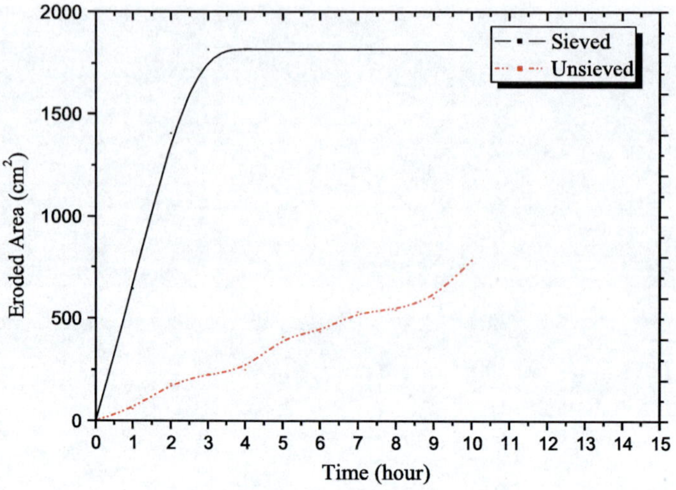

Fig. 28.5 Eroded area–time graph of sieved and unsieved soil

Table 28.2 Soil loss rate for different soil conditions

Experimental soil	Erosion pattern	Rainfall intensity mm h^{-1}	Amount of sediment collected (kg)	Soil loss rate kg m^{-2} h^{-1}
Sieved	Rill erosion dominated	90	2.53	1.38
Unsieved	Sheet erosion dominated		1.608	0.88

where M = mass of sediment collected (kg), A = area of flume box (m^2) and t = observation time in hour

From Table 28.2, it can be concluded that the total amount of sediment collected at the end of the 10-h rainfall event is larger for sieved soil as compared to the unsieved soil as rill erosion occurred in the case of sieved soil. Since the soil loss rate depends on many factors, viz. rainfall intensity, rainfall duration, slope gradient, topography, and area of the flume, a direct co-relation does not exist between the values obtained for soil loss rate. Previous researchers indicated that the rill erosion dominated the water erosion pattern larger as compared to the sheet erosion dominated water erosion pattern [4, 18]. Many experimental results revealed that when rill erosion dominates, soil erosion rate increases because soil structure was degraded by rill channel and flow in the rill can erode more soil.

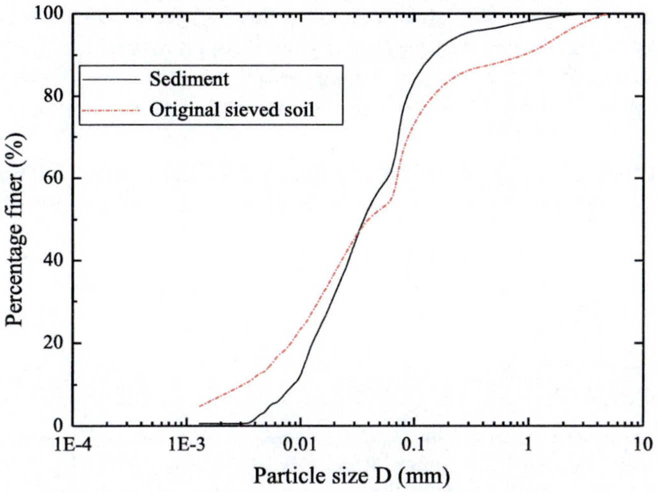

Fig. 28.6 Particle size distribution curve of sediment and original sieved soil

Comparison Between Particle Size Distribution of Sediment and Original Sieved Soil

The particle size of eroded soil sediment could give information about the erosion process and rills formation. Figure 28.6 shows the particle size distribution curve of the original sieved soil and the sediments. Form particle distribution curves D10, D30, D60, and D90 were obtained (Table 28.3). From the graph and table, it was observed that eroded soil has a high silt fraction followed by sand and low clay content. Previous studies also revealed that inter-rill or sheet erosion and rill erosion transport both finer and coarser particles.

Conclusion

In the current study, erosion process was studied on a small-scale soil surface by using an inclined flume, rainfall simulator, and camera. Based on the results obtained the following conclusion can be made.

Table 28.3 Characteristics diameters of sediment and original sieved soil

Type	D10 (mm)	D30 (mm)	D60 (mm)	D90 (mm)
Original sieved soil	0.0032	0.014	0.072	0.97
Sediment	0.009	0.019	0.06	0.14

1. In the case of sieved soil, sheet erosion followed by rill erosion was observed, and most of the eroded particles are carried through rills whereas in the case of unsieved soil only sheet erosion was observed.
2. During sheet or inter-rill erosion, soil loss rates were less as compared to the rill erosion.
3. When rills are formed during the erosion process, the sample contained coarser particles than fine particles as compared to the soil eroded during the sheet erosion phase.

References

1. Lei TW, Zhang QW, Zhao J, Tang ZJ (2001) Laboratory study on sediment transport capacity in the dynamic process of rill erosion. Trans ASAE 44(6):1537–1542
2. Chen XY, Zhao Y, Mi HX, Mo B (2016) Estimating rill erosion process from eroded morphology in flume experiments by volume replacement method. Catena. ISSN: 0341-8162
3. Hao H, Wang X, Guang J, Guo Z, Hua L (2019) Water erosion processes and dynamic changes of sediment size distribution under the combined effects of rainfall and overland flow. CATENA 173:494–504
4. Ni SM, Zhang DQ, Wen H, Cai CF, Wilson GV, Wang JG (2020) Erosion processes and features for a coarse-textured soil with different horizons: a laboratory simulation. J Soils Sedim 20:2997–3012
5. Zhang QW, Lei TW, Zhao J (2008) Estimation of the detachment rate in eroding rills in flume experiments using an REE tracing method. Geoderma 147:8–15
6. Sun L, Fang H, Qi D (2013) A review on rill erosion process and its influencing factors. Chin Georg Sci 23:389–402
7. Shen H, Zheng F, Wen L, Han Y, Hu W (2016) Impacts of rainfall intensity and slope gradient on rill erosion processes at loessial hillslope. Soil Tillage Res 155:429–436
8. Berger C, Schulze M, Rieke-Zapp D (2010) Rill development and soil erosion: a laboratory study of slope and rainfall intensity. Earth Surf Process Land 35:1456–1467
9. Shen H, Zheng F, Wen L, Lu J, Jiang Y (2015) An experimental study of rill erosion and morphology. Geomorphology 231:193–201
10. Seeger M (2007) Uncertainty of factors determining runoff and erosion processes as quantified by rainfall simulations. Physical Geography, University of Trier, 54286 Trier, Germany
11. Wang L, Shi ZH (2015) Size selectivity of eroded sediment associated with soil texture on steep slopes. Soil Sci Soc Am J 79(3):917–929
12. Aksoy H, Unal NE, Cokgor S, Gedikli A, Yoon J, Koca K, Inci SB, Eris E (2012) A rainfall simulator for laboratory-scale assessment of rainfall-runoff-sediment transport processes over a two-dimensional flume. CATENA 98:63–72
13. Jing X, Chen Y, Pan C, Yin T, Wang W, Fan X (2019) Erosion failure of a soil slope by heavy rain: laboratory investigation and modified GA model of soil slope failure. Int J Environ Res Public Health 16:1075
14. Balaguer-Puig M, Marqués-Mateu Á, Lerma JL, Ibáñez-Asensio S (2017) Estimation of small-scale soil erosion in laboratory experiments with structure from motion photogrammetry. Geomorphology 295:285–296. https://doi.org/10.1016/j.geomorph.2017.04.035
15. Schiettecatte W, Gabriels D, Cornelis WM, Hofman G (2008) Enrichment of organic carbon in sediment transport by inter-rill and rill erosion processes. Soil Sci Soc Am J 72(1):50–55
16. Zhang P, Tang H, Yao W, Zhang N, Xizhi LV (2016) Experimental investigation of morphological characteristics of rill evolution on loess slope. CATENA 137:536–544

17. Shinohara Y, Otani S, Kubota T, Otsuki K, Nanko K (2016) Effects of plant roots on the soil erosion rate under simulated rainfall with high kinetic energy. Hydrol Sci J 61:2435–2442
18. Jiang YL, Zheng FL, Wen LL, Shen HO (2019) Effects of sheet and rill erosion on soil aggregates and organic carbon losses for a Mollisol hillslope under rainfall simulation. J Soils Sediments 19:467–477

Chapter 29
Impact of Biochar Fraction on Thermal Characteristics of Soil-Biochar Composite

Deepak Patwa, K. Ravi, and S. Sreedeep

Introduction

The thermal backfill material is engineered geomaterials castoff as a replacement of locally available soil for efficiently prohibiting the heat movement through the source (such as underground heavy crude oil pipeline and crude oil storage tank). The thermal characteristics such as thermal conductivity and volumetric heat capacity of soil that is locally available at the site might not be suitable for thermal backfill material. As a result, it is necessary to supplement the locally accessible soil using appropriate geomaterials. The kind of soil and its mineral composition, particle size gradation, porosity, moisture content, and organic matter content all influence the thermal characteristics of the soil ([1, 20]). To increase the performance of backfill material, locally available soil might be combined with adequate porous organic matter content to reduce the thermal characteristics of soil-biochar composite (SBC). Because of its good physicochemical features and microstructural characteristics, soil-biochar composite (SBC) could be used as a backfill material [18].

Biochar is a porous carbonaceous organic product made from biomass by pyrolysis or thermochemical breakdown in a confined compartment under oxygen deficiency [13]. It is a porous, low thermal conductivity, and chemically stable substance [12]. As a result of the combined effect of changes in soil composition and porosity, the biochar amendment in the soil would influence the thermal characteristics (Glab et al. 2016; [8]). Patwa et al. [15] found that biochar amendments with three distinct particle size fractions reduced soil thermal characteristics. However, in the previous study, the thermal characteristics of the soil-biochar composite were measured at maximum possible compaction states [i.e., at the optimum water content and maximum dry density. In the above studies, the optimum water content (OMC) and maximum dry

D. Patwa (✉) · K. Ravi · S. Sreedeep
Department of Civil Engineering, Indian Institute of Technology Guwahati, Guwahati 781039, India
e-mail: deepak21@iitg.ac.in

K. Muthukkumaran et al. (eds.), *Transportation and Environmental Geotechnics*, Lecture Notes in Civil Engineering 298, https://doi.org/10.1007/978-981-19-6774-0_29

Table 29.1 Basic physical properties of the studied soil [15]

Properties	Value
Specific gravity (I)	2.7 ± 0.1
Grain distribution (%)	
Fine sand (0.075–0.425 mm)	9.1
Medium sand (0.425–2 mm)	12.1
Coarse sand (2–4.75 mm)	0.8
Silt fraction (0.075–0.002 mm)	39.74
Clay fraction (< 0.002 mm)	38.26
Liquid limit, LL (%)	54.42 ± 0.9
Plastic limit, PL (%)	35.25 ± 0.7
Shrinkage limit, SL (%)	30.12 ± 0.8
Optimum moisture content (%) Maximum dry density (Mg m^{-3})	22.72 1.579
Unified soil classification system	MH
BET surface area (m^2 g^{-1})	41.45 ± 1.5

MH: Inorganic slit of high plasticity

density (MDD) of SBC were first found as per the ASTM standard explained in [15]. Thereafter, the thermal test was done by compacting the samples in an acrylic mold. Thus, the above results are the combination of the influence of maximum possible compaction states and biochar particle size fractions.

Therefore, the current study will focus solely on the effect of different biochar fractions and amendment rates on the thermal characteristics of the soil. Three distinct biochar fractions obtained from Prosopis juliflora (hardwood of mesquite) are supplemented with locally available soil. The effects of the addition on SBC's thermal characteristics are discussed (Table 29.1)

Materials and Methodology

Soil and Biochar

This study used locally accessible residual soil gathered from the IIT Guwahati campus in Amingaon, NE state, Assam, India. The basic parameters of soil, such as particle size distribution, specific gravity, consistency limit, and compaction behavior of locally accessible soil, were estimated according to American Society for Testing and Materials (ASTM) recommendations (ASTM D422-63, 2007; ASTM D854-14, 2014; ASTM D4318, 2010 and ASTM D698-12, 2012). The locally available soil is classified as high plastic inorganic silt (MH) by following the standard grain size distribution analysis method mentioned in ASTM D2487-11. Biochar prepared from the hardwood of mesquite (Prosopis juliflora) was obtained from Greenfield Eco Solutions (An indigenous company). The biomass pyrolysis temperature was set to 500 °C. Initially, commercial biochar came in the form of big granules. For the thermal characteristics test, the biochar was grounded using an industrial-grade blender and sieved via a sieve with a 4.75 mm pore opening. Following that, the biochar was sieved into three distinct fractions for thermal characteristics testing, namely coarse biochar (CBC) with size 4.75–2 mm, medium biochar (MBC) with size 2–0.425 mm, and fine biochar (FBC) with size 0.425–0.075 mm, as shown in Fig. 29.1b. The ASTM D422-63, 2007 and ASTM D854-14, 2014 standards were used to determine the sieve analysis in the dry state and specific gravity (G) of biochar. The MBC and CBC fractions had G values of 1.43 and 1.26, respectively, while the finer biochar fraction had a value of 1.65. Table 29.2 lists the main physical parameters of the biochar that was employed.

Thermal Characteristics Measurement

The thermal characteristics (namely thermal conductivity, K and volumetric heat capacity, C) of biochar, soil, and soil-biochar composite (SBC) were studied using a standard KD2 Pro SH-1 sensor. The biochar percentage in SBC was selected as 5, 10, and 15% by dry mass. The thermal characteristics test of different biochar fractions, soil, and SBC was done in an acrylic container with an 80 mm interior diameter and 60 mm height (refer to Fig. 29.1c). The details of the mold dimension for the thermal properties' measurement have been reported in [15]. The bare soil and SBC samples were compacted under two different conditions. In the first case, the soil and SBC samples were compacted in three layers at a fixed load (2kN), while in the second case, it was compacted at a fixed density (1.3 Mg m^{-3}) using static loading equipment with the support of loading frame, proving ring, and plunger. In the first case, the mass of the soil sample and load are fixed, while the sample volume is not fixed. Comparatively, in the second case, both mass of soil and volume is fixed. The water content (25%) of bare soil and SBC was kept the same for both cases. Three duplicate samples were tested, and their average value was stated to check the repeatability in thermal properties measurement.

Results and Discussion

Thermal Characteristics of Biochar

The thermal conductivity (K) of the finer biochar fraction employed in this investigation was 0.127 ± 0.01 W/mK. The coarser and medium biochar fractions, respectively, have a K value of 0.121 ± 0.02 and 0.124 ± 0.02 W/mK (refer to

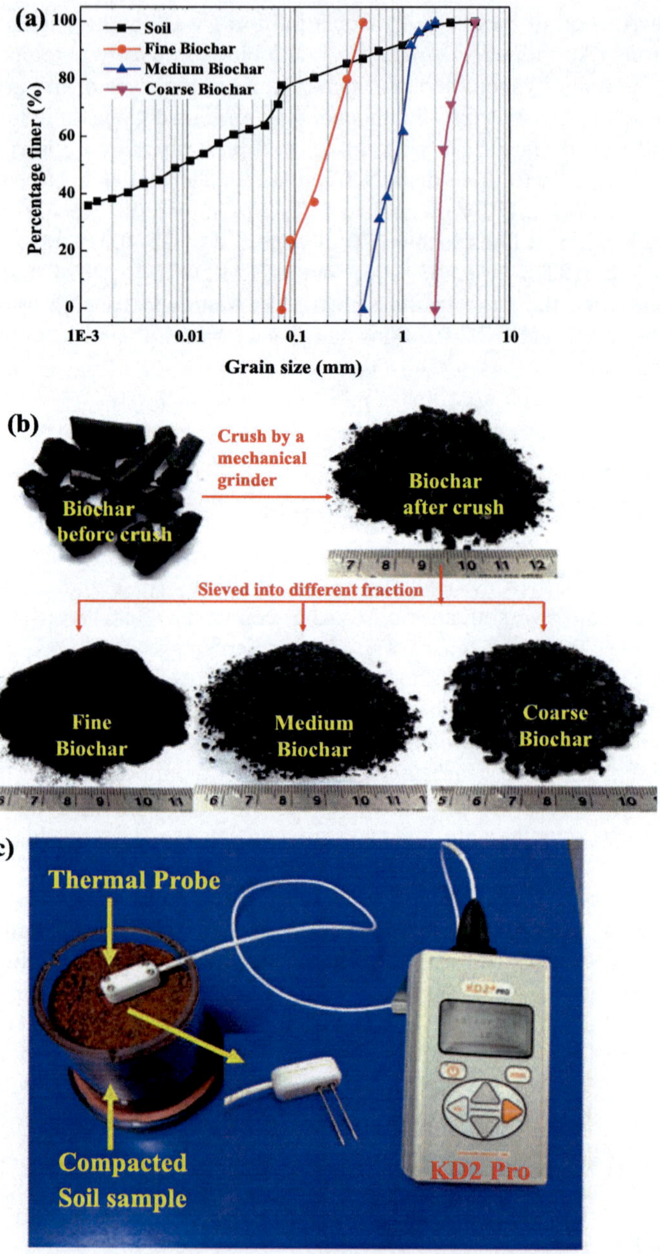

Fig. 29.1 **a** Gradation of used soil and biochar, **b** preparations of different biochar fractions, **c** thermal characteristics measurement [15]

Table 29.2 Basic properties of the biochar [15]

Parameter	Value				
Volatile matter (%)	12.84	Hydrogen(H) (%)	3.05	Molar ratios	
Moisture content (%)	10.12	Carbon (C) (%)	74.01	O/C	0.21
Ash content (%)	2.12	Nitrogen(N) (%)	0.18	H/C	0.49
Fixed carbon[a] (%)	74.92	Oxygen[a] (O) (%)	20.64	C/N	480
Properties		Coarse	Medium	Fine	
Mean diameter (mm)		2.542	0.800	0.20	
pH		7.75 ± 0.021	8.15 ± 0.010	8.65 ± 0.015	
Specific gravity		1.26 ± 0.008	1.43 ± 0.012	1.65 ± 0.012	
EC (μS cm^{-1})		244.7 ± 3.06	320.3 ± 6.81	492.7 ± 2.08	

[a] By Difference

Table 29.3 K, C, and DD of various biochar gradations [15]

Biochar fraction (mm)	Thermal conductivity (W/mK)	Heat capacity (MJ/m^3K)	Dry density, DD (Mg/m^3)
Fine biochar (0.075–0.425)	0.127 ± 0.01	0.729 ± 0.08	0.457 ± 0.004
Medium biochar (0.425–2)	0.124 ± 0.02	0.684 ± 0.11	0.386 ± 0.005
Coarse biochar (2–4.75)	0.121 ± 0.02	0.723 ± 0.12	0.363 ± 0.006

Table 29.3). The *K* value of biochar was significantly smaller than that of quartz mineral (7.72 Wm-1 K-1), water (0.596 W/mK, at 20 °C), and clay (1.52 W/mK) (Liu et al. 2018). The volumetric heat capacity (*C*) of the finer biochar fraction was 0.73 MJ/m^3K, whereas the coarser and medium biochar fractions had values of 0.72 MJ/m^3K and 0.684 MJ/m^3K, respectively. Similar results have been observed by other researchers [16, 18]. The above-stated biochar value from the reported literature and the present study was observed to be lesser than the reported soil constituent *C* value (1.91 MJ/m^3K), water (4.18 MJ/m^3K), and soil organic content (2.5 MJ/m^3K) (De Vries 1966). The presence of trapped air in the intrapores of biochar's skeleton is principally responsible for its reduced thermal conductivity and volumetric heat capacity. As a result, the addition of biochar was projected to lower the soil's thermal conductivity (*K*) and volumetric heat capacity (*C*).

Thermal Characteristics of Control Soil and SBC

Figure 29..2 presents the variation and percentage diminution in thermal characteristics of soil admixed with different gradations of biochar with respect to the control (bare) at a fixed load (2kN). The conductivity and heat capacity values were found to be 1.428 W/mK and 2.957 MJ/m^{3}K. The thermal conductivity of locally available soil declined in a range from 18 to 43% when the soil was admixed with fine fraction biochar (FBC), varying from 5 to 15% by dry mass of soil at the fixed load (see Fig. 29.2). Similarly, the decrement in volumetric heat capacity values was found to be 12, 17, and 25%. The trend was noted to be consistent with the incorporation of medium and coarser biochar fractions. A probable reason for the lessening in thermal properties could be the lessening in the density of the soil-biochar composite (SBC) due to biochar amendment. The heavier soil particles are replaced by lighter biochar particles, which increase the composite volume, consequently decreasing the composite density. The decrement in the density of soil-biochar composite leads to the enhancement in the total porosity of SBC, which results in more loss of contact between soil particles and a consequent decrease in the thermal properties.

The thermal characteristics of soil are reported to be closely dependent on porosity rather than the volume fraction of water [21]. Air has a thermal conductivity 25

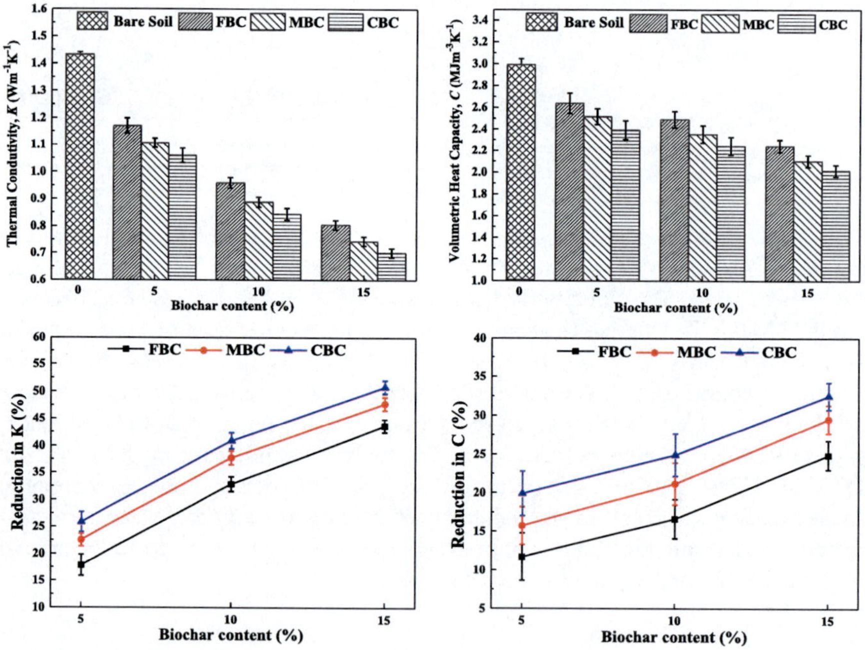

Fig. 29.2 Thermal characteristics variation and percentage reductions of bare soil with biochar percentage and fractions at a constant load

times that of water and nearly 100 times that of other soil constituents, decreasing heat transfer between soil solids and liquids molecules [14]. Furthermore, the biochar amendment's increase in *C*-rich organic components supports the decrease in soil heat conductivity. The volumetric heat capacity of air, on the other hand, is 0.00125 MJ/m^3K, which is significantly smaller than that of water (4.18 MJ/m^3K), soil mineral constituents (1.9 MJ/m^3K), and organic matter (0.00125 MJ/m^3K) (De Vries, 1966). As a result, the biochar amendment caused a decrease in thermal characteristics value. Moreover, it is noted from Fig. 29.2 that the soil added with the coarser biochar fractions revealed a higher reduction in thermal properties for the same biochar percentage at a fixed load. This is primarily due to the coarser biochar fraction having a lower specific gravity than the medium and finer biochar fractions. The coarser biochar fraction's lower particle density raises the total volume of the composite more than the medium and finer biochar fractions, consequently lowering the thermal characteristics.

The variation and percentage decrease in thermal characteristics of bare soil incorporated with various biochar percentages and biochar gradations at fixed density (1.3 Mg m^{-3}) are presented in Fig. 29.3. The thermal characteristics of control (bare) soil were detected to be 1.391 W/mK and 2.443 MJ/m^3K, respectively. The SBC was discovered to have lower K and C than bare soil, which is attributed to biochar's reduced *K* and *C* value. With the addition of 5, 10, and 15% finer biochar fractions, the thermal conductivity value of soil was found to be reduced by 21, 25, and 27%, respectively. Similarly, volumetric heat capacity was lowered by 7, 11, and 13%, respectively. The tendency was discovered to be consistent with the addition of medium and coarser biochar fractions.

Moreover, it is notable to point out from Fig. 29.3 that the rate of thermal properties decrement was noted to be diminished with a rise in biochar percentage. This might be accredited to the diminution in overall porosity of soil-biochar composite (SBC) as compared to bare soil. The thermal properties of the SBC at a particular dry density are governed by two aspects. The first one is the thermal characteristics of control or bare soil and biochar itself, and the second one is the total porosity of the soil-biochar composite. The diminution effect is mainly associated with the lower thermal characteristics of biochar itself, while the enhancement effect is due to the reduction in total porosity of soil-biochar mixes. Since the volume of mold and the weight of the soil sample are fixed and at a particular dry density. As the biochar content increases, the heavy soil particles are replaced with lighter biochar particles, which increases the total volume of the soil-biochar composite, leading to a decrease in the total porosity of the composite. Thus, at a higher biochar content (15%), the enhancement effect in thermal properties of the soil-biochar composite due to its reduced porosity is more dominant than the diminution effect due to biochar itself. Therefore, a reduction in a relative decrement in thermal properties of the soil-biochar composite was found at higher biochar amendment than the lower one.

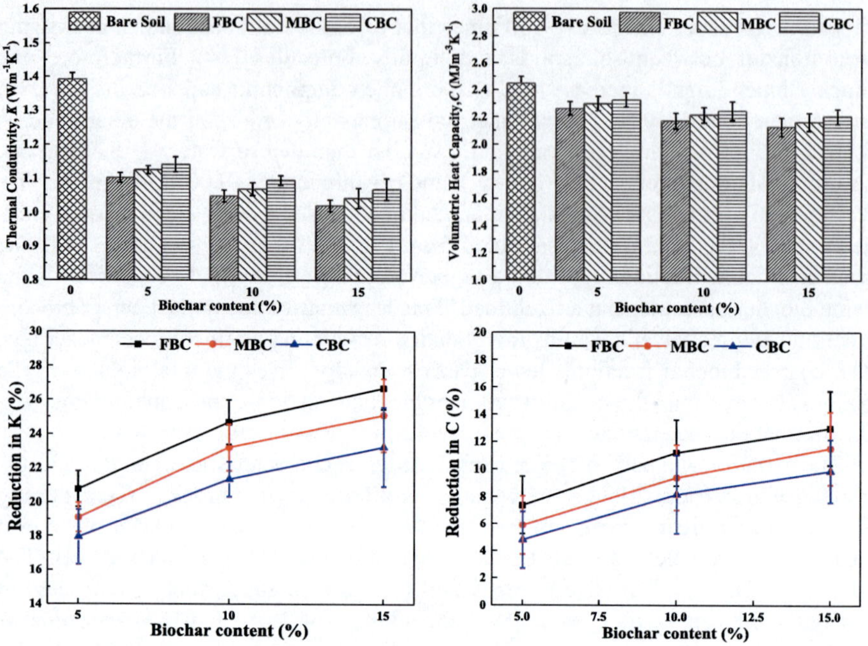

Fig. 29.3 Thermal characteristics variation and percentage reductions of bare soil with biochar percentage and fractions at fixed density

Conclusion

The impact of biochar fractions, namely coarse fraction (4.7–2 mm), medium (2–0.425 mm) fraction, and a finer fraction (0.425–0.075 mm) and amendment rates of 5, 10, and 15% on dry weight basis on the thermal characteristics of soil at constant load and fixed density is investigated in this study. At constant load, the thermal characteristics of bare soils added with coarser biochar (CBC) fractions are reduced more than those of soils incorporated with finer and medium biochar (FBC and MBC) fractions. At fixed density, the reduction is greater in the finer biochar fraction, followed by the medium and coarse fractions. According to the findings, the pore size distribution of the soil-biochar composite is predominantly responsible for the decrease in soil thermal characteristics when distinct biochar fractions are added to the bare soil.

References

1. Abu-Hamdeh NH, Reeder RC (2000) Soil thermal conductivity effects of density, moisture, salt concentration, and organic matter. Soil Sci Soc Am J 64(4):1285–1290
2. Abu-Hamdeh NH (2003) Thermal properties of soils as affected by density and water content. Biosys Eng 86(1):97–102

3. ASTM, D422–63 (2007). Standard Test Method for Particle-Size Analysis of Soils. *ASTM* International, *West Conshohocken, PA. doi*, https://doi.org/10.1520/D0422-63R07.
4. ASTM, D. 854 (2014) Standard test methods for specific gravity of soil solids by water pycnometer. *West Conshohocken, PA; ASTM International, 2014.*
5. ASTM, D4318–10. (2010). Standard test methods for liquid limit, plastic limit, and plasticity index of soils. *West Conshohocken, PA; ASTM International, 2010.*
6. ASTM D698–12 (2012) Standard test methods for laboratory compaction characteristics of soil using standard effort. *Annual Book of ASTM Standards.*
7. ASTM, D. 2487-11 (2011) Standard practice for classification of soils for engineering purposes (unified soil classification system) Annual Book of ASTM Standard. *ASTM International, 4, 08*
8. Burrell LD, Zehetner F, Rampazzo N, Wimmer B, Soja G (2016) Long-term effects of biochar on soil physical properties. Geoderma 282:96–102
9. Cuthbertson D, Berardi U, Briens C, Berruti F (2019) Biochar from residual biomass as a concrete filler for improved thermal and acoustic properties. Biomass Bioenerg 120:77–83
10. De Vries, D. A., 1966, Physics of Plant Environment, 2nd Ed., W. R. Van Wijk. Amsterdam: 487 North-Holland.
11. Głąb T, Palmowska J, Zaleski T, Gondek K (2016) Effect of biochar application on soil hydrological properties and physical quality of sandy soil. Geoderma 281:11–20
12. Gupta S, Kua HW (2017) Factors determining the potential of biochar as a carbon capturing and sequestering construction material: critical review. J Mater Civ Eng 29(9):04017086
13. Lehmann J, Joseph S (eds) (2015) Biochar for environmental management: science, technology and implementation. Routledge, New York, USA
14. Lu S, Ren T, Gong Y, Horton R (2007) An improved model for predicting soil thermal conductivity from water content at room temperature. Soil Sci Soc Am J 71(1):8–14
15. Patwa D, Chandra A, Ravi K, Sreedeep S (2021) Influence of Biochar Particle Size Fractions on Thermal and Mechanical Properties of Biochar-Amended Soil. J Mater Civ Eng 33(9):04021236
16. Patwa D, Muigai HH, Ravi K, Sreedeep S, Kalita P (2022) A novel application of biochar produced from invasive weeds and industrial waste in thermal backfill for crude oil industries. Waste Biomass Valorization 1–18. https://doi.org/10.1007/s12649-022-01694-0
17. Reddy KR, Yaghoubi P, Yukselen-Aksoy Y (2015) Effects of biochar amendment on geotechnical properties of landfill cover soil. Waste Manage Res 33(6):524–532
18. Liu Z, Xu J, Li X, Wang J (2018) Mechanisms of biochar effects on thermal properties of red soil in south China. Geoderma 323:41–51
19. Tong B, Kool D, Heitman JL, Sauer TJ, Gao Z, Horton R (2020) Thermal property values of a central Iowa soil as functions of soil water content and bulk density or of soil air content. Eur J Soil Sci 71(2):169–178
20. Usowicz B, Lipiec J, Usowicz JB, Marczewski W (2013) Effects of aggregate size on soil thermal conductivity: comparison of measured and model-predicted data. Int J Heat Mass Transf 57(2):536–541
21. Xie X, Lu Y, Ren T, Horton R (2020) Thermal conductivity of mineral soils relates linearly to air—filled porosity. Soil Sci Soc Am J 84(1):53–56

Chapter 30
Experimental Study on Volumetric Shrinkage Behavior of Indian Clays

Ankti Srivastava and Tadikonda Venkata Bharat

Introduction

Plastic clays such as bentonite are used as an engineering material in the nuclear waste repository [2, 9] as backfill and buffer material, earthen dam, road embankments, barriers in a cover system and landfill liner, etc. The civil engineering structures build on plastic clays are subjected to large settlement and uplifting due to change in moisture content with climatic and environmental conditions. Volume change of bentonite is directly related to the change in moisture content of the soil.

The volumetric shrinkage characteristic in plastic clays is due to the presence of montmorillonite mineral. When this clay comes in contact with the moisture, it forms a diffuse double layer around the clay particles which is responsible for swelling in the soil, and when the moisture is reduced due to evaporation, it causes a reduction in volume or shrinkage in the soil mass.

Understanding of volume change behavior, flow behavior, and shear strength behavior of unsaturated plastic clays requires a constitutive relationship which is soil water characteristics curve (SWCC, the relationship between moisture content and matric suction). Data of volume of soil specimen with moisture content defined as volumetric shrinkage curve (VSC) Fig. 30.1 during the evaporation process [1]. VSC is required to estimate SWCC with respect to the degree of saturation and water content since measurement of SWCC directly in the laboratory is time-consuming and gives inaccurate data as the change in the volume of soil is neglected [4, 10]. Structures built on plastic clays are subjected to excessive settlement due to a reduction in soil volume caused by the evaporation process. Hence, the study of volumetric shrinkage characteristics is very important to address settlement problems [8].

To establish volumetric shrinkage curve of plastic clays, mercury displacement method [3], and resin method were used in the laboratory. Accurate measurement

A. Srivastava (✉) · T. V. Bharat
Indian Institute of Technology Guwahati, Assam 781039, India
e-mail: ankit.srivastav@iitg.ac.in

K. Muthukkumaran et al. (eds.), *Transportation and Environmental Geotechnics*, Lecture Notes in Civil Engineering 298, https://doi.org/10.1007/978-981-19-6774-0_30

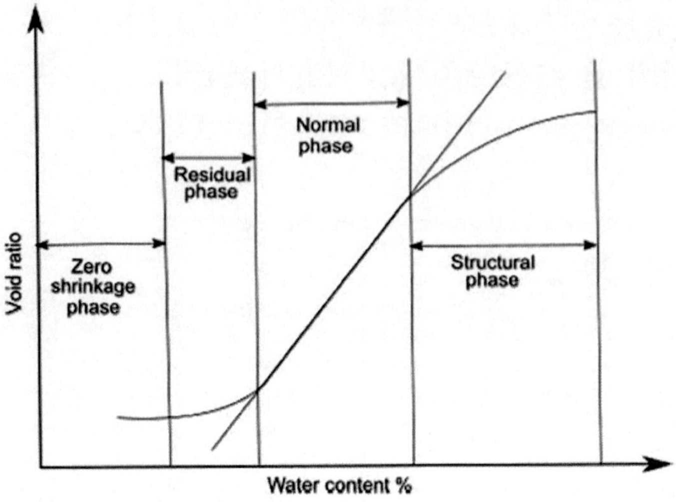

Fig. 30.1 Volumetric shrinkage curve [7]

of soil specimen volume is a very challenging task in the laboratory. Measurement of soil specimen volume was done by submerging soil clod into the kerosene and using application of Archimedes principle in resin test. Volume measurement was done from the slurry state (saturated) to dry state. Saturated soil sample dried in the atmosphere due to evaporation process. Soil specimen losses its moisture very fast in the initial stage of drying but after some time, the rate of decrease in moisture was decreases but shrinkage takes place continuously, and after some time, there was no significant change in volume was observed, the moisture content at which this state is reached is called shrinkage limit of a soil. Shrinkage limit of the soil is useful parameter when soil is subjected to large settlement or volume change through wetting and drying cycle, for example, the embankments, earth dam are subjected to wet and dry due to rainfall, changes in height of ground water table, excessive arid environment, etc. The VSC is very important characteristics of high to low plastic clays to understand flow, volume change, and strength behavior in unsaturated state and hence need to be established with different laboratory methods. The moisture content determination has been done by oven-dry method in the laboratory. Shrinkage limits for all type of clays were obtained from VSC curve and from standard shrinkage limit tests and then compared. The applicability of the resin method and mercury displacement technique was discussed.

Table 30.1 Geotechnical properties of soils used in the test [7]

Properties	*B*1	*B*2	Red soil
Particle size distribution (%)			
Coarse sand (4.75–2 mm)	0	0	0.034
Medium sand (2–0.425 mm)	0	0	18.59
Fine sand (0.425–0.075 mm)	2.35	3.79	65.046
Silt (0.075–0.002 mm)	35.33	27.48	10.32
Clay (<0.002 mm)	62.32	68.73	6.01
Atterberg limits (%)			
Liquid limit	175	296	41
Plastic limit	27	46	22
Shrinkage limit	12	14	18
Specific gravity, G_S	2.76	2.76	2.68

Materials and Methodology

Materials

Commercially available bentonites were taken which are representing as B1 and B2. These bentonites were obtained from Barmer district of Rajasthan. Naturally occurred red soil from the IIT Guwahati campus was taken and sieved through 0.425 mm sieve in the laboratory. The basic geotechnical soil properties of bentonites and red soil are given in Table 30.1.

Resin Test

Oven-dried samples of soil were taken and mixed with water, nearly 1.5 times its liquid limit. Mixing was done to make a slurry of desired consistency. Duplicate soil specimens were placed in shrinkage dish and kept in an atmosphere for 24 h to gain some stiffness. The specimens were coated with Fevicol glue (synthetic resin). Glue was diluted by adding with the ratio of 1 part of water and 10 parts of glue. All specimens were suspended with the help of thread as shown in Fig. 30.2 and kept it in the laboratory atmospheric condition for drying by an evaporation process. Glue will allow vapor transfer from the soil specimen to the atmosphere but it does not allow moisture from the atmosphere into the soil specimen. Measurement of specimen clod volume was done by submerging it in the non-polar liquid kerosene having density 0.8 g/cc. Archimedes principle was used to estimate specimen clod volume from the submerged weight and density of kerosene. Clod volume measurements were done from saturated state to air dry state until no shrinkage was observed in the soil specimens. Measurement of initial volume of the clod in normal shrinkage phase, just after the formation of clod specimens in shrinkage dish was not possible because it requires sufficient time to gain some stiffness so that glue can be applied. After the application of Fevicol glue (synthetic resin) to the clod specimens, it was kept in the laboratory at the atmospheric condition for some time to solidify the glue. When glue

solidifies, then volume measurement was started. One of the main advantages of this method is that the shrinkage test on single clod specimen provides entire shrinkage curve for the particular soil.

Results and Discussion

The volumetric shrinkage curve (VSC) is represented as the constitutive relationship between the state variables volume versus moisture content or void ratio vs. moisture content [5]. The plot of VSC data for bentonites $B1$, $B2$, and red soil was shown in Fig. 30.3a–c in the form of void ratio vs. water content. Initially, the VSC data follows 1:1 straight line representing the change in void ratio is equal to the change in water content with 100% degree of saturation. As of drying proceeds, there was further loss of moisture from the soil and at some water content, air starts entering into the soil pores resulted in partially saturated soil having a degree of saturation less than 100%. The VSC data deviates from the point of air entry into the soil specimen and not follow 1:1 straight line. The water content at which there is no reduction in the volume of the soil is defined as shrinkage limit of the soil. The value of shrinkage limits of low plastic clay $B1$, high plastic clay $B2$, and red soil was obtained from the VSC plots, and from the standard shrinkage, limit test data is reported in Table 30.2.

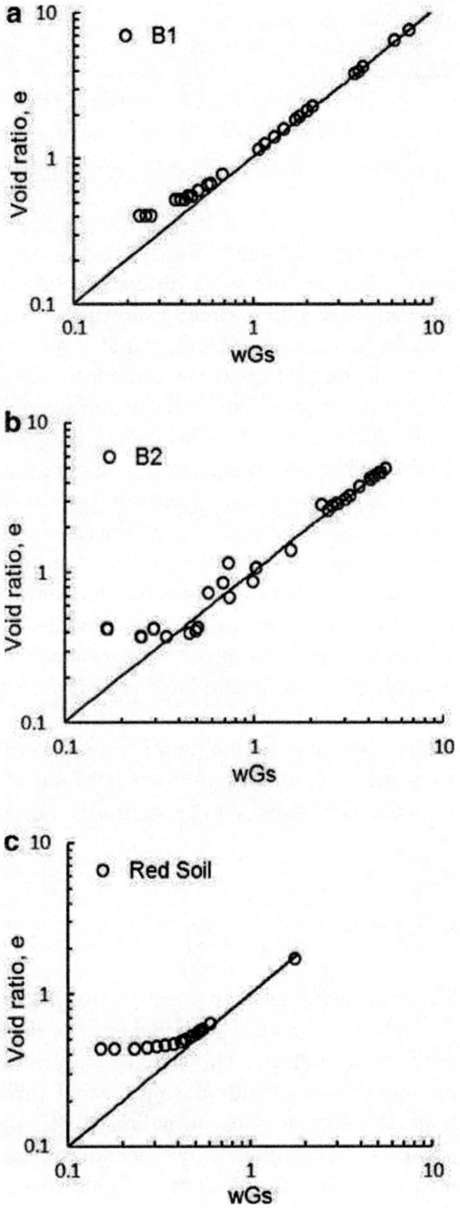

Fig. 30.3 **a** VSC of *B*1 by resin test, **b** VSC of *B*2 by resin test, **c** VSC of red soil by resin test

Table 30.2 Shrinkage limits of the soils from two different test methods

	Soil type	Clod test	Mercury displacement
Shrinkage limit, %	*B*2	10	12
	*B*2	11	14
	Red soil	13	18

Measured shrinkage limits for this soil from two different techniques were not matching and show 2–5% differences between their respective values for a particular type of soil [7] have already shown that the shrinkage limit obtained from resin test data for bentonites was well matched with the direct volume measurement technique. Thus, shrinkage limits calculated from the standard method (mercury displacement) may have an error. This error in shrinkage limit can be related to assumptions made in the derivation of the shrinkage limit formula. The assumption is that the volume of the soil at shrinkage limit is same as the volume of the soil in oven-dry state. Clays have very smaller pores that can hold moisture due to large matric suction at very low moisture content near the shrinkage limit. This moisture cannot be removed by drying the soil in the laboratory atmospheric condition and requires a massive amount of energy to remove this moisture. The oven drying process removes water from the smaller pores of the bentonites. After oven drying decrease in soil volume was observed because of moisture loss from the smaller pores of the soil. Hence, the volume at shrinkage limit and volume at fully dry state was not the same which consequence in error in shrinkage limit determined from mercury displacement technique. This may be the major limitation of shrinkage limit determination from the standard method of shrinkage limit laboratory test. Therefore, the clod test is more accurate as compared to the standard mercury displacement method.

Conclusions

The main advantage of resin test is that the single soil specimen is enough to determine the entire volumetric shrinkage curve of fine-grained soils which saves time and required less effort. The initial measurement of clod specimen volume, just after the application of glue during normal shrinkage phase is difficult because of glue requires sufficient time to solidify. The initial VSC data during normal shrinkage phase can be established by other volume measurement techniques but not all volume measurement techniques are reliable showed by [6, 7]. Limitations are involved in shrinkage limit measurement technique using standard shrinkage limit test for the plastic clays as the volume at shrinkage limit was not same as volume at oven-dry state which is the assumption involved in the derivation of shrinkage limit formula as a function of oven-dry volume. The water content at air entry point was found to be much higher than at the shrinkage limit. It means the soil becomes partially saturated at air entry point before reaching the constant volume (boundary of solid and semi-solid state) at the shrinkage limit point.

References

1. Bensallam S, Bahi L, Ejjaaouani H, Shakhirev V (2012) Shrinkage curve: experimental study and modelling. Int J Eng 25(3):203–210
2. Bharat TV, Sridharan A (2015) Prediction of compressibility data for highly plastic clays using diffuse double-layer theory. Clays Clay Miner 63(1):30–42
3. BIS I (1972) 2720 Methods of test for soils: part 6 determination of shrinkage factors
4. Fredlund DG, Houston SL (2013) Interpretation of soil-water characteristic curves when volume change occurs as soil suction is changed. Adv Unsat Soils 1:15
5. Fredlund DG, Rahardjo H, Rahardjo H (1993) Soil mechanics for unsaturated soils. John Wiley & Sons
6. Gapak Y, Das G, Yerramshetty U, Bharat TV (2017) Laboratory determination of volumetric shrinkage behavior of bentonites: a critical appraisal. Appl Clay Sci 135:554–566
7. Gapak Y, Yerramshetty U, Bharat TV (2015) Laboratory determination of volumetric shrinkage behaviour of plastic clay. In: 50th Indian geotechnical conference.
8. Nelson JD, Miller DJ (1992) Expansive soils. Prob Pract Found Pavement Eng 260
9. Pusch R (1992) Use of bentonite for isolation of radioactive waste products. Clay Miner 27(3):353–361
10. Wijaya M, Leong EC, Rahardjo H (2015) Effect of shrinkage on air-entry value of soils. Soils Found 55(1):166–180

Chapter 31 Thermo-Mechanical Behavior of Geothermal Energy Pile in Indian Sand

Shachi Yadav, Md Azhar, and Somenath Mondal

Introduction

During last two or three decades, excessive exploitation of natural resources as well as their depletion is occurring at an alarming rate. All these take place due to excessive use of fossil fuels and other anthropogenic activities which results in emission of greenhouse gases into the atmospheres [1, 2]. This activity results in increasing global carbon footprint. Due to India's tropical climate, a large amount of non-renewable energy has been utilized for space conditioning which can be reduced by geothermal energy. Geothermal energy is available in abundance and reliable i.e. constantly present at shallow depth of ground [3]. Additionally, the production of geothermal energy assured the continuation energy production for decades at lower cost and thus, it would not shrink as fossil fuel [4, 5]. With the implementation of geothermal energy piles, energy can be harnessed from the shallow earth surface [4]. Therefore, installation of geothermal energy pile reduces the electricity consumption in comparison to traditional method.

Over the decades, several studies have been conducted on the thermo-mechanical behavior of geothermal pile installation in various soils [5–8]. Saggu and Chakarborty [7] have analyzed the behavior of geothermal energy pile in loose and dense sand under thermal loading. Afterward, to improve the heat exchanger capacity and performance of energy pile, thermal conductive filler for piles such as steel fiber and graphite has been used [8]. Moreover, the laboratory investigation has been performed long-term behavior of thermo-mechanical behavior of pile under different loading condition with 30 temperature cycles [6]. Sani and Singh [4] have analyzed that water content of soil and granular soil result better performance in energy system by inhibiting the temperature rise in surrounding soil. Furthermore, many studies have been analyzed to enhance the performance of geothermal energy pile by using

S. Yadav · M. Azhar (✉) · S. Mondal
Civil Engineering Department, National Institute of Technology, Jamshedpur 831014, India
e-mail: aazhar8603@gmail.com

K. Muthukkumaran et al. (eds.), *Transportation and Environmental Geotechnics*,
Lecture Notes in Civil Engineering 298, https://doi.org/10.1007/978-981-19-6774-0_31

FEM-based software [4, 9] and U-loop system [10] to harness greater amount of energy under different climatic condition.

In India, production and requirement of energy increases day by day, and it is mostly fulfilled by conventional methods, but utilization of geothermal energy has not been taken place to full extent. Nevertheless, these studies may not provide an appropriate representation of situations if energy piles are installed in Indian soil. With this in view, an attempt has been made in the present study to explore the thermo-mechanical behavior of pile in Indian soil and the influence of varied geometry on the soil pile interaction of shallow geothermal energy pile. Numerical simulation has been carried out using FEM-based software COMSOL Multiphysics under thermo-mechanical load on single floating pile under geological and climatic condition of Jamshedpur. Moreover, the performance of the geothermal energy piles has also been examined for different L/d ratio.

Methodology and Numerical Analysis

A hypothetical single pile is installed in sand and thermo-mechanical behavior of energy pile is simulated using FEM software COMSOL Multiphysics for Indian city Jamshedpur. Analysis has been performed for three different L/d ratio, i.e., 15, 18, and 20. In this study, models were analyzed as 2D axisymmetric, in which pile and soil geometries were created separately. Pile was considered as made of concrete and placed in single layer of dry sand. Pile was experienced 1000 kN constant mechanical load on pile head throughout the simulation period. Subsequently, a close U-loop for heat exchanger was created within pile, and temperature is applied for 2D axisymmetric analysis. Moreover, sand as well as concrete pile were modeled as linearly elastic material. A perfect contact has been assumed between pile and sand. Furthermore, bottom, side and top edge of sand was considered as fixed, roller and free respectively. The thermal properties of sand were taken from Mondal et al. [11], and Young's modulus and Poisson's ratio of sand were taken from Hegde and Sitaram [12]. Pile properties were adopted from Saggu and Chakarborty [7]. Table 31.1 presents the material properties of concrete and sand both. Figure 31.1 shows a typical finite element mesh of pile and surrounding soil with triangular elements. Diameter of pile was kept constant as 1 m and modeled for different L/d ratio. In few studies, it is recommended to adopt model dimensions 14–15 times of diameter as a width and more than two times length of pile as long [6, 7]. Therefore, in this study, dimension of surrounding soil was taken as 50 m long with 7 m diameter, which was approx. 14 times of diameter in width and more than 2.5 times to length of pile as depth.

According to climate condition of Jamshedpur, heating as well as cooling both actions has been experienced. The behavior of pile and surrounding soil was computed for coldest day, hottest day, and annual average ground ambient temperature. The average ambient temperature of Jamshedpur ground surface was obtained from Singh and Sharma [13]. The constant ground temperature is maintained at

Table 31.1 Properties of concrete and sand

Properties	Concrete	Sand
Young's modulus of elasticity (MPa)	33×10^3	15
Poisson's ratio	0.2	0.3
Coefficient of thermal expansion (/°C)	10^{-5}	1.62×10^{-4}
Thermal conductivity (W/m/°C)	2.1	1.87
Specific heat capacity (J/kg/°C)	800	613.49
Ratio of specific heat capacity	1.56	1.87
Density (kg/m^3)	2500	1630
Compressive strength (MPa)	45	–
Bulk modulus (MPa)	–	5.77
Shear modulus (MPa)	–	12.5

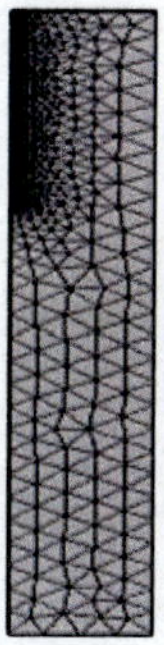

Fig. 31.1 Mesh generation of model

24 °C. Moreover, all boundaries of model were considered as adiabatic. Annual ground surface temperature variation of city in January, April, and December month is observed as 19 °C, 39 °C, 22 °C, respectively. Diurnal variation of temperature on hottest day in the month of April is 28, 56 and 33 °C. Diurnal variation of temperature on coolest day in the month of January is 18, 36 and 22 °C.

Results and Discussion

Validation of FE Model Under Thermo-Mechanical Loading

The thermo-mechanical analysis of geothermal energy piles was validated by comparing the numerical simulation results with the field pile load test data and the results of numerical simulations performed by Adinolfi et al. [14]. Adinolfi et al. [14] performed pile load test for 120 days of cooling and heating (as shown in Fig. 31.2).

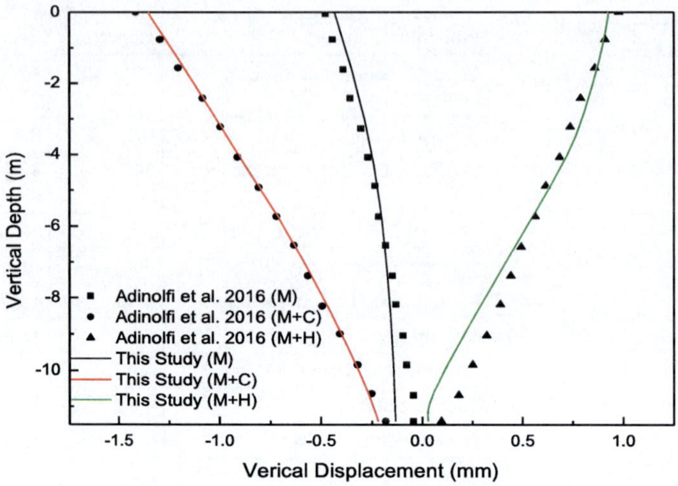

Fig. 31.2 Validation graph of energy pile with Adinolfi et al. [14]

The axial load was applied at pile head and was 1200kN. The length and diameter of pile were 11.4 m and 0.8 m, respectively, and installed in tuff and pozzolanas.

Vertical Displacement of Pile

Vertical displacement of pile along the length has been analyzed for coolest day, hottest day, and annual variation of the Jamshedpur against different L/d ratio. According to climate condition, hottest day, coolest day, and annual variation of temperature were considered as 18–36–22, 28–56–33, and 19–39–22 °C, respectively. For coolest and hottest day, simulation period was 24 h, and annual variation was modeled for entire year.

According to above figures, it can be concluded that vertical displacement of pile is almost similar for hottest and coolest day. Figures 31.3, 31.4 and 31.5 show the simulation of geothermal energy pile for hottest day, coolest day, and annual variation, respectively. For annual variation, displacement was more as compared to diurnal variation due time variation. Moreover, by increasing length of pile, vertical displacement reduced slightly due increment in surface area. Subsequently, during heating, expansion of soil observed and contraction was observed in cooling. However, the magnitude of expansion and contraction is negligible. Although obtained displacement is irreversible in nature, under cooling cycle any significant change has not been noticed. For the annual variation, the initial displacement is more as compare to diurnal variation. Further increment in temperature with time displacement slightly reduced with time. Therefore, it can be opined from above results that due to application of thermal loading, vertical displacement did not show significant change. So installation of geothermal energy pile did not hamper the stability of structure and could be utilized as a renewable energy source.

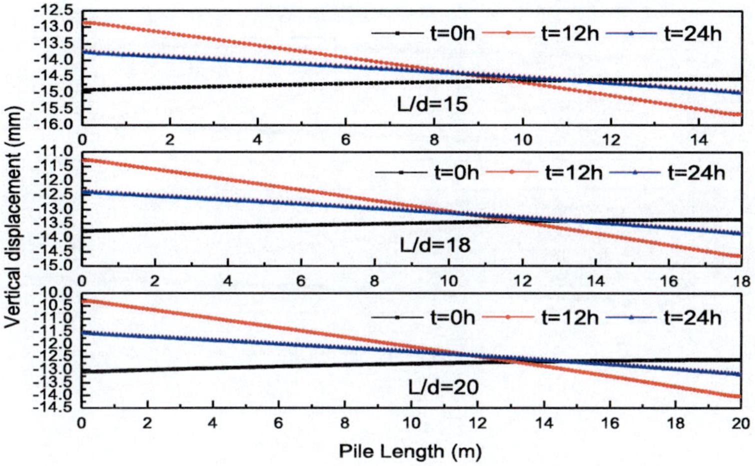

Fig. 31.3 Vertical displacement of energy pile for hottest day

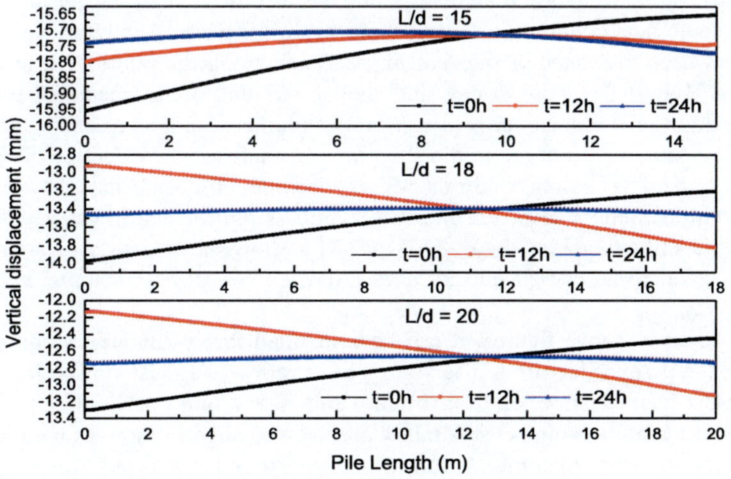

Fig. 31.4 Vertical displacement of energy pile for coolest day

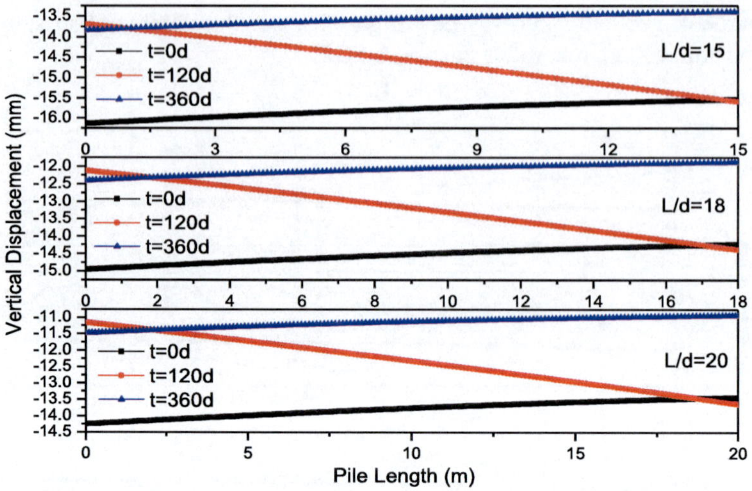

Fig. 31.5 Vertical displacement of energy pile for annual variation

Volumetric Strain of Pile

Volumetric strain of geothermal energy pile has also been analyzed for climatic as well as geological condition of Jamshedpur against different L/d ratio. Volumetric strain has been evaluated at pile–soil interference along the pile length at certain intervals. Hottest day, coolest day, and annual variation of Jamshedpur have been simulated for geothermal energy pile. Mainly, volumetric strain has been calculated due to temperature variation within the pile and analyzed the effect of thermal as well as mechanical loading condition. Moreover, volumetric strain has been observed at different location at pile–soil interface, which is represented in graphs as *(r, z)*. Hence, the following Figs. 31.6, 31.7 and 31.8 show the variation of volumetric strain of geothermal energy pile for hottest day, coolest day, and annual variation, respectively.

According to above figures, it can be concluded that volumetric strain of pile is under the permissible limit. The volumetric strain was almost similar for hottest and coolest day variation, and maximum strain was around 0.5% after $t = 18$ h. For annual variation, volumetric strain was observed slightly more as compared to diurnal variation and maximum strain was 0.6% around at 5000 h. The magnitude of volumetric strain was very less; hence, it did not affect the stability of structure under the climatic condition of Jamshedpur. Moreover, not any significant change in volumetric strain has been observed due changing the L/d ratio. Furthermore, a smooth curve has been obtained during entire simulation. Hence, it can be concluded that there is no any gap formation observed between pile and sand due to volumetric expansion and contraction during hearing and cooling process.

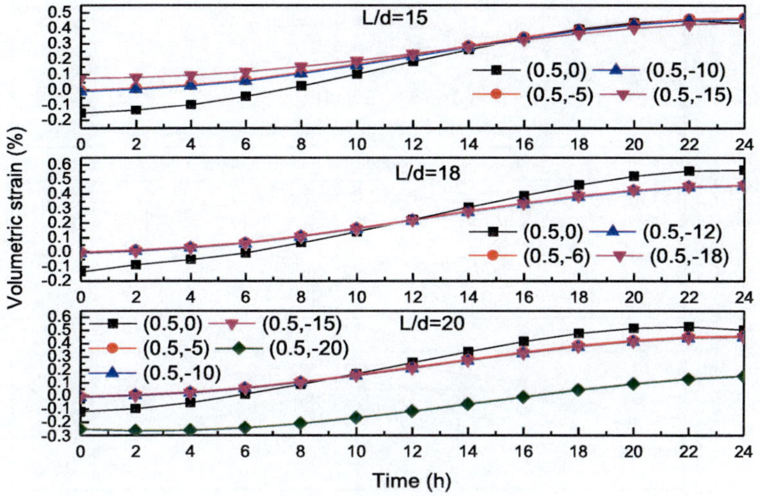

Fig. 31.6 Variation volumetric strain energy pile for hottest day

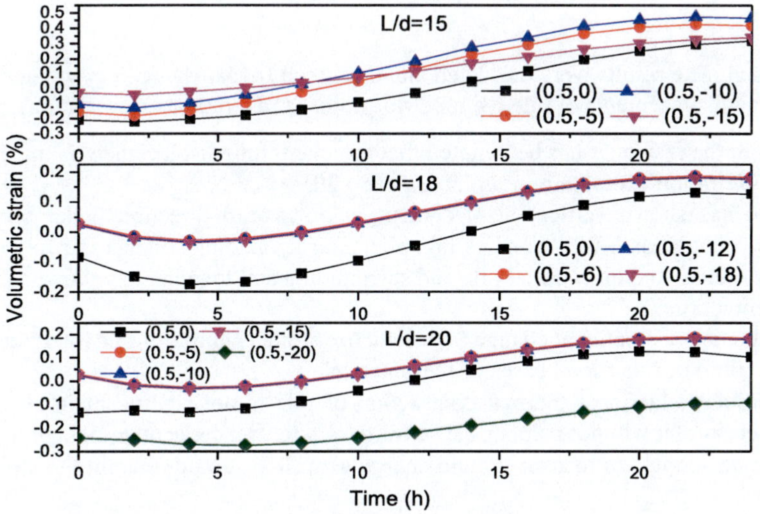

Fig. 31.7 Variation volumetric strain energy pile for coolest day

Conclusions

This research study has been presented the numerical analysis using finite element software. Numerical simulation has been performed with the aim of investigating the thermo-mechanical behavior of pile and resulting soil pile interaction in Indian

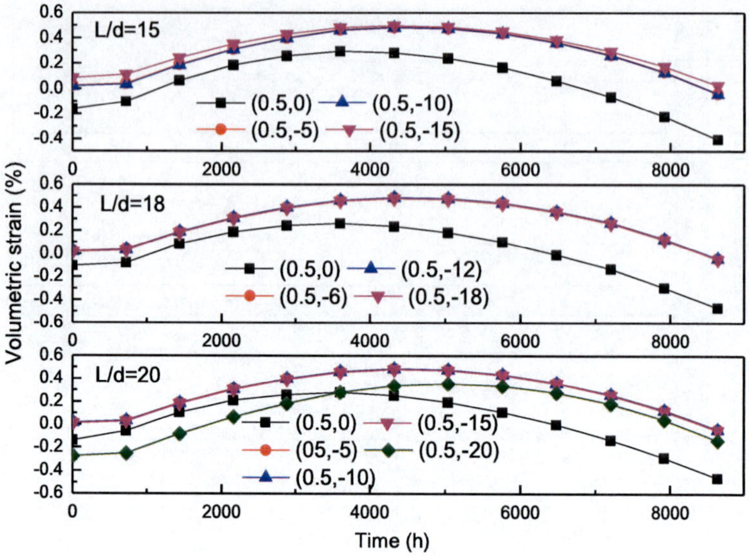

Fig. 31.8 Variation volumetric strain energy pile for annual variation

sand soil. The results were evaluated and discussed for Jamshedpur city in terms of vertical displacements of pile, volumetric strains, and temperature variations.

- From the results, it has been noted that the maximum displacement is maximum at *L*/*d* ration 15 and minimum at *L*/*d* ratio 20.
- Pile has shown vertical displacement in downward direction under constant mechanical load. While on heating, pile and surrounding soil has shown thermal expansion, and on cooling, pile and surrounding soil together has shown thermal contraction.
- There is no significant change in volumetric strains along soil pile interface with *L*/*d* ratio of pile for all cases of thermal cycle.
- Thermal dilation or thermal contraction of pile is not significant in all cases. Therefore, it will not hamper the stability of pile. Hence, geothermal energy pile can be adopted to recover ground energy as well as provide stability to structure both.
- Developed volumetric strain of lesser magnitude in all cases hence, no gap formation is observed between pile and surrounding soil due to volumetric expansion or volumetric contraction.
- GEPs provide cost-effective and long-term solution for meeting space heating and cooling requirements in buildings.

References

1. Rajamani (2015) The devilish details: key legal issues in the 2015 climate negotiations. Wiley Online Library
2. Möhner A (2018) The evolution of adaptation metrics under the UNFCCC and its Paris Agreement. Adapt Metrics Perspect Measure Aggregating Comp Adapt Results 15
3. Laloui L, Nuth M, Vulliet L (2006) Experimental and numerical investigations of the behaviour of a heat exchanger pile. Int J Numer Anal Methods Geomech 30(8):763–781
4. Sani AK, Singh RM (2020) Response of unsaturated soils to heating of geothermal energy pile. Renew Energy 147:2618–2632
5. Sani AK, Singh RM, Amis T, Cavarretta I (2019) A review on the performance of geothermal energy pile foundation, its design process and applications. Renew Sustain Energy Rev 106:54–78
6. Nguyen VT, Tang AM, Pereira J-M (2017) Long-term thermo-mechanical behavior of energy pile in dry sand. Acta Geotech 12(4):729–737
7. Saggu R, Chakraborty T (2015) Thermal analysis of energy piles in sand. Geomech Geoengin 10(1):10–29
8. Faizal M, Bouazza A, Singh RM (2016) Heat transfer enhancement of geothermal energy piles. Renew Sustain Energy Rev 57:16–33
9. Anongphouth A, Maghoul P, Alfaro M (2020) Geothermal energy pile-soil interaction under mechanical loading and thermal cycles. Int J 18(70):1–8
10. Lyu W, Pu H, Xiao H, Hu D, Ma Q (2021) Thermal performance of energy pile with deeply penetrating 1-U-shape heat exchanger. Geothermics 91:102023
11. Mondal S, Dangayach S, Singh DN (2018) Establishing heat-transfer mechanisms in dry sands. Int J Geomech 18(3):6017024
12. Hegde A, Sitharam TG (2015) 3-Dimensional numerical modelling of geocell reinforced sand beds. Geotext Geomembr 43(2):171–181
13. Singh RK, Sharma RV (2017) Numerical analysis for ground temperature variation. Geotherm Energy 5(1):1–10
14. Adinolfi M, Mauro A, Maiorano RMS, Massarotti N, Aversa S (2016) Thermo-mechanical behaviour of energy pile in underground railway construction site. In: Energy geotechnics: proceedings of the 1st international conference on energy geotechnics, ICEGT 2016, vol 29–31. Kiel, Germany, pp. 83–90

Chapter 32
A Review on Use of Biochar for Remediation of Heavy Metal Polluted Soils

Bhoomi A. Kamdar and Chandresh H. Solanki

Introduction

Due to rapid urbanization and industrialization, it is found that environment is recently getting polluted dramatically with heavy metals. Accumulation of heavy metals like lead, chromium, zinc, mercury, and arsenic is adversely affecting the soil biota and agro ecosystems along with ground water pollution. The main causes of heavy metal pollution are both geogenic and anthropogenic activities [1]. Heavy metals are persistent in the ecosystems and concentrate in the biosphere through trophic levels of the food web [2]. Thus, it is important to prevent and reduce the heavy metal contents in soil for future environmental safety.

It is critical to find a sustainable and effective method for remediation of heavy metal contaminated soils. There are various ex situ and is situ methods applied to contaminated soils like soil washing, stabilization/solidification, phytoremediation, electrokinetic remediation, etc. Use of organic and inorganic binders for remediation has also been explored. Li et al. (2019) compared two popular inorganic binders namely OPC and lime with reactive MgO to stabilize/solidify six individual heavy metal contaminated soil [3]. Due to constant CO_2 emissions related to the production of ordinary Portland cement, its use as a binder for immobilization is not feasible [4]. So, need arises for the use of alternative eco-friendly and cost-effective organic materials for treating the heavy metal contaminated soils. Organic binders like compost, microorganisms, and biopolymers have been applied to polluted soils and their effectiveness is measured. Leachate studies conducted by Anandha kumar et al. (2021) indicate that biopolymer like guar gum when amended in soil has good ability to remove lead, copper, zinc, and cadmium [5]. Anandha Kumar and Sujatha (2021) explored the use of different biopolymers for heavy metal attenuation in soils. It was

B. A. Kamdar (✉) · C. H. Solanki
Department of Civil Engineering, Sardar Vallabhbhai National Institute of Technology, Surat 395007, India
e-mail: bkamdar91@gmail.com

K. Muthukkumaran et al. (eds.), *Transportation and Environmental Geotechnics*, Lecture Notes in Civil Engineering 298, https://doi.org/10.1007/978-981-19-6774-0_32

concluded from their research that biopolymers have immense potential for various environmental applications due to their carbon neutral nature unlike conventional binder like cement [6].

Recently, biochar has caught attention for its use in many environmental and agricultural applications. It is an organic material formed by thermal decomposition of biomass in absence of oxygen. Biochar properties like SSA, pH, CEC, and surface functional groups play crucial role for remediation of heavy metals. This paper presents review of biochar use as an effective adsorbent for removal of different heavy metals from soil. Also, the review of articles showing the increased performance of pristine biochar with its further activation by different methods is done.

Effect of Biochar on Remediation of Heavy Metal Contaminated Soils

Immobilization of heavy metals in biochar amended soil is a combination of direct physical and chemical adsorption by biochar and indirect immobilization by changing the physicochemical properties of soil. Adsorption by biochar depends on the biochar characteristics which again depends on the biochar production temperature and other related conditions. Increase in soil pH due to addition of biochar also helps in converting heavy metals into more stable form.

Beesley and Marmiroli (2011) showed that biochar produced from hardwood proved to be more effective for sorption of Cd and Zn but had limited response for as [7]. Abdelhafez et al. (2014) studied effect of sugarcane bagasse and orange peel derived biochar on lead and arsenic contamination. It was concluded that addition of biochar increased soil properties like nutrients, CEC, and water holding capacity. Increased soil pH and organic matter due to biochar converted labile Pb into more stable form but increased the solubility of as [8]. Bian et al. (2014) amended wheat straw biochar with agricultural soil to immobilize Pb and Cd. Pollutants bonded with mineral phases of Al, Fe, and P available in biochar to convert into more stable forms [9]. Ehsan et al. (2011) showed that heavy metals content was reduced to a great extent after addition of biochar produced from unfertilized dates. After incubation, the water-soluble Ni and NH4NO3-extractable soil Cd and Ni contents were significantly lower in all the biochar treatments than the control [10].

The main mechanisms of heavy metal adsorption through biochar are (i) physical sorption on biochar surface and (ii) chemical sorption like precipitation, surface complexation, cation exchange, and electrostatic attraction. Four different crop straws derived biochar were used in two variable charged soils for lead adsorption by Jiang et al. (2014) [11]. Increased pH, electrostatic adsorption, cation exchange capacity, and precipitation were the main mechanisms for lead adsorption. Application of biochar to enhance properties of agricultural soils has been widely accepted. Adsorption of heavy metals from soils helps in reduced uptake by plants and thus preventing the health of organisms. Zhao et al. (2016) added chicken manure and rape straw biochar which turned Cd into more stable form due to increase in soil pH and CEC. It was concluded that 5% dose can be considered as optimum as addition of more than 5% biochar restricted the growth of maize plant due to high soil pH [12]. Alaboudi et al. (2019) presented agricultural waste biochar succeeded in immobilizing Pb and Cd into more stable form by binding with organic matter and improving soil properties. In contrast, it increased Cr solubility due to high pH formation and its uptake also increased in maize plant [13].

Summary provided in Table 32.1 shows that all biochars are not equally effective for adsorption of all heavy metal type. So proper care should be taken before using biochar for field application. It is also seen that mobilization of heavy metals occur after few years of application of biochar in soil. This may be due to biochar's contact with microorganisms present in soil and other environmental factors like acid rain, wind erosion, and soil pH [14]. So, it is necessary to study the long-term effects of biochar addition in soil for removal of heavy metals.

Effect of Biochar Modification on Remediation of Heavy Metal Contaminated Soils

Use of biochar for remediation of heavy metal contamination is proved to be efficient and eco-friendly. But its efficiency can be increased through modifications. Several researches has shown that modified biochars perform better than pristine biochar because of enhanced biochar properties like pH, SSA, porosity and surface functional groups. Hence, recent attention has been focused on modification of biochar. Modification of biochar is obtained by various physical and chemical methods such as activation, magnetization, oxidation, and digestion. Steam activation provides biochar surface with more complex pore structure and a larger pore volume. With magnetization of biochar, added advantage is that the spent biochar can be easily separated and collected by magnetic separation. Gong et al. (2020) used magnetic wheat straw biochar for lead adsorption in agriculture soil. Bonding of Pb particles with iron oxide coated on biochar particles was observed. Removal efficiency for both polluted soils was high. Also, it was found that uptake of Pb in ryegrass decreased by 30% [15]. Fan et al. (2020) observed that removal capacity was higher after thiol modified biochar was amended in individual heavy metal polluted soil. For binary metal system, Cd was adsorbed selectively over Pb. Main mechanism involved was complexation due to functional groups [16]. Zubair et al. (2021) used chitosan-coated textile biochar for Cd adsorption in soil. Results concluded that chitosan coating on textile biochar could effectively adsorb Cd from soil along with low uptake by the roots and shoots of moringa plant. Plant productivity and nutrition values also got enhanced due to addition of biochar [17].

Table 32.2 summarizes different modifications applied to biochar. Modifications show positive changes in physicochemical properties of biochar. These modifications are closely related to the adsorption efficacy of heavy metal contaminated soils.

Table 32.1 Effect of biochar on heavy metal removal from soil

Biochar used	Heavy metal type	observations	References
Hardwood	Cd, Zn and As	Biochar showed encouraging results for Cd sorption as leachate concentration dropped by 300 folds	[7]
Sugarcane bagasse and orange peel	Pb and As	Pb and As smelter soil stabilized and improvement in physicochemical properties of soil determined after addition of biochar	[8]
Wheat straw	Pb and Cd	Mechanism for removal of contaminants is cation exchange capacity and precipitation with mineral phases present in biochar	[9]
Unfertilised dates	Cd and Ni	Addition of biochar improved soil physical properties and successfully immobilized Cd and Ni	[10]
Crop Straw	Pb	Surface complexation, high pH, CEC and precipitation between biochar and lead helped in reduction of lead from soils	[11]
Chicken manure and rape straw	Cd	Addition of biochar increased soil pH and CEC values which helped in turning the speciation of cadmium into more stable state	[12]
Agricultural waste	Pb, Cd and Cr	Soil quality improved with biochar amendment along with good Pb and Cd removal efficiency	[13]

Table 32.2 Different modifications applied to biochar for heavy metal removal from soil

Biochar used	Heavy metal type	Modification	References
Wheat straw	Pb	Magnetization	[15]
Rice straw	Cd and Pb	Addition of thiol functional group	[16]
Textile biochar	Cd	Chitosan coated on biochar	[17]

Conclusion

Heavy metals in soil do not easily degrade and pose high environmental risks. They occur in the environment via many sources like disposal from industries, use of fertilizers and pesticides, coal combustion, and other natural sources. Research for removal of heavy metals from the soil has found effectiveness of many organic and inorganic binders. Among all the binders, biochar has been proved more effective due to its unique physical and chemical properties along with long-term stability in soils.

This review summarizes the use of different types of biochar for converting different heavy metals into less soluble forms. It is seen from the articles reviewed that any type of agricultural waste, manure, or industrial waste can be used to produce biochar. Effectiveness of biochar depends on production temperature and residence time as well as feedstock. It can be concluded that biochar is useful for heavy metal adsorption but its long-term stability in soils should be deeply investigated [18]. Also, presence of heavy metals in biochar should be studied before its application. Apart from this, it is seen that activation of biochar enhances the performance of pristine biochar. Use of different methods for biochar activation should be adopted to improve its properties thus making biochar most suitable and environmentally friendly material for immobilization of heavy metals in contaminated soils.

References

1. Reddy VA, Solanki CH, Kumar S, Reddy KR (2020) Stabilization/Solidification of Zinc- and Lead-Contaminated soil using limestone calcined clay cement (LC 3): an environmentally friendly alternative (2020)
2. Bandara T, Franks A, Xu J, Bolan N, Wang H, Tang C (2020) Chemical and biological immobilization mechanisms of potentially toxic elements in biochar-amended soils. Crit Rev Environ Sci Technol 50(9):903–978
3. Li W, Ni P, Yi Y (2019) Science of the total environment comparison of reactive magnesia, quick lime, and ordinary Portland cement for stabilization/solidi fi cation of heavy metal-contaminated soils. Sci Total Environ 671:741–753
4. Wang L et al (2019) Green remediation of As and Pb contaminated soil using cement-free clay-based stabilization /solidi fi cation. Environ Int 126(Nov):336–345
5. Anandha S, Evangelin K, Sujatha R, Pugazhendi A, Jamal MT () Guar gum—stabilized soil : a clean, sustainable and economic alternative liner material for landfills. Clean Technol Environ Policy 0123456789
6. Kumar SA, Sujatha ER (2021) An appraisal of the hydro-mechanical behaviour of polysaccharides, xanthan gum, guar gum and β -glucan amended soil. Carbohydr Polym 265(Mar):118083
7. Beesley L, Marmiroli M (2011) The immobilisation and retention of soluble arsenic, cadmium and zinc by biochar. Environ Pollut 159(2):474–480
8. Abdelhafez AA, Li J, Abbas MHH (2014) Feasibility of biochar manufactured from organic wastes on the stabilization of heavy metals in a metal smelter contaminated soil. Chemosphere 117(1):66–71
9. Bian R, Joseph S, Cui L, Pan G, Li L, Liu X, Zhang A (2014) A three-year experiment confirms continuous immobilization of cadmium and lead in contaminated paddy field with biochar amendment. J Hazard Mater 272:121–128
10. Ehsan M, Barakat MA, Husein DZ, Ismail SM (2014) Immobilization of Ni and Cd in soil by biochar derived from unfertilized dates. Water Air Soil Pollut 225(11)
11. yu Jiang T, kou Xu R, xia Gu T, Jiang J (2014) Effect of crop-straw derived biochars on pb(II) adsorption in two variable charge soils. J Integr Agric 13(3):507–516

12. Zhao B, Xu R, Ma F, Li Y, Wang L (2016) Effects of biochars derived from chicken manure and rape straw on speciation and phytoavailability of Cd to maize in artificially contaminated loess soil. J Environ Manage 184:569–574
13. Alaboudi KA, Ahmed B, Brodie G (2019) Effect of biochar on Pb, Cd and Cr availability and maize growth in artificial contaminated soil. Annal Agric Sci 64(1):95–102
14. Wang J et al (2021) Ecotoxicology and environmental safety analysis of the long-term effectiveness of biochar immobilization remediation on heavy metal contaminated soil and the potential environmental factors weakening the remediation effect : a review. Ecotoxicol Environ Saf 207(Sep):111261
15. Gong H, Chi J, Ding Z, Zhang F, Huang J (2020) Removal of lead from two polluted soils by magnetic wheat straw biochars. Ecotoxicol Environ Saf 205(Dec):111132
16. Fan J et al (2020) Remediation of cadmium and lead polluted soil using thiol-modified biochar. J Hazard Mater 388(Oct):122037
17. Zubair M et al. (2021) Efficacy of chitosan-coated textile waste biochar applied to Cd-polluted soil for reducing Cd mobility in soil and its distribution in moringa (Moringa oleifera L.). J Environ Manage 284(Feb)
18. Ahmad M et al (2014) Biochar as a sorbent for contaminant management in soil and water: a review. Chemosphere 99:19–33

Chapter 33
Influence of Slope Geometry on Stability of Bioreactor Landfills—A Numerical Analysis

Dharmendra Kumar Gurjar and Arvind Kumar Jha

Introduction

Several methods have been adopted to dispose the municipal solid waste (MSW) which includes an open dumping, shredding and pulverization, incineration and thermal pyrolysis, sanitary land filling, composting, disposal by barging out in sea. In India, landfilling method is used in the large scale for managing the MSW. Currently, development and construction of engineered or modern landfill has been given more priority in India. Modern landfills are well engineered which provides an atmosphere for the decomposition of solid waste. During decomposition, lot of contaminants release into the environment can lead the harmful effects which need to be considered during designing of landfill. However, the slower decomposition rate due to the absence of moisture is a major drawback of landfill system. Hence, bioreactor landfill is becoming popular worldwide currently due to its faster decomposition rate to handle MSW in effective and efficient manner.

The major concept of bioreactor landfill is to discompose and stabilize the solid waste by the recirculating the generated leachate, which creates a more suitable environment for the biological decomposition of organic substance in the landfill. The decomposition proceeds within 5–10 years only as compared to conventional landfill (i.e., 30–100 years) [1]. In this process, decomposition can be occurred in any or combination of processes, i.e., aerobic or anaerobic. Based on the decomposition process, it is classified into aerobic, anaerobic and hybrid (aerobic–anaerobic) bioreactor landfills. The key advantages of bioreactor landfills are to increase effective waste density and landfill capacity, increase the rate of gas production, in-situ leachate treatment, which may make an energy recovery more promising, reduce the risk of damage to the final cover due to settlement before placement of the final

D. K. Gurjar · A. K. Jha (✉)
Department of Civil and Environmental Engineering, Indian Institute of Technology Patna, Patna 801103, India
e-mail: jhaarvind@iitp.ac.in

K. Muthukkumaran et al. (eds.), *Transportation and Environmental Geotechnics*, Lecture Notes in Civil Engineering 298, https://doi.org/10.1007/978-981-19-6774-0_33

cover and an acceleration of waste decomposition. This may decrease the regulated post closure monitoring period as well as the overall cost of the landfill [2]. Despite several advantages, it is not so popular in India, except only few cities like Bangalore. The bioreactor landfill needs to be constructed in order to solve the several problems of waste disposal system and to overcome the limitations of conventional liner.

However, the acceleration in decomposition also leads to change the geotechnical characteristics of the waste considerably which has a great concern over stability and deformation behavior of landfill. So, there is a need to explain and quantify such impact on strength parameters. Previous studies related to stability analysis of bioreactor landfill using several techniques are summarized in Table 33.1. It is reported to be difficulties in performing tests on MSW materials due to their heterogeneity behavior and the presence of wide range of particle sizes. Based on the in-situ tests, Matasović and Kavazanjian [3] observed that the values of Poisson's ratio located closer to the surface are lower than deeper portion of the landfill. Most of the analyses were performed previously by assuming homogenous properties of materials throughout with entire age of MSW in landfills. However, it is not the same case in reality and differs with the type of materials such as organic and inorganic wastes. Also, uniform shear strength parameters have been adopted throughout the depth of landfill in many of the previous studies, whereas it varies with the age of decomposition and due to the overburden pressure caused with filling of the wastes. Further, stability analysis has not performed to optimize the slope geometry of bioreactor landfill which can provide a desire FoS.

The present study is aimed to analyze the slope stability of bioreactor landfills considering time and rate of decomposition. Further, effect of slope geometries such as angles, bench, and its width and combinations of these are examined using PLAXIS2D to examine their influence on the stability of bioreactor landfill. Influence of shear strength characteristics is also brought out clearly.

Material Model and Validation of Model

Material Model

Finite element method (FEM) is adopted to analyze the slope stability of bioreactor landfills considering several parameters such as slope geometries, material properties. In order to achieve the objectives, PLAXIS 2D, a sophisticated FE software is used. Mohr–Coulomb method in PLAXIS is used to analyze the slope stability analyses. Mohr–Coulomb method is based on mainly five characteristics of materials. These are Young's modulus (E), cohesion (c), Poisson's ratio (ν), frictional angle (ϕ), and di-latency angle (ψ).

Table 33.1 Summary of literature review on stability analysis of bioreactor landfill

Author(s)	Objectives	Key observations
Warith [4]	To determine the effect of solid waste size, nutrient balance on the frequency of MSW biodegradation and leachate recirculation	Smaller MSW in size has ability to degrade the waste in faster rate. Leachate recirculation reduces the organic load (i.e., measured in term of BOD and COD)
Zekkos et al. [5]	Used large-scale laboratory data and field data to develop hyperbolic relationship of MSW	Unit weight of MSW is controlled mainly by waste composition and practices of landfill operational, like cover soil placement, compaction effort, and liquids management during waste placement
Hossain et al. [6]	Performed seismic stability of bioreactor landfill with decomposition using PLAXIS and GSTABLE	FoS reduces with an increase in decomposition time for both static and dynamic cases as the decomposition increased with time
Haque [7]	Analyzed the dynamic characteristics and stability analysis using PLAXIS and GSTABLE of MSW in bioreactor landfills	The bioreactor landfill stability should be assessed using the strength characteristics determined as a function of time and decomposition rather than by using average values
Hossain et al. [8]	Determined the deformation of bioreactor landfills using PLAXIS	The settlement of wastes should be assessed as function decomposition rather than taking average values using the compressibility parameters
Hossain et al. [9]	Resonant Column (RC) test to examine dynamic behavior of MSW in bioreactor landfill	Degree of decomposition affects the damping curves and shear modulus reduction significantly
Srivastava et al. [10]	Reliability approach to find out the stability of bioreactor landfills slopes due to leachate recirculation and degradation	The stability of landfill slope increases with age elapsed. The landfill slopes changes with the value of reliability index between 10 and 15 years compared to the fresh MSW landfills

(continued)

Table 33.1 (continued)

Author(s)	Objectives	Key observations
Hossain and Penmethsa [11]	Laboratory test to understand the alteration in geotechnical behavior of MSW with degradation in bioreactor landfill	Degradation of waste increases the moisture content of the MSW subsequent reduction in pore spaces. Also, hydraulic conductivity reduces with an increase in stages of decomposition
Giri and Reddy [12]	Effect of unsaturated hydraulic properties on the bioreactor landfill slope stability using FLAC program	The unsaturated hydraulic behaviors of MSW influence the moisture distribution, capillary pressures, and pore water, and the landfill slope stability during leachate recirculation
Babu et al. [13]	Effect of spatially variable geotechnical properties stability of MSW landfill slope	FoS is obtained to be lower for spatially varied geotechnical properties than that of the same with uniformly constant geotechnical properties of MSW
Rajesh et al. [14]	Slope stability of MSW landfill using PLAXIS	FoS reduces with the rate of decomposition progress with time and hence, strength characteristics need to be considered for the stability analysis
Ismail et al. [15]	Stability of landfill slope using PLAXIS with waste degradation	FoS reduces correspondingly with increase in the slope angles
Rezwan et al. [16]	Monitor an amount of moisture variations in bioreactor landfill due to leachate recirculation using resistivity imaging technique	Resistivity of the solid waste reduced up to 80% by adding leachate/water using the recirculation pipe

Mesh Generation and Boundary Conditions

PLAXIS offers two types of elements, i.e., 6 and 15 nodes triangular element. 15 nodes triangular elements are very accurate which produces high quality results. Hence, 15 nodes triangular element is selected to use for further analyses (Fig. 33.1). Boundary conditions are selected in such that both side are allowed to move freely in vertical direction, whereas the bottom of the model is fixed in both the horizontal as well as vertical direction (Fig. 33.1). The foundation soil was assumed to be stiff

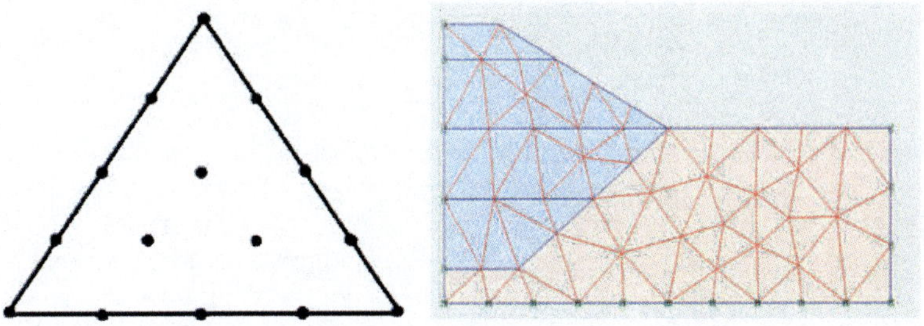

Fig. 33.1 Illustration triangular element (left) and cross section of generated mesh (right)

soil. The stability of foundation soil does not include in the present study, and hence, the boundary at bottom is fixed.

Analysis and Model Validation

The FoS was calculated using c-Phi reduction method in each cases. The incremental multiplier (i.e., M_{sf}) is used to stipulate the increase in the strength reduction. Successive reduction in the strength parameters can be observed in each step till completion of each the steps. After development of the failure mechanism, the PLAXIS gives the FoS as follows:

$$\text{FoS} = \text{Value of sum of } M_{sf} \text{ at failure} = \text{Strength available/Strength at failure.}$$

Experimental analysis on slope stability of bioreactor landfill has not found in the literatures. Further, the validation of model is carried out by considering simple geometrical model (Fig. 33.2) and comparing the result (FoS) obtained from manual calculation (i.e., Swedish circle approach) and numerical analysis using PLAXIS. The material properties used in the model are presented in Table 33.2. It is observed that the FoS values from manual calculation and PLXIS2D are obtained to be 2.26 and 2.06, respectively. The difference in results is 8.84% which is less than 10% and hence can be considered within permissible limit for further analysis.

Model Geometry

In the present study, geometrical models of bioreactor slopes are shown in Figs. 33.3–33.5. The bioreactor landfill geometry is conceptualized from the published works of Hossain and Haque [17] and Ismail et al. [15]. The step in slope is introduced which divides the regular slope into two parts, i.e., upper and lower slope (Fig. 33.3). The slope is divided into 4 phases of decomposition as similar to Hossain and Haque

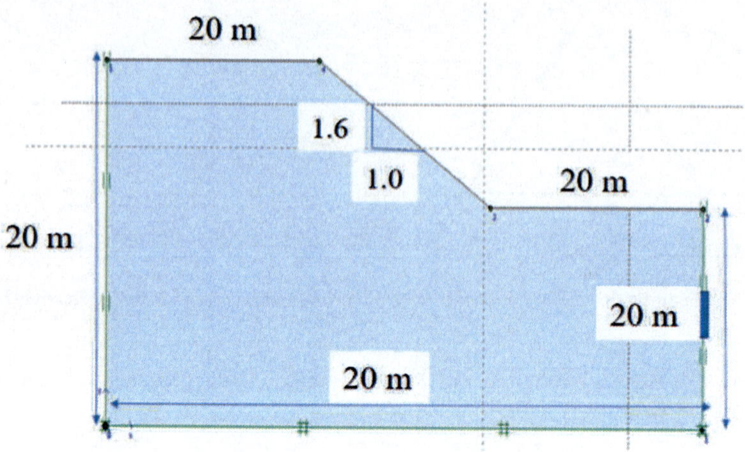

Fig. 33.2 Geometrical model of slope

Table 33.2 Properties of material used in geometrical model for validation

Parameters	Values
Soil type	Sandy clay
Cohesion	30 kN/m^2
Frictional angle	20°
Unit weight	18 kN/m^3
Young's modulus of elasticity (*E*)	1260 kN/m^2
Dilatency angle	0

[17]. The two other different geometrical models have been selected to examine the stability analyses and are shown in Figs. 33.4 and 33.5, in which two steps in slope are provided instead of uniform slope. Figure 33.4 shows the landfill consisting of four layers, out of which three layers are above natural ground level. The thickness of top layer, i.e., layer 1 is kept smaller compared to other layers, i.e., 2, 3, and 4 subsequently. Further, slope consists of two steps having different widths of 6 m and 9 m. However, angles of slope in each layer are kept uniform, i.e., 1:1 (Fig. 33.4). Similarly, landfill geometry as shown in Fig. 33.5 is also similar to Fig. 33.4; however, slopes in each layer are varying from 1:1, 2:1, 3:1, and 4:1.

Material Properties

All the properties of materials used in the FEM analysis of slope geometry (Fig. 33.3) for the different phases of decomposition are selected from Hossain and Haque [17]

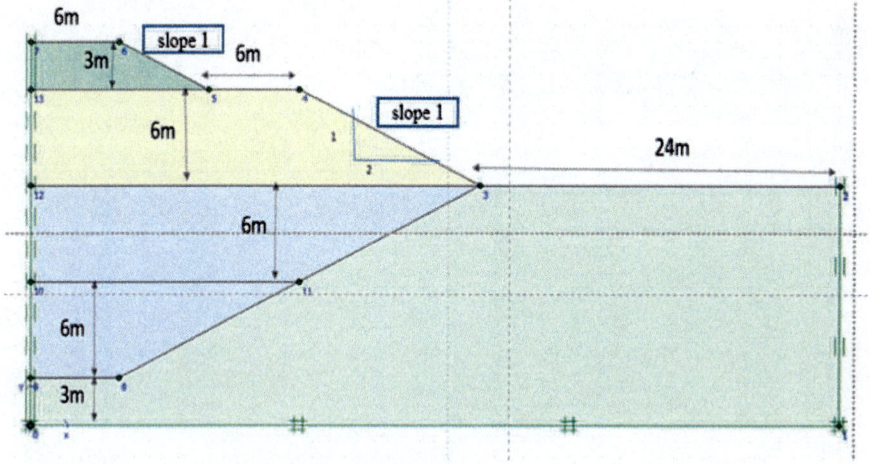

Fig. 33.3 Bioreactor landfill model with step in slope

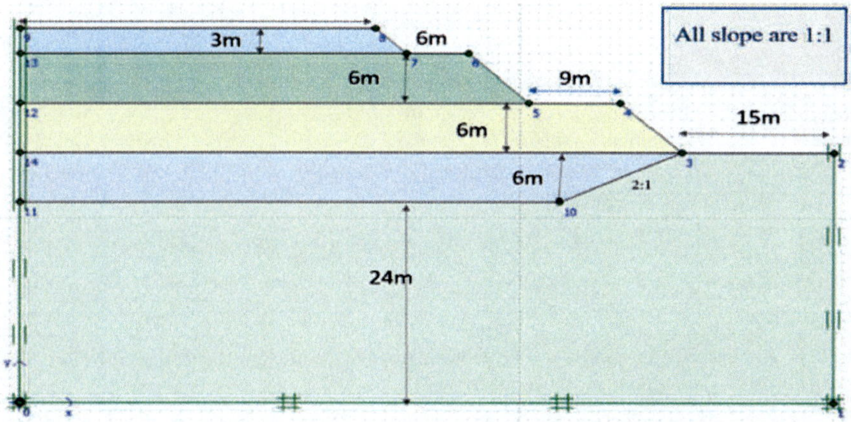

Fig. 33.4 Geometrical model of bioreactor landfill consisting of two unequal steps

and are presented in Table 33.3. Materials of each layers of bioreactor landfill that are used in analysis of models as shown in Figs. 33.4 and 33.5 are taken from previous research works [13–15]. The details of material properties of each phases are listed in Table 33.4.

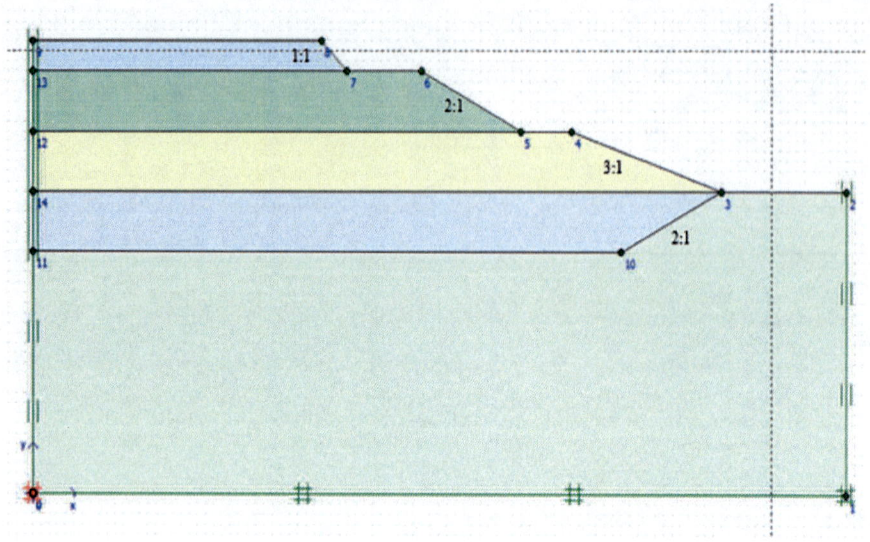

Fig. 33.5 Geometrical model of bioreactor landfill consisting of two unequal steps and different slope angles

Table 33.3 Properties of MSW for M–C model in present analysis [17]

Material set phase	Drainage condition	γ_{moist} (kN/m^3)	K (10^{-3}) (m/day)	υ	v_s (m/s)	c (kN/m^2)	$\Phi°$
I	Drained	11.00	860.0	0.25	23.70	–	33.0
II		11.80	86.0	0.40	30.30	–	27.0
III		12.60	86.0	0.42	71.70	–	25.0
IV		14.10	8.60	0.45	73.60	–	24.0
Natural ground		15.90	1.00	0.35	11.80	9.60	30.0

[***Note:*** γ_{moist} = Moist unit weight; υ = Poisson's ratio; K = coefficient of permeability; v_s = shear wave velocity; c = cohesion; φ = angle of friction]

Table 33.4 Characteristic properties of MSW for M–C model in present analysis [13–15]

Material set phase	Drainage condition	γ_{moist} (kN/m^3)	K (10^{-3}) (m/day)	υ	v_s (m/s)	c (kN/m^2)	$\Phi°$
I	Drained	7.5	190	0.25	25,000	10.0	30.0
II		9.3	173	0.40	25,000	23.8	27.5
III		10.6	130	0.42	25,000	32.9	25.8
IV		11.8	86	0.45	25,000	42.1	24.1
Natural ground		15.9	1	0.35	25,000	9.6	30.0

Table 33.5 Decomposition phases at each stages [17]

Cases/layers	1	2	3	4
Case 1	I	II	III	IV
Case 2	II	III	IV	IV
Case 3	III	IV	IV	IV
Case 4	IV	IV	IV	IV

Analysis Cases

Similar to Hossain and Haque [17], the cores section of the MSW is separated into four equal layers. The uniform rate of decomposition is maintained at each layer. Table 33.5 presents the several cases of decomposition situations adopted in the analysis. Case 1 represents a situation immediately after landfill closure. Layer 1 is taken at initial phase of decomposition (Phase I), whereas other layers of 2, 3, and 4 are considered as the second, third, and fourth decomposition phases, respectively. To understand the effect of layered homogeneous rate of decomposition on slope stability, it was presumed that each layers entered the next phase concurrently and consequently, other three cases of 2, 3, and 4 layers are considered for analysis of all slope geometries.

Result and Discussion

Effect of Varying Slope Angles and Constant Step Width on the Bioreactor Landfill Stability

In this problem, the FoS is determined by keeping constant width of step in slope (i.e., 6 m) (Fig. 33.3) but, slope angles are varying such as 2:1, 3:1, and 4:1. However, the slope above and below the step is kept same. Further, the FoS of different slopes (i.e., 2:1, 3:1 and 4:1) without/with providing steps is compared as shown in Fig. 33.6.

It is observed that FoS increases drastically with reduction in slope angles. However, FoS reduces with cases from case 1 to case 2, irrespective of change in slope angles. This is due to change in the properties of different phases in landfill. It is of interesting to note that FoS reduces continuously with cases, irrespective of change in slope angles for both conditions (i.e., with and without step). However, FoS reduces drastically with reduction in the slope angle in both conditions, irrespective of change in cases. This may be due to the increase in self-weight of landfill waste with change in phases from cases 1 to 4. Comparing the both conditions, it is observed that influence of providing step in the slope on FoS has pronounced beneficial in cases of 1 and 2, generally for higher slope angles.

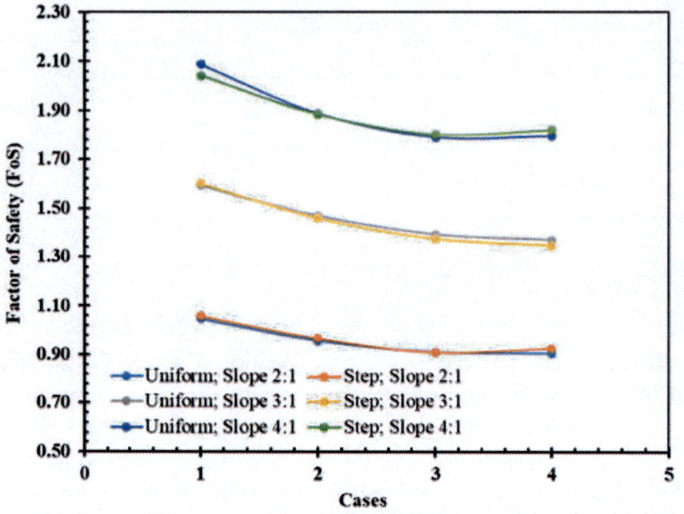

Fig. 33.6 Comparison of FoS of bioreactor landfills with varying slopes and cases

Effect of Varying Width of Step and Constant Slope Angle on the Bioreactor Landfill Stability

In this analysis, the slope of bioreactor landfill (Fig. 33.3) is kept constant to 2:1, while width of step in the slope is varying from 3 to 9 m. However, other materials properties for different phases are kept similar to the previous problem solved in Section "Effect of Varying Slope Angles and Constant Step Width on the Bioreactor Landfill Stability". The result of FoS for different conditions and cases is shown in Fig. 33.7. It is of interesting to note that FoS of bioreactor landfill slope reduces continuously with increase in width of step. However, variation in FoS is seemed to be constant with an increase in the width of step beyond 4 m for all cases. It is again observed that FoS reduces drastically with change in material properties of landfill, i.e., from cases 1 to 4 (Fig. 33.7).

Effect of Constant Step Width and Non-uniform Slope Angles on the Bioreactor Landfill Stability

The geometrical model of the bioreactor landfill slope is shown in Fig. 33.3. The width of step is kept constant (i.e., 6 m), but slope angles above (i.e., designated as Slope 1) and below (i.e., designated as Slope 2) of step (Fig. 33.3) are varying. The results are presented in Fig. 33.8a–c. When slope 1 is kept constant to 1:1 and slope 2 is varying from 1:1 to 4:1, the FoS is observed to less than 1 in all cases (Fig. 33.8a).

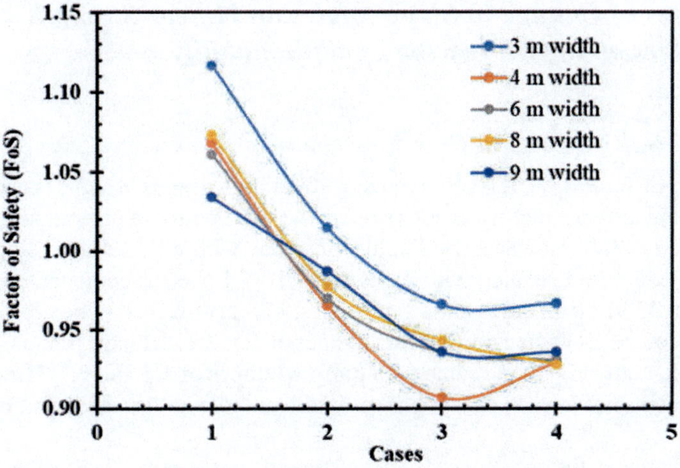

Fig. 33.7 Effect of different cases on FoS of bioreactor landfill having varying width of step

Although, FoS increases with reduction in angles (from 1:1 to 4:1) of slope 2 in all cases (Fig. 33.8a). When slope 1 is kept constant to 2:1 and slope 2 is varying from 1:1 to 4:1 (Fig. 33.8b), FoS is observed to be more than 1, particularly for slope angles of 3:1 and 4:1 in cases from 1 to 3. However, maximum FoS is seemed to be 1.41 when angles of slope 1 and 2 are 2:1 and 4:1 in case 1, respectively. When slope 1 is kept constant to 3:1 and slope 2 is varying from 1:1 to 4:1 (Fig. 33.8c), FoS is observed to be more than 1, particularly for slope 2 angles of 3:1 and 4:1 in case of 1 to 4. However, maximum FoS is appeared to be 1.97 when angles of slope 1 and 2 are 3:1 and 4:1 in case 1, respectively. On comparison (Fig. 33.8a–c), it is revealed that the stability of bioreactor slope could be improved by providing bench with proper combinations of slope angles.

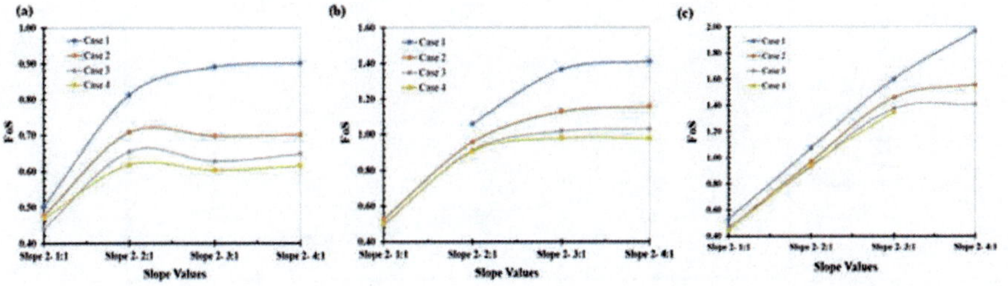

Fig. 33.8 Effect of varying angle of Slope 2 on stability of different cases of bioreactor landfill having constant step width and Slope 1 of **a** 1:1, **b** 2:1, and **c** 3:1

Influences of Change in Slope Angles and Shear Strength Parameters on the Bioreactor Landfill Stability

In this study, effect of two steps in slope with varying angle and shear strength parameters (Figs. 33.4 and 33.5) is analyzed. The material properties are taken as mentioned in Table 33.4. It is observed that effect of two steps in slope is pronounced significant in term of stability as compared to the same with one or, non-stepped slope landfill (Fig. 33.9). However, FoS values increase with a decrease in slope angles from 1:1 to 3:1 and, reduce with cases from 1 to 4 (i.e., decomposition rate and time) (Fig. 33.9). In all cases and slope angles, FoS is observed to be greater than 3. Landfill geometries having non-uniform slopes of 1:1, 2:1, 3:1 have pronounced to be more stable compared to the same with uniform slopes of 1:1 and 2:1. However, the material properties in each phases play a significant role in the stability of bioreactor landfill.

To understand the influence of shear strength parameters, FoS values of both geometrical models (Figs. 33.4 and 33.5) are examined for Case 1 (Table 33.5) by considering two extreme conditions of shear strength characteristics of landfills, i.e., (a) Cohesion = 0 and (b) Angle of friction = 0 in material properties (Table 33.4). It is revealed (Table 33.6) that effect of cohesion value of waste on the stability of bioreactor landfill is observed to be more in comparison of the same with an angle of friction. However, the stability of bioreactor landfill has observed to be reduced in both extreme conditions of shear strength parameters as compared to the same with materials having both c-phi (Table 33.6).

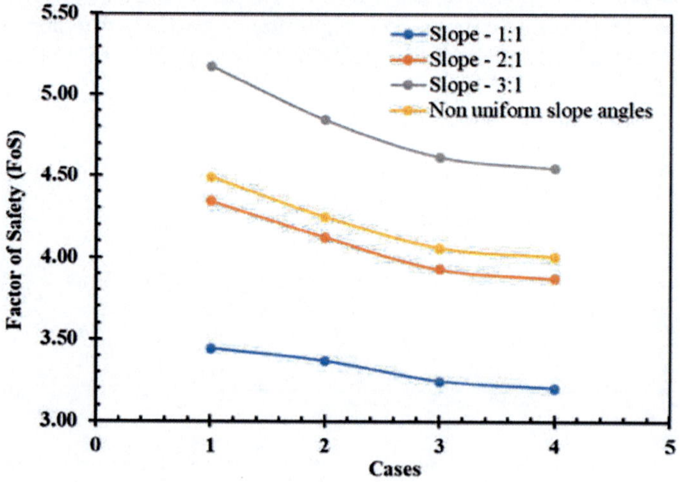

Fig. 33.9 FoS of stepped bioreactor landfill slope with varying slope angles

Table 33.6 FoS of stepped landfill slopes with varying shear strength parameters

Parameters		Both c and φ	$c = 0$	$\varphi = 0$
Uniform slope angles (Fig. 33.4)	1:1	3.445	0.826	1.863
	2:1	4.345	1.010	2.128
	3:1	5.177	1.457	2.419
Non-uniform slopes (Fig. 33.5) of 1:1, 2:1, 3:1		4.493	1.062	2.154

Conclusion

The key conclusions that can be brought out from present studies are as follows:

1. Reduction in slope angle increases the stability of bioreactor landfill. However, integration of one step in slope is more beneficial in terms of stability on bioreactor landfill having different stage of decomposition (i.e., gradually increase in rate and time of decomposition in each layers from top to bottom) and gentler uniform slope angles. Further, increasing the width of step/step in slope of bioreactor landfill has pronounced to be ineffective for any slope angles and particularly for cases where stages of decomposition in each layers are not gradual.
2. Stability analyses of bioreactor landfill consist of one step and varying slope angles in each layers are revealed that the bioreactor landfill system having combinations of steeper angles in slope with step are observed to be unstable compared to the same with gentler slope angles.
3. Provision of two non-uniform width stepped slope has better performance in terms of stability as compared to the same with one or, non-stepped/stepped slope landfill, particularly with gentler and uniform slopes. Effect of cohesion value of waste on stability is more as compared to the same with angle of friction.

Overall, it has been brought out clearly that slope stability of bioreactor landfill system is not only controlled by decomposition rate and time, and related change in geotechnical properties of wastes but also, influenced significantly by slope geometrical conditions. However, investigation of optimum geometry of bioreactor landfill system still needs further comprehensive studies by considering several parameters such as dynamic loading conditions, microbial activities in decomposition of wastes, etc.

References

1. Barlaz MA, Bareither CA, Hossain A, Saquing J, Mezzari I, Benson CH, Yazdani R (2010) Performance of north american bioreactor landfills. II: Chemical and biological characteristics. J Environ Eng 136(8):839–853
2. Kumar S, Chiemchaisri C, Mudhoo A (2011) Bioreactor landfill technology in municipal solid waste treatment: an overview. Crit Rev Biotechnol 31(1):77–97

3. Matasović N, Kavazanjian E Jr (1998) Cyclic characterization of OII landfill solid waste. J Geotech Geoenviron Eng 124(3):197–210
4. Warith M (2002) Bioreactor landfills: experimental and field results. Waste Manage 22(1):7–17
5. Zekkos D, Bray JD, Kavazanjian E Jr, Matasovic N, Rathje EM, Riemer MF, Stokoe KH (2006) Unit weight of municipal solid waste. J Geotech Geoenviron Eng 132(10):1250–1261
6. Hossain M, Haque M, Qasim S, Hoyos L (2007) Seismic stability of bioreactor landfill with decomposition—a numerical modeling. In: Proceedings of the 4th international conference on earthquake geotechnical engineering, Thessaloniki, Greece, Balkans, pp 25–28
7. Haque MA (2007) Dynamic characteristics and stability analysis of municipal solid waste in bioreactor landfills. The University of Texas at Arlington
8. Hossain MS, Gabr MA, Haque MA (2008) Deformation of MSW bioreactor landfills: properties and analysis approach. In: GeoCongress 2008: geotechnics of waste management and remediation, pp 216–223
9. Hossain MS, Haque MA, Hoyos LR (2010) Dynamic properties of municipal solid waste in bioreactor landfills with degradation. Geotech Geol Eng 28(4):391–403
10. Srivastava A, Babu GS, Reddy KR, Kulkarni HS (2011) Effect of leachate recirculation and extent of degradation on the stability of bioreactor landfill slopes. In: Geo-Risk 2011: risk assessment and management, pp 720–727
11. Hossain MS, Penmethsa KK (2011) Changes in geotechnical properties of municipal solid waste (MSW) in bioreactor landfill with degradation. Int J Environ Eng 3(3–4):349–370
12. Giri RK, Reddy KR (2014) Slope stability of bioreactor landfills during leachate injection: effects of unsaturated hydraulic properties of municipal solid waste. Int J Geotech Eng 8(2):144–156
13. Babu GS, Reddy KR, Srivastava A (2014) Influence of spatially variable geotechnical properties of MSW on stability of landfill slopes. J Hazard Toxic Radioact Waste 18(1):27–37
14. Rajesh S, Babel K, Mishra SK (2017) Reliability-based assessment of municipal solid waste landfill slope. J Hazard Toxic Radioact Waste 21(2):04016016
15. Ismail S, Taib AM, Rahman NA, Hasbollah DZA, Ramli AB (2019) Slope stability of landfill with waste degradation. Int J Innovat Technol Explor Eng 9:393–398
16. Rizwan M, Saif Y, Almansoori A, Elkamel A (2019) Environmental performance of municipal solid waste processing pathways. Energy Procedia 158:3363–3368
17. Hossain MS, Haque MA (2009) Stability analyses of municipal solid waste landfills with decomposition. Geotech Geol Eng 27(6):659–666

Chapter 34
Interfacial Mechanical Properties and Microstructural Properties of Lead Contaminated Soil

Rima M. Shah and Manas K. Bhoi

Introduction

Industrial development is necessary for the growth of any country, but it is also responsible for various land pollution. In many areas, groundwater is a source of drinking water that is being contaminated due to leaching of land pollutants. Large industrial activities cause the hectares of land is being contaminated with different chemicals and heavy metals which is responsible for a shortage of uncontaminated land for future development. In recent scenarios, geotechnical engineers have to use contaminated land for the development of infrastructure. Different heavy metals have different reactions to soil properties. Also, the reaction on the soil is different from changes in the type of soil.

In the past few decades, a great deal of research has been carried out on contamination of soil and their remediation techniques. A few example of such studies are [1–4] (Sridharan et al. 2010). Rather most of the studies are based on agriculture or environmental purpose. Soil is being contaminated when hazardous materials are present or other alteration has been made to its natural state. This is often caused by accidentally or intentionally by release of chemicals, oils, Petrocarbons, or the improper disposal of hazardous waste. Leakage of crude oil by damaged pipelines, oil drilling procedures, petroleum transportation services, tanker accidents, oxidized tanks, and natural seepage cause oil pollution in the surrounding soil. Effects on the quality of groundwater and human health are always a prime concern but soil contamination also affects the engineering properties of soils. Soil-waste contact affects the soil properties and the effects of different pollutants on soil are complex which is the reason that they can be better understood if various issues are isolated and consider independently [5]. In developing countries like India, the availability of

R. M. Shah (✉) · M. K. Bhoi
Civil Department, SOT, PDPU, Gandhinagar, Gujarat, India
e-mail: rima15shah.rs@gmail.com

K. Muthukkumaran et al. (eds.), *Transportation and Environmental Geotechnics*,
Lecture Notes in Civil Engineering 298, https://doi.org/10.1007/978-981-19-6774-0_34

land is a major issue so that an engineer has to be reliable on remediation or stabilization/solidification methods. S/S methods have been extensively used in the remedy of contaminated soil internationally as a consequence of their low cost, efficiency, and ease to handle. Soil remediate with S/S method could also be applied as road subgrade material or filling material.

There is a need to find the effect of different contaminants on different types of soil. There are so many case studies of structural damage to residential and industrial buildings due to different contamination of soils attend to highlight the importance to be given to the problem of variation in engineering properties of soil due to different contamination. There was a distress in the floors, beam and upheaval of foundations at number of places in a fertilizer plant in Kerala State since soil contained high phosphate content in an acidic environment which resulted in the heaving of soil (Sridharan et al. 1981). A case of unintentional spilling of highly concentrated caustic soda solution through a cracked drain in industrial area in Tema (Ghana) produced significant structural damage to an industrial building. (Kumapley and Ishola 1985). In this case, it was reported that the undrained shear strength of the contaminated soil had drastically reduced making the failure imminents.

Lead is a major source of pollution and it is found in industrial activities, electronic wastes, dyes, and paints, etc. Geotechnical properties of soil are affected when a lead is adsorbed by soil. Changes in the geotechnical parameters of soil like Atterberg's limits, permeability, strength, and swelling, etc., may lead to failure of soil as pavement or sub-base material. So geotechnical engineer needs to get the idea about the outcome of Pb contamination on soil properties. The major thrust of the study is to examine the effect of heavy metal (lead) contamination in soil concerning their engineering behavior.

Experimental Program

Soil collected from the Ahmedabad location was used in the present investigation for artificial contamination which is classified as silty sand Representative disturbed samples were collected and kept into the laboratory. Experiments were conducted to find out different micro and macro properties of virgin soil such as physical properties, soil classification. Proctor test to find out the optimum moisture content and maximum dry density were conducted. Strength characteristics of both types of soil were established using GDS triaxial system.

In the current investigation, a known quantity of lead nitrate solution was mixed in to the soil. The lead nitrate solution was chosen due to its high solubility and it does not inhibit stabilization reaction. The soil was artificially contaminated with lead nitrate solution of concentration 500, 1000, and 5000 mg/kg. The first solution contained a mixture of 1 g/l Pb. The second solution contained 2 g/l Pb, and the third solution contained 10 g/l Pb. 5 ml concentrated nitric acid was added to each solution and stir. Dried soil samples of 1 kg were separately added with each solution

Table 34.1 Index properties of soil

Properties	Soil
Grain size distribution	
Gravel size (%)	0
Sand fraction (%)	93.96
Silt fraction (%)	6.04
Clay fraction (%)	0
Soil classification	Silty sand (SM)
Consistency limits	
Liquid limit (%)	14.11
Plastic limit (%)	NP
Plasticity index (%)	NIL
IS classification of soil	SM
Specific gravity	2.63
Compaction characteristics	
Optimum moisture content (%)	8.5
Maximum dry density (g/cm^3)	1.637

in plastic containers. Samples kept in container with lead nitrate solution for 30 days. After 30 days, samples were dried, powdered and sieved for further investigation.

The results obtained for virgin soils are presented in Table 34.1. Soil classification is done according to the unified soil classification system. It shows that soil samples were collected from Ahmedabad region is called silty sand.

Results and Analysis

Specific Gravity Test

The specific gravity of virgin and contaminated soils has been determined in a laboratory by density bottle method. The 50 ml capacity density bottle was used (as per IS 2720_3). It can be seen that specific gravity of soil reduces with increasing concentration of lead in soil. Results obtained show that specific gravity is decreased with increasing metal adsorption. The effect of lead (Pb) on specific gravity of soil is shown in Fig. 34.1.

Results indicate that some modification in the morphology and mineralogy has taken place on contaminated with lead nitrate solution so that the new mineral formation may have more volume and less mass, which leads to decrease in specific gravity of soil [6].

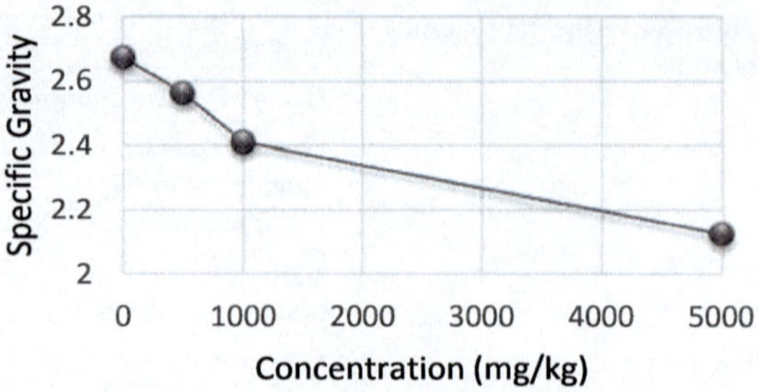

Fig. 34.1 Specific gravity of silty soils with different concentration of lead

Liquid Limit Test

Geotechnical engineering properties can reflect by Atterberg's limits values. Atterberg's limits have been associated with several factors, such as the specific surface area, soil mineralogy, permeability, swelling and shrinkage behavior, compressibility and strength of soil [7]. The Atterberg's limits were significantly affected by the susceptibility of minerals of clay in to the soil to liquid [8]. In current paper, Atterberg's Limits of soil were measured along with IS methods for soil (IS 2720-1991).

Results of liquid limit show that, proportion of contamination in soil increases, soil's liquid limit reduces. When concentration of contamination at its top, liquid limit was decreased by almost 36%. The variation in the liquid limit values of soils is summarized in Fig. 34.2. This may be taken as the typical behavior of soils as the liquid limit of the soils fluctuates directly with the thickness of the double diffuse layer [9]. Reduction in the values of liquid limit may be the reason of the reduction in the thickness of double diffuse layer. Liquid limit of the soil reduces as the inter particle repulsion forces decreased by increasing the concentration solution and because of that particles moving more easily in reduced water contents [10].

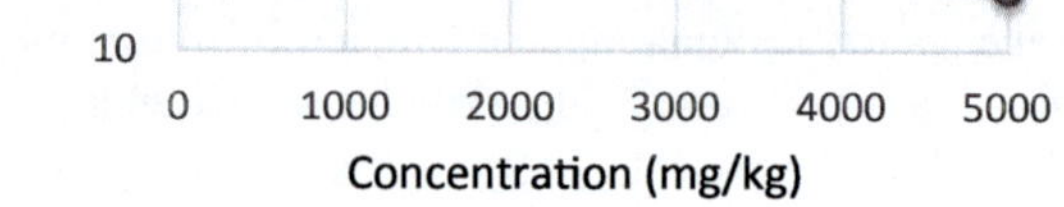

Fig. 34.2 Liquid limits of silty soils with different concentration of lead

Compaction Characteristics

Standard proctor tests have been performed to determine the optimum moisture content (OMC) and maximum dry density (MDD) are summarized in Figs. 34.3 and 34.4. Results obtained shows that dry density decreases and water content increases with increase in effect of contamination. The optimum moisture content was observed to decrease by around 7% and maximum dry density was observed to decrease by around 6% for SM soils with the maximum addition of Pb solution. Reducing the thickness of the double diffuse layers brought the particles closer to the same number of dense forces when the thickness of the diffused layer decreases and as a result of that MDD increases [11]. Decreased water holding capacity by particles becoming closer, the optimum moisture content decreased (Fig. 34.5).

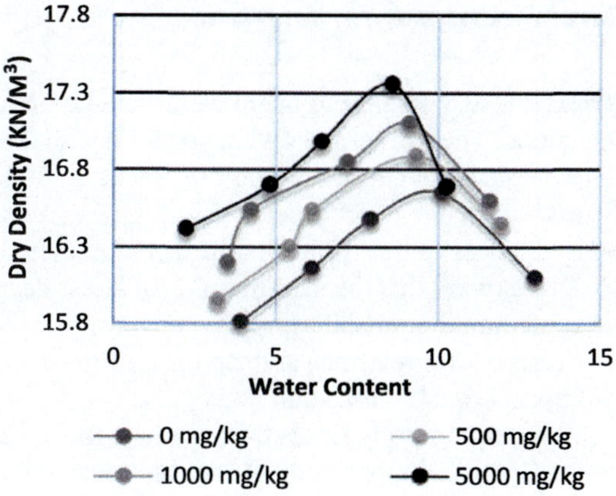

Fig. 34.3 Dry density versus water content of silty soils with different concentration of lead

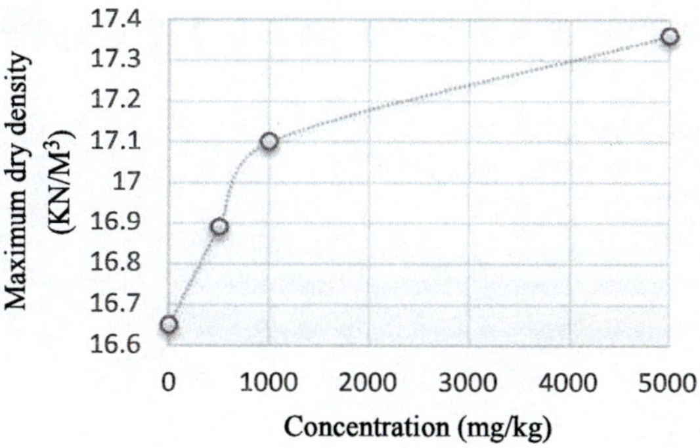

Fig. 34.4 Maximum dry density of silty soils with different concentration of lead

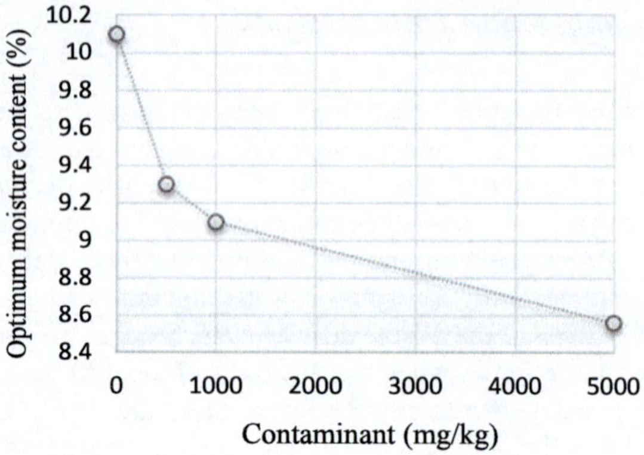

Fig. 34.5 Optimum water content of silty soils with different concentration of lead

Stress–Strain Characteristics

Triaxial tests with GDS system were carried out using virgin and contaminated soil specimen. The soil samples were prepared with their corresponding OMC and MDD. Three tests were carried out on 50, 100 and 150 kPa cell pressure for each sample. Stress–strain behavior for each soil samples are presented in figures. Typical curves show the stress–strain relationship of contaminated and uncontaminated soil samples. Results express that the maximum shear stress decrease with increased concentration of contamination in soil samples.

Stress–strain relations at different cell pressures for different soils are presented in Figs. 34.6, 34.7, 34.8 and 34.9.

A maximum strain is the strain at which the maximum deviatoric stress is achieved. From Graphical representation, It can be seen that maximum strain is achieved in virgin soil at 150 kPa normal stress. When concentration of contamination (Pb) is increased maximum strain is reduced and it is at a low level for 50 kPa normal stress at the highest concentration of contamination. Relations between maximum

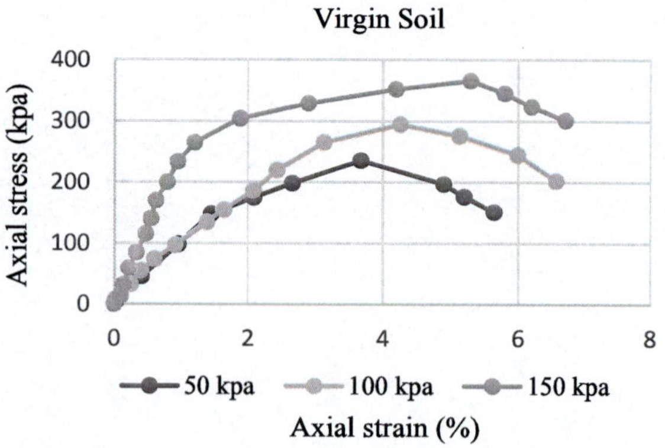

Fig. 34.6 Stress–strain relation of virgin silty soil at different cell pressure

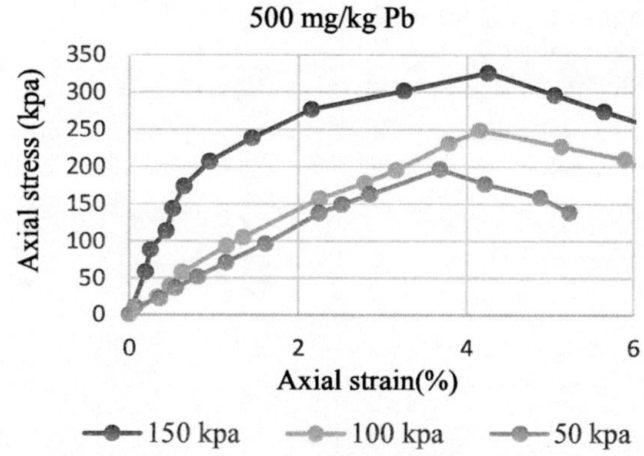

Fig. 34.7 Stress–strain relation of silty soil contaminated with 500 mg/kg Pb at different cell pressure

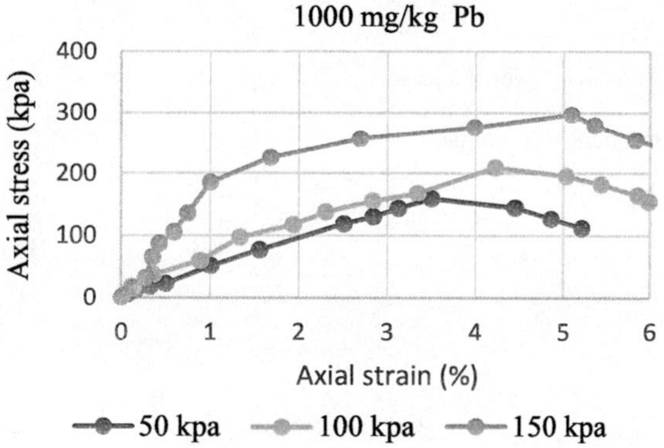

Fig. 34.8 Stress–strain relation of silty soil contaminated with 1000 mg/kg Pb at different cell pressure

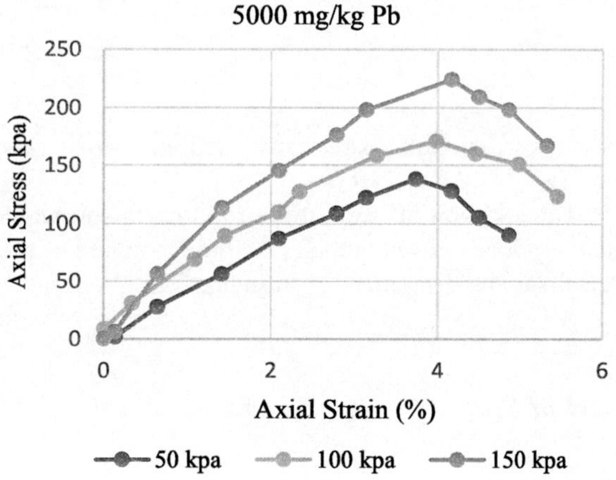

Fig. 34.9 Stress–strain relation of silty soil contaminated with 1000 mg/kg Pb at different cell pressure

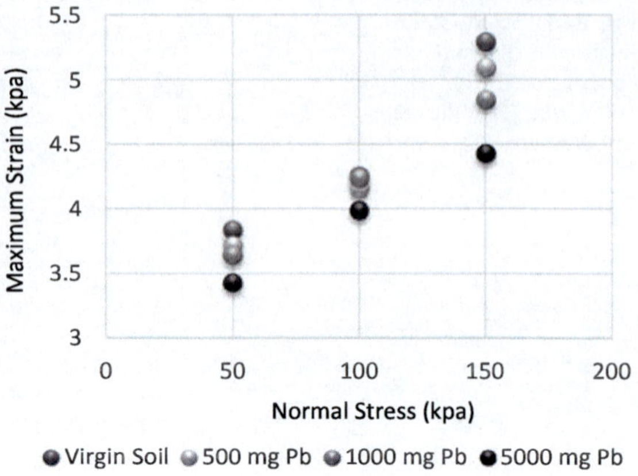

Fig. 34.10 Normal stress versus maximum strain for different soil samples

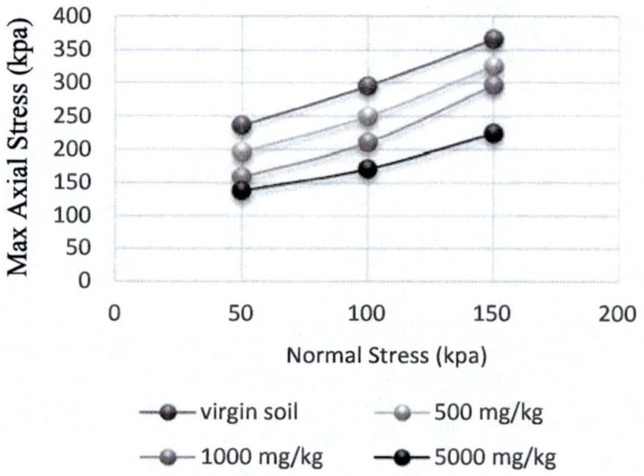

Fig. 34.11 Normal stress versus maximum stress for different soil samples

strain versus normal stress and maximum stress versus normal stress are presented in Figs. 34.10 and 34.11.

The addition of lead nitrate solution in soil would delay the hydration reaction and produce defects in the crystalline structure or breakage of internal bonds which results in the decrement of strength.

PH of Soil

When hydrogen ion activity increases, soil pH decreases. Change in salt concentration and reduction in organic matter might affect the pH. Concentration of hydrogen ion in the soil changes all biological activities and chemical reaction [12]. Precipitation having pH value less than 5.6 has resulted from atmospheric pollution which resulted

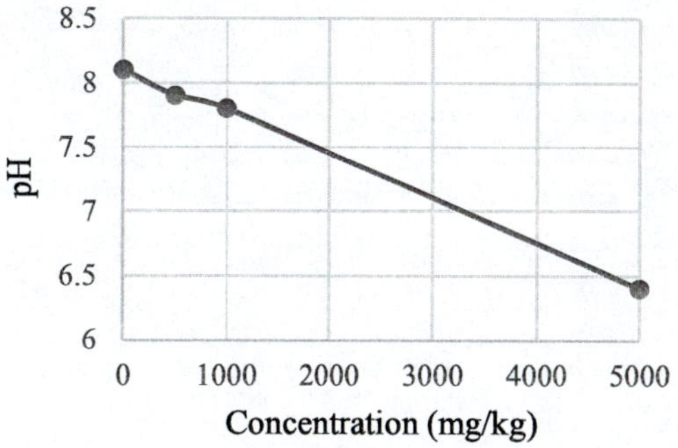

Fig. 34.12 Values of pH of soil at different concentration of lead

in acidification of soils and affects the engineering and mineralogical properties of soil (Stalin and Muthukumar 2002).

It is concluded from Fig. 34.12 that the pH of soil decreases with increasing the amount of concentration of lead nitrate solution in soil. It reacts with nitrates (NO_3^{2-}) to form nitrogen dioxide (NO_2) and water. This might be the reason for decreasing the pH of the soil.

X-Ray Diffraction Analysis

X-ray diffraction analysis of virgin soil and soil contaminated with 1000 mg/kg was carried out to evaluate the mineralogical changes due to lead contamination as shown in Figs. 34.13 and 34.14. Considerable changes can be observed due to lead nitrate solution. Interlayer spacing of minerals increased with lead nitrate solution. Interlayer spacing of quartz increased from 4.218 to 4.226 Å, for calcite 3.010–3.023 Å, for dolomite 1.538–1.540 Å. The reason may be attributed to the lead nitrate solution adsorption in inter layers increases with decreasing cation size and increasing interlayer spacing. It might be credited to the ion exchange reactions between silica, calcium, sodium, aluminum, and lead ions [13].

Conclusion

Current research has been done on to evaluate the outcome of lead contamination on physical and microscopic parameters of silty sand. The following conclusion can be given based on the test results.

- The special gravitational force of the soil has been reduced by the steady growth of lead solution. The formation of new minerals may have less mass and more volume, leading to a decrease in specific gravity.

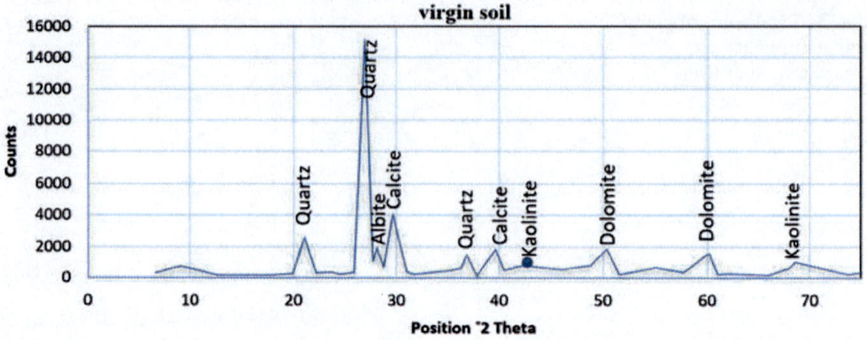

Fig. 34.13 X-ray diffraction analysis for virgin soil

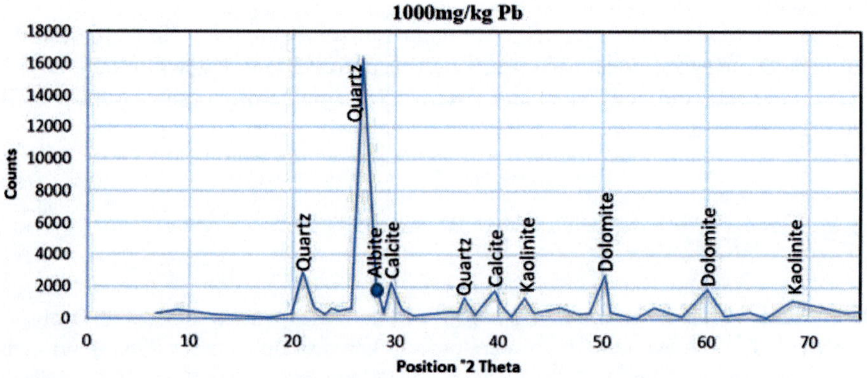

Fig. 34.14 X-ray diffraction analysis for soil contaminated with 1000 mg/kg Pb

- The impact of lead pollution on geotechnical soils can be explained by the theory of diffuse double layer (DDL). Lead solution may affect the DDL and float the clay particles leading to a change in mechanical behavior of the soil. With the growth of a mixture of lead nitrate solution in the soil, the liquid limit decreases. As lead concentration grows, the size of the soil particles increases, leading to an increase in silt content.
- Lead nitrate solution reduced moisture content and increased the density of the soil. When the double-layer thickness decreases, the soil particles come closer together under the same amount of bonding force and the dry density increases. Water holding capacity decreased when particles becoming closer which results in the moisture content decreased.
- From a stress–strain relationship, it is evident that maximum shear stress is reduced by increasing the contamination level of soil samples. Heavy metal addition to the soil can prevent hydration reactions and produce errors in crystal matrix or breakage of internal bonds which results into lower strength.

- pH of soil decreases with increasing the amount of concentration of lead nitrate solution increases in soil. The reason behind the typical behavior might be hydrogen ion concentration increases.

Remediation of contaminated soils is becoming essential practice concerning for environmental and geotechnical aspects. Long-term effect of individual heavy metal used in dyes and pigment industrial activity should be determined.

Acknowledgements The above work is done in a geotechnical and environmental engineering laboratory of the Pandit Deendayal Petroleum University, Gandhinagar, India. Institute was greatly acknowledged for technical support.

References

1. Khan MI, Irfan M, Aziz M, Khan AH (2017) Geotechnical characteristics of effluent contaminated cohesive soils. J Environ Eng Landscape Manag 25(1):75–82
2. Krishna AK, Govil PK (2007) Soil contamination due to heavy metals from an industrial area of Surat, Gujarat, Western India. Environ Monit Assess 124(1–3):263–75
3. Sunil BM, Shrihari S, Nayak S (2009) Shear strength characteristics and chemical characteristics of leachate-contaminated lateritic soil. Eng Geol 106(1–2):20–25, Elsevier B.V
4. Wong CSC, Li X (2003) Analysis of heavy metal contaminated soils. Pract Periodic Hazard Toxic Radioact Waste Manag 7(1):12–18
5. Oluremi JR, Adedokun SI, Olaoye RA, Ajamu SO, Eng M (2012) Assessment of cassava wastewater on the geotechnical properties of lateritic soil. Pacific J Sci Technol 13(1):631–39
6. Sivapullaiah P V, Manju (2005) Kaolinite–Alkali interaction and effects on basic properties. Geotech Geol Eng 23(5):601–14
7. Yang Y Wen Q, Hu L (2012) Experimental study on application of industrial waste in landfill liner. In: Geotechnical special publication, no 225, GSP, pp 3881–3890
8. Schmitz RM, Schroeder C, Charlier R (2004) Chemo-mechanical interactions in clay: a correlation between clay mineralogy and atterberg limits 351–58
9. Mitchell JK (1993) Fundamentals of soil behavior. Wiley, USA
10. Warkentin BP (1961) Interpretation of the upper plastic limit of clays. Nature 190:287–288
11. Singh S, Prasad A (2007) Effects of chemicals on compacted clay liner. Electron J Geotech Eng 12(D) (2007):1–15
12. Gupta AK, Singh DN (1997) Interaction of contaminants with soils: a review. Indian Geotech J 75–87
13. Wu H, Hu LM (2014) Microfabric changes of electro-osmotic stabilizedv bentonite. Appl Clay Sci 101:503–509
14. Sunil, BM, Nayak S, Shrihari S (2006) Effect of PH on the geotechnical properties of laterite. Eng Geol 85(1–2):197–203
15. Chu Y, Liu S, Wang F, Bian H, Cai G (2018) Electric conductance response on engineering properties of heavy metal polluted soils. J Environ Chem Eng

Chapter 35
Impact of Aquaculture Solid Waste on Environment in the Delta Region of Andhra Pradesh: A Case Study

T. V. Nagaraju, B. M. Sunil, and Babloo Chaudhary

Introduction

From plastic waste to e-waste, from household garbage to municipal solid waste, from industrial waste to medical waste, environmental concern has found an important aspect in the future sustainable development world over. Waste is threat for the ecosystem in recent and past [1, 2]. Due to lack of awareness, proper management, and advanced equipment, many large-scale landfills or dump yards came into existence. The Indian waste management has been no exception in large-scale dump yards. For example, dump yard in Hyderabad, India, having radius of 10 km (over 4700 metric tons of garbage waste was dumping every day) [3]. Municipal solid waste (MSW) landfill, which comprise a variety of wastes such as garbage, plastic, wood, glass bottles, and other wastes which are threat to environment if dumped in the open places. Last few years, many revolutionary changes in research and practice have been taken place in MSW landfills. Apart from MSW, hazardous wastes such as chemical waste, medical waste and e-waste were most concern in many parts of the world [4].

Nowadays, due to rapid increase in global population with an average annual rate of 1.6% from 1960 to 2018, needs were increased. To fulfill food demand, aquaculture production is playing vital role and intensively expanding in all parts of the world with an increased rate of average annual production of 3% for the same period [5]. In developing countries, particularly in India, expanding aquaculture creates employment for local community and export earnings for country. In India, Andhra Pradesh is the state which produces the highest production of 34.5 lakh metric tons annually. Aquaculture has been traditionally practices in both fresh water and brackish water from time immemorial [6]. Around 10% demand is increased in production

T. V. Nagaraju (✉) · B. M. Sunil · B. Chaudhary
Department of Civil Engineering, National Institute of Technology, Mangaluru, Karnataka 575025, India
e-mail: varshith.varma@gmail.com

K. Muthukkumaran et al. (eds.), *Transportation and Environmental Geotechnics*, Lecture Notes in Civil Engineering 298, https://doi.org/10.1007/978-981-19-6774-0_35

per year, and for 2020, the target was setup to 15 million tons by supporting blue revolution. However, due to intensive aquaculture farming worldwide, the environment has been negatively impacted by high-density cultivation, high-loading feed, high-concentration disinfectants, chemicals, minerals, and probiotics, and the subsequent pollution of nearby canals and groundwater bodies [6]. Further, aquaculture-contaminated water having high concentration of ammonia and nitrates has become a major concern on ecosystem and human health [7].

Mainly aquaculture ponds or marine aquaculture generates waste in the form of aquaculture sludge (settled feeds and fish or shrimp feces) and fish or shrimp processing waste. Due to intensive farming, aquaculture sludge and processing waste significantly increased, particularly in underdeveloped and developing nations. Aquaculture sludge which is suspended solids and settled solids depends on the feed coefficient and continuous monitoring. It is estimated that about 10% of the total feed was wasted because of uneaten. Moreover, many parts of the world still traditional depend on traditional practice where aquaculture sludge was not collected at the end of the crop. Instead, aquaculture sludge was discharged into nearby canals periodically. Among the aquaculture contaminant constituents, ammonia and nitrates are major concern because of its higher concentration [7]. Several studies have reported that treatment or control of ammonia in landfill leachate using air stripping and adsorption. Nonetheless, only a few researchers reported the application of air stripping in aquaculture effluent treatment [8].

In addition, recirculating systems (RAS) are in vogue in developed nations to treat the aquaculture effluents and only in small scale [9]. RAS consists of filters work by mechanically and bio-absorbers to treat aquaculture effluent water and reuse the filter water for next crop. Moreover, RAS needs expensive setup, so, undeveloped nations still not looking for RAS systems in their practice even though intensive farming [9, 10]. Moreover, geotextile bags containing flocculant-aid for suspended or settled solids are an emerging alternative to traditional treatment process. Geotextile bags could be helpful for reducing total suspended solids, chemical oxygen demand (COD), and total nitrates [11].

In this study, investigation was carried out on aquaculture leachate and nearby groundwater bodies. Field surveys were carried to know the source and disposal of ASW, and to identify the lacunae of practice. This study also explores the intensity of aquaculture practices in the delta region of Andhra Pradesh and its impact on environment.

Aquaculture Solid Waste Management at Global Scenario

Currently, the top five countries which are generating inland aquaculture production are China, India, Bangladesh, Myanmar, and Cambodia. Moreover, Asia aquaculture production consistently accounted for 75% of global inland aquaculture production since 2000 [12]. Recent years, China inland aquaculture production decreased from 2.18 million tons to 1.96 million tons from 2017 to 2018. Cambodia and Myanmar showing same production trend in both 2017 and 2018. India and Bangladesh production was rapidly increasing every year since 2000 [12]. However, due to intensive farming, many negative consequences may occur [13].

In aquaculture ponds, shrimps and fishes which are unable to consume the substantial content of nitrogen and phosphorus, which are main ingredients (nutrients) presence in the feed, may acts as a key source of water pollution. Hence, in many countries, it was considered as industrial waste.

Globally, government authorities have regulated the management of aquaculture waste. Particularly, developed nations that have stringent rules on aquaculture practices and waste management include China (fisheries law of the people's republic of China, 1986); the USA (procedures for the safety and sanitary processing and importing of fish and fish products, 1995); Canada (aquaculture regulation, 2002); and Australia (fisheries management regulation, 1995). In developing countries, most of the ponds the aquaculture effluents were directly released to nearby canals, which lead to environmental pollution. In most of the countries, no treatment of effluent takes place at the source. As a result, effluents (nitrogen and phosphorus) may form ammonia and total nitrates, over the time due to biological activity their concentrations will be increased and behave as toxic compounds [14].

Fish or shrimp production waste which includes shrimp shell, scale, fish fins, fish tail and bones is about 70% before sold or imported. Based on the United Nations Food and Agriculture Organization (FAO 2020) reports, nearly 13 million tons of processing waste generated globally which is equal to the annual municipal solid waste of Australia in the year 2020. Surprisingly, China has reduced the intensity of farming and production in recent years and in few provinces inland aquaculture was banned due to environmental concern [15].

Aquaculture Practice in the Delta Region of Andhra Pradesh

Delta region of Andhra Pradesh, India, indicates that the aquaculture farming is now the heartland of India's inland aquaculture production. In particular, Undi constituency in Andhra Pradesh, land cover is dedicated to intensive aquaculture ponds with an area ranges from 5 to 100 acres. Early 2000's traditional aquaculture practices include species of tiger shrimp with lower density of shrimps did not affect the environment [6].

From 2010s, after the state bifurcation newly formed state Andhra Pradesh, the government relaxed their management and expansion of aquaculture, and species vannamei shrimp culture boosts the state economy with the highest aquaculture production. Lake-based aquaculture and inland aquaculture ponds considered as promising economic growth for the state development. So, the delta region of Andhra Pradesh ranks largest producer of shrimp in India [6, 16].

Table 35.1 Physicochemical characteristics of aquaculture leachate

Property	Average values of tested samples	
	Pre-monsoon	Post-monsoon
pH	8.4	7.8
Dissolved oxygen, DO (ppm)	3.0	5.0
Electrical conductivity, EC (μs/cm)	7084	4522
Total dissolved solids, TDS (ppm)	2655	1280
Chemical oxygen demand, COD (ppm)	64	28
Biological oxygen demand, BOD (5 days) (ppm)	14.8	6.4
Ammonium, NH_4 (ppm)	3.56	1.24
Nitrate, NO_3 (ppm)	8.2	3.15
Nitrite, NO_2 (ppm)	1.48	0.23
Sulfate, SO_4 (ppm)	842	324
Sodium, Na (ppm)	215	160
Magnesium, Mg (kg/acre)	94	74
Sulfur, S (kg/acre)	168	160
Potassium, K (kg/acre)	688	515
Phosphate, P (kg/acre)	490	385

Aquaculture Sludge Leachate Characteristics

Aquaculture pond sludge samples were collected from 35 locations in the delta region of Andhra Pradesh during the pre-monsoon and post monsoon between 2019 and 2021; no rains are occurred during the sample collection. The sample locations were represented the different zones based on the intensity of the farming and proximity of potential-contaminated zones. The physicochemical characteristics of the aquaculture sludge leachate samples were illustrated in Table 35.1.

The physicochemical characteristics of the samples before and after monsoon shows noticeably decreased after the post-monsoon. This can be due to the dilution of the sludge with the rainwater. The values of ammonia and nitrates exhibit higher concentration during the pre-monsoon, which leads to eutrophication process.

Effect of Aquaculture Waste Sludge on Clay Mineralogy

In Table 35.2, clays mineral behavior before and after exposed with aquaculture leachate for 20 days was observed. The test results were revealed that the cation exchange takes place between the clays and leachate. The main ions were NH_4^+, Na^+, and K^+ are significantly increased with leachate exposure. Moreover, NH_4^+ was observed highest amount due to the favorable monovalent cation to effect clays

Table 35.2 Effect of expansive clay before and after aquaculture sludge leachate exposure

Properties	pH	Calcium (mEq/100 g)	Sodium (mEq/100 g)	Ammonia (mEq/100 g)	Potassium (mEq/100 g)	Magnesium (mEq/100 g)
Aquaculture leachate before exposed values	7.6	40.21	0.21	0.14	3.12	1.32
Aquaculture leachate after exposed values	8.2	43.28	34.23	42.82	26.24	2.34

and further influence the structure and volume change behavior of clays. Therefore, the surface forces of the clay particles reduced with the aquaculture leachate exposure and cause an improvement in the hydraulic conductivity by ion exchange and decrease in the double diffusion layer in expansive clays.

Conclusions

In this study, the effect of aquaculture contaminants on the expansive clays and overview of the emerging aquaculture contaminants were presented. The following conclusions were drawn from this case study.

Nowadays, intensity of aquaculture practices has evolved rapidly in the delta region of Andhra Pradesh, India, which may contribute negative geoenvironmental consequences due to the unengineered culture.

Sampling events were conducted in pre-monsoon and post-monsoon to identify the potential problems with the aquaculture contaminants. Physicochemical characteristics of the aquaculture leachate clear that contamination occurred in the aquaculture ponds and the surrounding water bodies. Ammonia and nitrates are the key compounds with higher concentration found in the aquaculture systems.

Expansive clays exposed with aquaculture leachate show that cation exchange between clays and leachate. Major cation exchange takes place found in the NH_4^+, Na^+, and K^+. Moreover, the ammonia concentration in the aquaculture leachate may cause replacement of monovalent ions, which leads to reduction in the thickness of double diffusion layers and surface forces.

References

1. Karak T, Bhagat RM, Bhattacharyya P (2012) Municipal solid waste generation, composition, and management: the world scenario. Crit Rev Environ Sci Technol 42(15):1509–1630
2. Paterok, K. (2020). The race for waste-the evolution and implementation of India's municipal solid waste management agenda over the past three decades until 2016

3. Kamble BS, Saxena PR, Kurakalva RM, Shankar K (2020) Evaluation of seasonal and temporal variations of groundwater quality around Jawaharnagar municipal solid waste dumpsite of Hyderabad city, India. SN Appl Sci 2(3):1-22
4. Misra V, Pandey SD (2005) Hazardous waste, impact on health and environment for development of better waste management strategies in future in India. Environ Int 31(3):417–431
5. Tacon AG (2020) Trends in global aquaculture and aqua feed production: 2000–2017. Rev Fish Sci Aquaculture 28(1):43–56
6. Nagaraju TV, Malegole SB, Chaudhary B, Ravindran G (2022) Assessment of environmental impact of aquaculture ponds in the Western Delta Region of Andhra Pradesh. Sustainability 14(20):13035
7. Dauda AB, Ajadi A, Tola-Fabunmi AS, Akinwole AO (2019) Waste production in aquaculture: sources, components and managements in different culture systems. Aquaculture Fish 4(3):81–88
8. Ulu F, Kobya M (2020) Ammonia removal from wastewater by air stripping and recovery struvite and calcium sulphate precipitations from anesthetic gases manufacturing wastewater. J Water Process Eng 38:101641
9. Badiola M, Mendiola D, Bostock J (2012) Recirculating aquaculture systems (RAS) analysis: main issues on management and future challenges. Aquacult Eng 51:26–35
10. Van Rijn J (2013) Waste treatment in recirculating aquaculture systems. Aquacult Eng 53:49–56
11. Guerdat TC, Losordo TM, DeLong DP, Jones RD (2013) An evaluation of solid waste capture from recirculating aquaculture systems using a geotextile bag system with a flocculant-aid. Aquacult Eng 54:1–8
12. Food and Agriculture Organization (2020) The State of World fisheries and aquaculture sustainability in action. Food and Agricultulture Organization of the United Nations. Rome Italy. https://doi.org/10.4060/ca9229en
13. Mzula A, Wambura PN, Mdegela RH, Shirima GM (2021) Present status of aquaculture and the challenge of bacterial diseases in freshwater farmed fish in Tanzania; a call for sustainable strategies. Aquac Fish 6(3):247–253
14. Robles-Porchas GR, Gollas-Galván T, Martínez-Porchas M, Martínez-Cordova LR, Miranda-Baeza A, Vargas-Albores F (2020) The nitrification process for nitrogen removal in biofloc system aquaculture. Rev Aquac 12(4):2228–2249
15. Newton R, Zhang W, Xian Z, McAdam B, Little, DC (2021) Intensification, regulation and diversification: the changing face of inland aquaculture in China. Ambio 1–18.
16. Jayanthi M, Duraisamy M, Thirumurthy S, Samynathan M, Kabiraj S, Manimaran K, Muralidhar M (2020) Ecosystem characteristics and environmental regulations based geospatial planning for sustainable aquaculture development. Land Degrad Dev 31(16):2430–2445

Chapter 36
Performance of Permeable Reactive Barrier by Using Rubber Particles

G. Venkatesan, M. Renganathan, and L. Kavin Raj

Introduction

These days, the nature of the ground water is getting contaminated by different sources including enterprises, mining, and landfill destinations. Groundwater is moreover one of the most widespread wellsprings of water for irrigation system and furthermore for special applications. Lamentably, groundwater is defenseless to pollutants. Groundwater tainting occurs when man-made items, for example, gas, oil, street salts, and synthetic concoctions get into the groundwater and causes it to get hazardous and unfit for human use. Materials from the land's surface can travel through the grime and cease up in the groundwater. For instance, pesticides and manures can find out their way into groundwater supplies after some time. Harmful resources from mining locales and utilized engine oil likewise may additionally saturate groundwater. Moreover, it is conceivable for untreated waste from septic tanks and hazardous artificial compounds from underground stockpiling tanks and improper landfills to contaminate groundwater.

A PRB is a penetrable responsive treatment zone and is a creating innovation that has been perceived just like a financially savvy innovation for in situ (at the site) groundwater remediation. PRBs are boundaries which permit a few yet not all materials to go through. PRB would offer cost motivating forces because of aloof nature for the innovation that additionally lessen the activity, and upkeep expenses and subsequently, vows to give a powerful and conservative treatment for groundwater at numerous destinations the nation over like India. PRB is exceptionally feasible innovation for since quite a while ago run as a result of its capability of preferred treatment execution over customary techniques. Creating nations like India needs its

G. Venkatesan (✉) · M. Renganathan · L. Kavin Raj
Department of Civil Engineering, University College of Engineering (BIT Campus), Anna University, Trichy, India
e-mail: gvenkat1972@gmail.com

K. Muthukkumaran et al. (eds.), *Transportation and Environmental Geotechnics*, Lecture Notes in Civil Engineering 298, https://doi.org/10.1007/978-981-19-6774-0_36

acknowledgment to remediate the sullied groundwater as fast as conceivable to arrive at its spotless crucial. After the establishment of the PRB, the outside of the ground will have the option to re-use for different purposes. Right now, vitality would not be required, as a characteristic angle of ground water stream will be utilized to bring contaminants through the receptive zone.

Reactive Materials for PRBs

Removal of contaminants in PRBs takes place normally in the area of reactive medium. There are a number of types of the reactive substances to take away contaminants through physical contact whereas others work by means of various biogeochemical procedures in the treatment zone. At present, there is an extensive range of reactive substances available. The most frequent of them is zero valent iron. Others are the activated carbon, zeolites, peat, noticed dust, oxygen releasing compounds, etc. In this study, rubber particle which was obtained from the waste tire is taken as the reactive material. Along with rubber red soil and M-sand is used. These materials are selected based on various environmental aspects.

Leachate Sample

In this study area, leachate samples were collected from Ariyamangalam dump yard as shown in Fig. 36.1 which is located in Trichy. The leachate sample was collected from the surfaces of the stagnant leaching water as shown in Fig. 36.2 in the dumpsite. The samples are gathered in the polythene cover (soaked in a single day in 2% nitric acid and washed with distilled water) from every zone. The accumulated samples have been appropriately labeled which shows the supply of the samples time and date of collection. The samples had been taken to the laboratory, and assessments for the parameters had been taken within the forty-eight hours of collection (Table 36.1).

Preparation of Reactive Medium

The filtration is done by two methods: One is by acrylic box and the other is done by glass tube. The reactive medium is made of acrylic material which is a transparent material made of plastic with outstanding strength, optical clarity, and stiffness. It is easy to fabricate, and also, it bonds well with adhesive and solvents. Acrylic sheet has superior weathering while comparing with other type of transparent plastics. The acrylic sheet is made into box shape in which the dimensions are 30 cm × 20 cm

Fig. 36.1 Ariyamangalam dump yard

Fig. 36.2 Leachate

Table 36.1 Rubber characteristics

Test	Sample 1	Sample 2
Specific gravity	0.973	1.13
pH	6.2	6
Particle size (mm)	0.5	0.7

× 20 cm. The reactive materials are packed as shown in Fig. 36.3. In this reactive medium, the bottom layer is filled with 10 mm coarse aggregate up to 2 cm height. Next 5 cm is filled with M-sand mixed with rubber fiber in the ratio of 1:2. Then red soil is mixed with rubber at the ratio of 2:1, and it is filled for 5 cm. Next to that only rubber is filled for 5 cm. Next to rubber again red soil is mixed with rubber at the ratio of 2:1 and filled for 5 cm. And finally, M-sand is mixed to rubber at the ratio of

2:1 and filled for 5 cm. M-sand and red soil were sieved by 1.4 mm sieve, and rubber is sieved by 2.36 mm sieve. The characteristics of rubber, red soil, and M-sand are tabulated in Tables 36.2, 36.3, and 36.4 (Fig. 36.4).

Fig. 36.3 Reactive medium

Table 36.2 Red soil characteristics

Test	Sample 1	Sample 2
Specific gravity	2.55	2.9
pH	5.8	6.2
Organic impurities	Nil	Nil
Silt content (%)	6.25	7.1
Bulkage factor (%)	11	13

Table 36.3 M-sand characteristics

Test	Sample 1	Sample 2
Specific gravity	2.5	3.2
pH	9.8	8.2
Organic impurities	Nil	Nil
Silt content (%)	Nil	Nil
Bulkage factor (%)	14.3	15

Table 36.4 Leachate analysis

Characteristics	Sample 1	Sample 2	Average
pH	8.3	7.9	8.1
Temperature (°C)	29	27	28
Alkalinity	11,250	12,346	11,798
TDS (mg/l)	2589	3201	2895
BOD (mg/l)	1496	1854	1675
COD (mg/l)	11,582	19,835	15,708.5
Ca (mg/l)	734	542	638
Hardness (mg/l)	21,345	19,660	20,502.5
Iron (mg/l)	11.96	14.76	13.36
Zinc (mg/l)	2.3	3.6	2.95
Lead (mg/l)	0.18	0.27	0.225
Copper (mg/l)	0.23	0.17	0.2
Cadmium (mg/l)	0.06	0.043	0.0515
Chromium (mg/l)	0.06	0.03	0.045

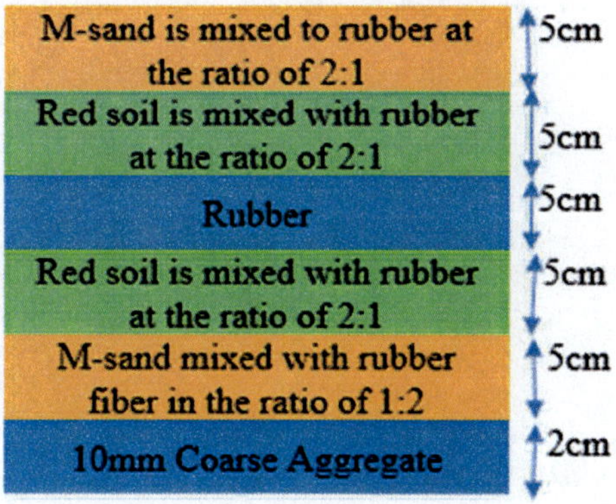

Fig. 36.4 Schematic representation of reactive medium

Result and Discussion

The leachate collected from the landfill site is analyzed, and the results of the analysis are shown in Table 36.4.

Then, the leachate is allowed to filter in the reactive medium so that the difference in the parameter between fresh leachate and treated leachate will show the reactive materials effect in the leachate. The treated leachate quality is analyzed, and the reports are mentioned in Table 36.5.

Table 36.5 Analysis of treated leachate

Characteristics	Sample 1
pH	7.6
Temperature	27
TDS (mg/l)	1564
Alkalinity	7354
BOD (mg/l)	1245
COD (mg/l)	9800
Calcium (mg/l)	592
Iron (mg/l)	10.94
Copper (mg/l)	0.14
Zinc (mg/l)	2.8
Lead (mg/l)	0.24
Chromium (mg/l)	0.015
Cadmium (mg/l)	0.4

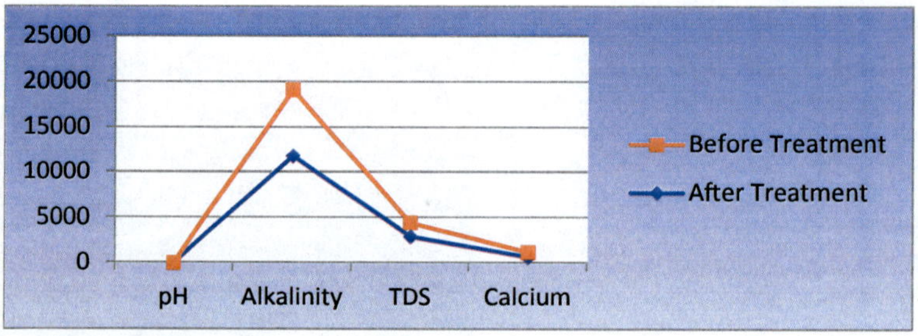

Fig. 36.5 Comparison study

Comparison Study

The comparison is made between the fresh leachate characters and the treated leachate characters. Figure 36.5 shows the variation of parameters.

Conclusion

In this study, the rubber is selected as the reactive material for the permeable reactive barrier. The rubber is obtained from the automobile wastes. The rubber fiber is produced by Ambient grinding, which is conducive to any size particle, consisting of entire tires. It can be executed in two approaches granulation or cracker mills. Along with the rubber fiber, the pink soil and M-sand are used to enhance the effectivity of the treatment. The characteristics of the rubber fiber, red soil, and M-sand are studied. Then, the leachate water is allowed to filter in reactive medium, and then, the filtered water's characters are analyzed. And then, the comparison between the leachate water and the treated water is studied.

From this study, it is concluded that the reactive medium which is packed with rubber, M-sand, and red soil alters the parameters (pH, TDS, alkalinity, and calcium) with huge variation in the leachate. And there are slight modifications in other parameters like iron and copper.

References

1. Tang Z, Butkus MA, Xie YF (2009) Enhanced performance of crumb rubber filtration for ballast water treatment. Chemosphere 74(10):1396–1399
2. Xie Y (2007) Filter media: Crumb rubber for wastewater filtration. Filtr Sep 44(5):30–32
3. Park J, Evans EA, Ellis TG (2011) Development of a biofilter with tire-derived rubber particle media for hydrogen sulfide odor removal. Water Air Soil Pollut 215(1–4):145–153
4. Thiruvenkatachari R, Vigneswaran S, Naidu R (2008) Permeable reactive barrier for groundwater remediation. J Ind Eng Chem 14(2):145–156
5. Wilkin RT, Su C, Ford RG, Paul CJ (2005) Chromium-removal processes during groundwater remediation by a zerovalent iron permeable reactive barrier. Environ Sci Technol 39(12):4599–4605
6. Jun D, Yongsheng Z, Weihong Z, Mei H (2009) Laboratory study on sequenced permeable reactive barrier remediation for landfill leachate-contaminated groundwater. J Hazard Mater 161(1):224–230
7. Park J, Ellis TG, Lally M (2006) Evaluation of tire derived rubber particles for biofiltration media. Procs Water Environ Fed 9:3217–3230
8. Oh JH, Park J, Ellis TG (2014) Septic wastewater treatment using recycled rubber particles as biofiltration media. Environ Technol 35(5):637–644
9. Park J, Ellis TG (2008) Development of biofilter with TDRP (tire derived rubber particles) media for odor removal. Evaluation of tire derived rubber particles as biofilter media and scale-up and design considerations for the static granular bed reactor (SGBR), pp 35
10. Ludwig RD, McGregor RG, Blowes DW, Benner SG, Mountjoy K (2002) A permeable reactive barrier for treatment of heavy metals. Groundwater 40(1):59–66
11. Guerin TF, Horner S, McGovern T, Davey B (2002) An application of permeable reactive barrier technology to petroleum hydrocarbon contaminated groundwater. Water Res 36(1):15–24
12. Benner SG, Blowes DW, Ptacek CJ, Mayer KU (2002) Rates of sulfate reduction and metal sulfide precipitation in a permeable reactive barrier. Appl Geochem 17(3):301–320

Chapter 37
Influence of Freezing–Thawing Cycles on Biotreated Sand Using MICP

Meghna Sharma and Neelima Satyam

Introduction

The geotechnical engineers and biotechnologists inspected the feasibility of soil cementation through the bacterial phase of soil. The living phase produces calcium carbonate in soil media; hence, the technique is called as microbially induced calcite precipitation (MICP) [1]. The common components of the MICP technique include urease-producing microbe, carbamide (urea), and calcium source [2]. The urease enzyme is secreted by bacteria which expedites the urea hydrolysis to produce ions of ammonia and carbonate. The carbonates react with calcium ions and produce calcium carbonate [3].

It is stated in literature that cementation increases the strength of soil and simultaneously reduces the permeability and compressibility of soil matrix which mitigate the liquefaction of soil effectively [4–18]. Nonetheless, the endurance of biologically cemented soil in terms of freezing–thawing is not fully researched yet.

The present study includes the investigation of permeability and calcite formation variation of biologically cemented soil after 18 days of treatment. Different injection intervals and pore volumes were adopted to optimize the cementation solution. The soil precipitated with different percentages of calcite subjected to freezing–thawing cycles to determine the endurance with respect to ultrasonic pulse velocity and unconfined compressive strength.

M. Sharma (✉) · N. Satyam
Department of Civil Engineering, Indian Institute of Technology, Indore, India
e-mail: meghnasharma586@gmail.com

K. Muthukkumaran et al. (eds.), *Transportation and Environmental Geotechnics*, Lecture Notes in Civil Engineering 298, https://doi.org/10.1007/978-981-19-6774-0_37

Materials and Methods

Soil Properties and Bacterial Solution

MICP technique was applied to poorly graded Narmada River sand. The values of D_{50}, C_u, and C_c were 0.32, 2.60, and 0.72, respectively [19].

Sporosarcina pasteurii bacterial strain was injected in soil using bioaugmentation methodology. The bacteria cultivation was carried out in non-sterile conditions defined in previous study [15].

Biotreatment and Test Procedures

The sand was filled in cylindrical PVC pipes at 40% relative density. The sand was augmented with *S. pasteurii* solution and treated with different pore volumes and injection interval frequencies, as quoted in Table 37.1. The molarity of carbamide/urea and calcium chloride dihydrate was kept 0.50 M in cementation solution [20, 21].

The permeability (constant head method) of biologically cemented sand was determined in PVC molds only which were used to prepare cylindrically specimens for shear wave velocity and unconfined compressive strength testing.

The biologically cemented samples were exposed to freezing–thawing cycles and tested for shear wave/ultrasonic pulse velocity and unconfined compressive strength after every 5th cycle up to 20 cycles. A single freezing–thawing cycle of 24 h was a combination of freezing up to 12 h at −14 °C and thawing for next 12 h at 25 °C [22]. The specimens were saturated with water during the freezing–thawing cycles.

Table 37.1 Sample designations according to cementation solution pore volumes and injection frequencies

Sample designation	Pore volumes (PV)	Injection interval (h)
1 PV–12 h	1	12
0.75 PV–12 h	0.75	12
0.50 PV–12 h	0.50	12
1 PV–24 h	1	24
0.75 PV–24 h	0.75	24
0.50 PV–24 h	0.50	24

Results and Discussion

Permeability and Calcite Content of Biotreated Sand

The permeability of biologically cemented samples varied with the calcite precipitation percentage. The specimen treated with 1 PV–12 h combination of cementation media, resulted in 95.8% permeability reduction with 11.9% calcite content (Fig. 37.1). Nevertheless, the reduced pore volumes and increased injection interval of cementation media resulted in lesser reduction of permeability due to lesser percentage of calcite formation. Moreover, the permeability withholding is advantageous, as the biotreatment can be continued to achieve higher strength.

Ultrasonic Pulse Velocity (UPV)

The calcite crystal formation between soil grains densifies soil media and simultaneously influences the UPV [23]. As shown in Fig. 37.2, the UPV of "1 PV–12 h" sample at 0 FT (before exposure to freezing–thawing cycles) was observed maximum, i.e., 2670 m/s. The same sample was precipitated with maximum amount of calcite content. Thus, the UPV of biologically cemented specimens decreases with the reduction in calcite content. Moreover, the minimum calcite-precipitated sample resulted in UPV more than 1000 m/s and showed rock-like behavior complementary to sandstone.

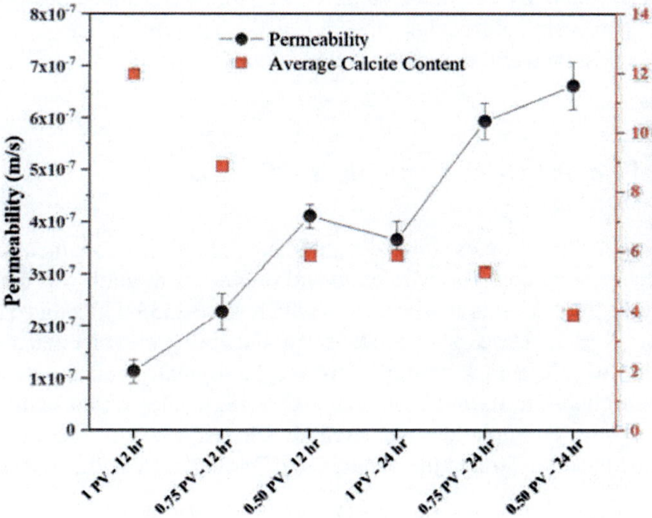

Fig. 37.1 Impact of treatment conditions on permeability and calcite precipitation

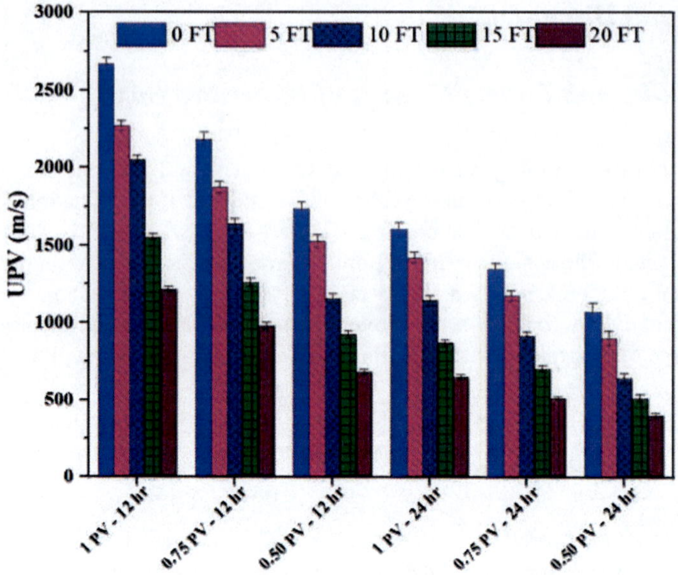

Fig. 37.2 UPV variation with biotreatment conditions and freeze–thaw cycles

The biologically cemented samples exposed to freezing–thawing cycles and the observation showed reduction in UPV up to 17, 43, 55, and 67% after 5, 10, 15, and 20 freezing–thawing cycles, consecutively. The decrease in UPV was related to generation of microcracks during freezing–thawing cycles. Moreover, the permeability retainment of biologically cemented samples was found advantageous to improve resistance to the exposure of freezing–thawing cycles. The similar behavior of porous soil media was observed and demonstrated by Litvan [24].

Unconfined Compressive Strength (UCS)

Biologically cemented samples showed significant strength improvement including less precipitated specimens. The maximum and minimum strength was observed in "1PV–12 h" and "0.50 PV–24 h" specimens which were 2333.3 kPa and 513.8 kPa, respectively (Fig. 37.3). The exposure to freezing–thawing cycles resulted in decrease of strength due to microcrack formation in the biologically cemented sand. The decrease in strength after 10 and 20 freezing–thawing cycles was less than 10 and 45%, consecutively. The upholding of strength after exposing to one of the major weathering agents demonstrated the endurance of biologically cemented sand.

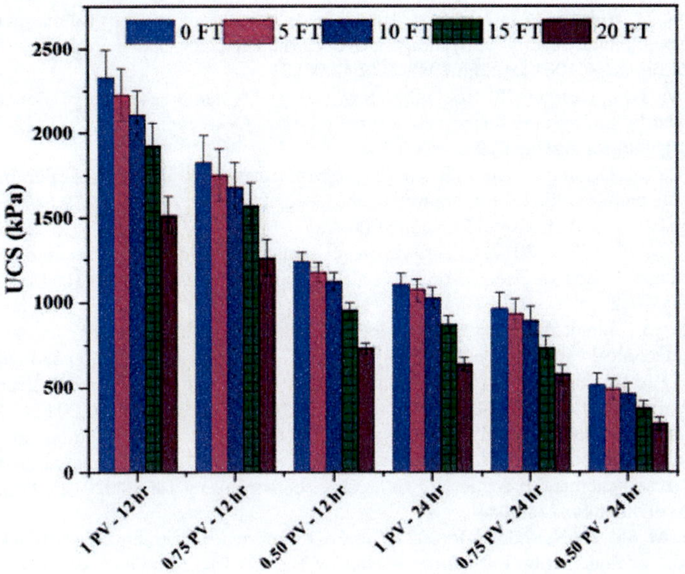

Fig. 37.3 UCS variation with different biotreatment condition and freeze–thaw cycles

Conclusion

The present study showed potential of MICP technique to increase the UPV and UCS of poorly graded sand. Moreover, the strength upholding even after 20 freezing–thawing cycle showed the endurance of biologically cemented sand.

References

1. Mitchell JK, Santamarina JC (2005) Biological considerations in geotechnical engineering. J Geotech Geoenviron Eng 131:1222–1233. https://doi.org/10.1061/(asce)1090-0241(2005)131:10(1222)
2. Cardoso R, Pedreira R, Duarte SOD, Monteiro GA (2020) About calcium carbonate precipitation on sand biocementation. Eng Geol 105612. https://doi.org/10.1016/j.enggeo.2020.105612
3. Stocks-Fischer S, Galinat JK, Bang SS (1999) Microbiological precipitation of CaCO3. Soil Biol Biochem 31:1563–1571. https://doi.org/10.1016/S0038-0717(99)00082-6
4. DeJong JT, Fritzges MB, Nüsslein K (2006) Microbially induced cementation to control sand response to undrained shear. J Geotech Geoenviron Eng 132:1381–1392. https://doi.org/10.1061/(asce)1090-0241(2006)132:11(1381)
5. Burbank MB, Weaver TJ, Green TL et al (2011) Precipitation of calcite by indigenous microorganisms to strengthen liquefiable soils. Geomicrobiol J 28:301–312. https://doi.org/10.1080/01490451.2010.499929

6. Burbank M, Weaver T, Lewis R et al (2013) Geotechnical tests of sands following bioinduced calcite precipitation catalyzed by indigenous bacteria. J Geotech Geoenviron Eng 139:928–936. https://doi.org/10.1061/(asce)gt.1943-5606.0000781
7. Montoya BM, DeJong JT, Boulanger RW (2013) Dynamic response of liquefiable sand improved by microbial-induced calcite precipitation. Géotechnique 63:302–312. https://doi.org/10.1680/geot.sip13.p.019
8. Martinez BC, DeJong JT, Ginn TR et al (2013) Experimental optimization of microbial-induced carbonate precipitation for soil improvement. J Geotech Geoenviron Eng 139:587–598. https://doi.org/10.1061/(ASCE)GT.1943-5606.0000787
9. Feng K, Montoya BM (2017) Quantifying level of microbial-induced cementation for cyclically loaded sand. J Geotech Geoenviron Eng 143:06017005. https://doi.org/10.1061/(asce)gt.1943-5606.0001682
10. Aishwarya T, Juneja A (2018) Sand bonded with calcite precipitation under cyclic simple shear. In: Geotechnical earthquake engineering and soil dynamics V GSP 290, pp 554–560
11. Xiao P, Liu H, Xiao Y et al (2018) Liquefaction resistance of bio-cemented calcareous sand. Soil Dyn Earthq Eng 107:9–19. https://doi.org/10.1016/j.soildyn.2018.01.008
12. Pablo ACMS, Lee M, Graddy CMR et al (2020) Meter-scale biocementation experiments to advance process control and reduce impacts: examining spatial control, ammonium by-product removal, and chemical reductions. J Geotech Geoenviron Eng 146:04020125. https://doi.org/10.1061/9780784482780.044
13. Sharma M, Satyam N, Reddy KR (2021) State of the art review of emerging and biogeotechnical methods for liquefaction mitigation in sands. J Hazard Toxic Radioact Waste 25:03120001. https://doi.org/10.1061/(ASCE)HZ.2153-5515.0000557
14. Sharma M, Satyam N, Reddy KR (2021) Comparison of improved shear strength of biotreated sand using different ureolytic strains and sterile conditions. Soil Use Manag. https://doi.org/10.1111/sum.12690
15. Sharma M, Satyam N, Reddy KR (2021) Effect of freeze-thaw cycles on engineering properties of biocemented sand under different treatment conditions. Eng Geol 284:106022. https://doi.org/10.1016/j.enggeo.2021.106022
16. Sharma M, Satyam N, Reddy KR (2021) Rock-like behavior of biocemented sand treated under non-sterile environment and various treatment conditions. J Rock Mech Geotech Eng 13:705–716. https://doi.org/10.1016/j.jrmge.2020.11.006
17. Sharma M, Satyam N, Reddy KR (2020) Strength enhancement and lead immobilization of sand using consortia of bacteria. J Hazard Toxic Radioact Waste 24:04020049. https://doi.org/10.1061/(ASCE)HZ.2153-5515.0000548
18. Tiwari N, Satyam N, Sharma M (2021) Micro-mechanical performance evaluation of expansive soil biotreated with indigenous bacteria using MICP method. Sci Rep 11:10324. https://doi.org/10.1038/s41598-021-89687-2
19. Sharma M, Satyam N (2021) Strength and durability of biocemented sands: wetting-drying cycles, ageing effects, and liquefaction resistance. Geoderma 402:115359. https://doi.org/10.1016/j.geoderma.2021.115359
20. Sharma M, Satyam N, Reddy KR (2019) Investigation of various gram-positive bacteria for MICP in narmada sand, India. Int J Geotech Eng 15:220–234. https://doi.org/10.1080/19386362.2019.1691322
21. Sharma M, Satyam N, Tiwari N et al (2021) Simplified biogeochemical numerical model to predict pore fluid chemistry and calcite precipitation during biocementation of soil. Arab J Geosci 14:807. https://doi.org/10.1007/s12517-021-07151-x
22. Cheng L, Shahin MA, Mujah D (2016) Influence of key environmental conditions on microbially induced cementation for soil stabilization. J Geotech Geoenviron Eng 143:04016083. https://doi.org/10.1061/(asce)gt.1943-5606.0001586

23. Sharma M, Satyam N, Reddy KR (2021) Hybrid bacteria mediated cemented sand: Microcharacterization, permeability, strength, shear wave velocity, stress-strain, and durability. Int J Damage Mech 30:1–28. https://doi.org/10.1177/1056789521991196
24. Litvan G (1980) Freeze-thaw durability of porous building materials. In: Sereda P (ed) Durability of building materials and components, vol 691. ASTM International, West Conshohocken, PA, pp 455–463. https://doi.org/10.1520/STP36080S